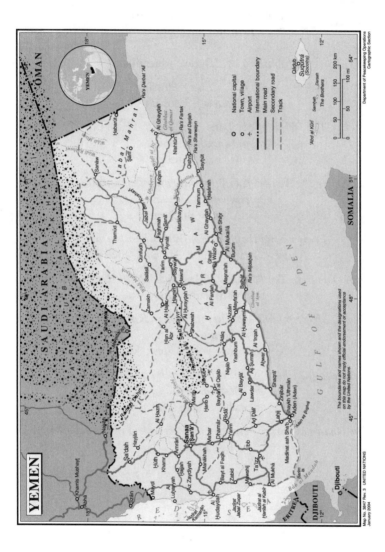

Yemen, no. 3847 Rev. 3, January 2004, courtesy of the UN Cartographic Section

CONTEMPORARY ANTHROPOLOGY OF RELIGION

*A series published with the Society for the
Anthropology of Religion*

Robert Hefner, Series Editor
Boston University

Published by Palgrave Macmillan

Body / Meaning / Healing
By Thomas J. Csordas

*The Weight of the Past: Living with History in
Mahajanga, Madagascar*
By Michael Lambek

*After the Rescue: Jewish Identity and Community in
Contemporary Denmark*
By Andrew Buckser

Empowering the Past, Confronting the Future
By Andrew Strathern and Pamela J. Stewart

Islam Obscured: The Rhetoric of Anthropological Representation
Daniel Martin Varisco

*Islam, Memory, and Morality in Yemen: Ruling Families in
Transition*
Gabriele vom Bruck

*A Peaceful Jihad: Negotiating Identity and Modernity in
Muslim Java*
Ronald Lukens-Bull

The Road to Clarity: Seventh-Day Adventism in Madagascar
Eva Keller

Islam, Memory, and Morality in Yemen

Ruling Families in Transition

Gabriele vom Bruck

palgrave
macmillan

First published in 2005 by
PALGRAVE MACMILLAN™
175 Fifth Avenue, New York, N.Y. 10010 and
Houndmills, Basingstoke, Hampshire, England RG21 6XS
Companies and representatives throughout the world.

PALGRAVE MACMILLAN is the global academic imprint of the Palgrave Macmillan division of St. Martin's Press, LLC and of Palgrave Macmillan Ltd. Macmillan® is a registered trademark in the United States, United Kingdom and other countries. Palgrave is a registered trademark in the European Union and other countries.

ISBN 1–4039–6664–8
ISBN 1–4039–6665–6

Library of Congress Cataloging-in-Publication Data

Vom Bruck, Gabriele.
    Islam, memory, and morality in Yemen : ruling families in transition / Gabriele vom Bruck.
        p. cm.—(Contemporary anthropology of religion)
    Includes bibliographical references (p.) and index.
    ISBN 1–4039–6664–8 (alk. paper)—
    ISBN 1–4039–6665–6 (pbk. : alk. paper)
        1. Yemen—History. 2. Yemen—Kings and rulers. 3. Zaydiyyah—Yemen—History. I. Title. II. Series.

DS247.Y45V66 2005
953.3—dc22   2004052433

First edition: November 2005

10 9 8 7 6 5 4 3 2 1

Printed in the United States of America.

Memory is not the instrument of exploring the past but its theatre. It is the medium of past experience, as the ground is the medium in which dead cities lie interred. He who seeks to approach his own buried past must conduct himself like a man digging. This confers the tone and bearing of genuine reminiscences. He must not be afraid to return again and again to the same matter; to scatter it as one scatters earth, to turn it over as one turns over soil.

Walter Benjamin, *Berlin Chronicle*

Contents

Part IV   Engaging Difference

Illustrations

Abbreviations

b. = ibn
bt. = bint

Acknowledgments

I owe the greatest debt to those whose names must go unmentioned. I must thank a great number of people for engaging with the subject matter over several years, and for never losing patience with an apparently never ending string of inquiries. Many committed themselves in ways that go well beyond reasonable expectations of collecting ethnographic data. Rather than expressing displeasure, those who invested an extraordinary amount of time and energy in the project would quote a saying "*Ihya' al-'ilm mudhakiratuhu*" (To keep knowledge alive, remember it). I was touched by those who extended kinship to me by insisting that I accept the *'asb* (gift) on *'id* days like their mothers, daughters, and sisters. In spite of the many demands that researchers make on the people they stay with, the hospitality, kindness, warmth, and respect that many of us were afforded in the Yemen made us feel treated like visiting notables.

A special obligation must be acknowledged to several 'ulama for nurturing me with their knowledge and good humor. I greatly benefited from their remarkable intellectual generosity. I should like to thank especially the late Mufti Ahmad b. Muhammad Zabarah who happily corresponded with me until he was well into his nineties; Sayyid Muhammad al-Mansur who, despite being approached by advice-seekers from 6 a.m. onwards, was always willing to give of his time; the late Sayyid 'Abd al-Qadir b. 'Abdullah 'Abd al-Qadir, Sayyid Hamud al-Mu'ayyad, the late Sayyid Hamud al-Mudwahi, Sayyid Isma'il al-Mukhtafi, and Qadi Muhammad b. Isma'il al-'Amrani. The late Sayyid Ahmad b. Muhammad al-Shami, who never tired explaining Yemeni history to me, held the generous view that teaching me would bring him *ajr* (merit).

A very special word of thanks is due to Maurice Bloch for his dedication to the project, his patience, and encouragement; and for helping me understand the nature of Shi'i authority. I am also greatly indebted to the late Ernest Gellner, who co-supervised the dissertation out of which this book grew, for his wise guidance and enthusiasm. Both were a constant source of inspiration. The late Alfred Gell, who is sadly missed, generously contributed to shaping the book through discussions of vital issues. His unconditional support, wittiness, intellectual acumen, and loyalty will never be forgotten by generations of students and friends. I must also thank him and his wife Simeran for their hospitality in London and Cambridge.

I appreciate the thoughtful readings given to the chapters in an earlier form by Fred Halliday, Michael Lambek, Wilferd Madelung, André Mazawi, Yitzhak Nakash, and Christina Toren. I am grateful to Shree Berke and Harold Schickler for help with preparing the manuscript. I also wish to thank Johnny Parry for allowing me to complete the book in the stimulating environment of the London School of Economics. As both teacher and neighbor, Johnny offered advice and assistance over many years. To Bernard Haykel, Wendy Levitt, Madawi al-Rasheed, Azam Torab, and Shelagh Weir I would like to express my appreciation for discussions and encouragement. Dale Eickelman deserves special thanks for his helpful advice and invaluable support. I also extend my gratitude to Robert Hefner who was committed to the project from the outset and guided me during the final stages, and Farideh Koohi-Kamali, chief editor, for confidently taking it on at the beginning of her new career.

The study could not have been completed without the help of Lieutenant 'Ali al-Anizi of the Presidential Office and his right hand, Qadi 'Ali Abu Rijal. I must also thank Dr. Husayn al-'Amri, former ambassador to London, Muhammad al-Radi of the Yemeni Center for Studies and Research, and Dr. 'Abd al-Karim al-Iryani, advisor to President 'Ali 'Abdullah Salih. The final phase of the work was overshadowed by the tragic death of Yahya al-Mutawakkil, Member of the Advisory Council, on January 13, 2003. His frankness and animated discussions over the years were much appreciated. Jamal Muhammad al-Sharjabi, who also died prematurely in 2004, provided wonderful companionship. His integrity and loyalty made a lasting impression on me.

The Studienstiftung des Deutschen Volkes in conjunction with the German Historical Institute in London (1984–87) financed research in San'a and a village in northern Yemen (referred to as Falih) on which this study draws. I gratefully acknowledge the support of these institutions and the LSE Malinowski Memorial Fund which helped during the later period. Since that time I have had ample opportunity to follow up the life histories of Yemenis who came to London on visits, some of whom stayed for several months. To the despair of one of my supervisors, who often told our friends that I had "brought the field back home," the book is the product of ongoing dialogues with Yemenis of diverse backgrounds and ideological proclivities who helped shape the arguments over the past years. The analytical claims put forward relate primarily to the period from 1982 (the twentieth anniversary of the September revolution) to 1990 (the unification of the two Yemeni states).

Parts of chapter 4 were published in slightly different form in "Ibrahim's childhood," *Middle Eastern Studies* 1999, 35(2). Earlier versions of chapters 6 and 7 appeared as "Enacting tradition: the legitimation of marriage practices amongst Yemeni *sadah*," *Cambridge Anthropology* 1992/93, 16(2). Chapter 11 builds on "Evacuating memory in post-revolutionary Yemen," in *Counter-histories: History, contemporary society, and politics in Saudi Arabia and Yemen*, edited by M. al-Rasheed and R. Vitalis. New York: Palgrave, 2004. Portions of chapter 12 appeared as "Disputing descent-based authority in the idiom of religion: the case of the Republic of Yemen," *Die Welt des Islams* 1998, 38(2).

Arabic terms have been transliterated in accordance with the guidelines given by the *International Journal of Middle East Studies*, but diacritics indicating long vowels and emphatic consonants have been omitted. Words such as 'ulama, shaykh, waqf, hadith, and hijrah are treated as English words and not placed in italics. The ta marbuta and some colloquial speech has been maintained where it deemed appropriate. All dates follow the Christian calendar.

Foreword

Few countries in the world can exercise such fascination, over visitor and social scientist alike, as Yemen, the site, along with China, Iran, and Egypt, of one of the only four states to have political and cultural continuity over three millennia. Located in the southwest corner of Arabia, and reunited after three centuries of division into North and South in 1990, Yemen is today a country of around 20 million people, half of the total population of Arabia, but, by lacking oil, poorer by many degrees. The larger part, the former North, was for many years a conservative society, ruled by an Imam, who sought to exclude the external world, and was one of the few third world states not to fall under European colonial rule. The Imamate itself, based on the Zaydi teachings of Islam, a form of moderate Sh'ism, had been a significant force in Yemen since the ninth century AD and the Imams had, in modern times, become political rulers, custodians of a modern if rudimentary state, and leaders of an early, some would say nationalist, resistance to the Turkish occupation that began in 1870.

Turkey never had it easy in Yemen. An hour from the capital San'a, on a spectacular escarpment that dominates the area around, is the town of Kawkaban, known as 'maqbarat al-atrak-, "the Graveyard of the Turks." In the period around 1900 Yemen was in effect the Vietnam of the Ottoman Empire. Hence the song declaimed to this day by recruits to the Turkish army "Kanli Yaman, Kanli" "Bloody Yemen, Bloody." After years of fighting, in 1911 the Imam and the Turks signed the Treaty of Da"an, which in effect gave sovereignty to the Imam in much of the land. When the Turks departed in 1918, following the collapse of the Ottoman Empire, Yemen became independent, the first Arab country to do so. The two Imams, Yahya and Ahmad, then ruled Yemen successively, keeping the outside world at bay. Ahmad famously declared, "If I have to choose between being rich but dependent and poor but independent, I will choose the latter," something he lined by.

Change could not, however, be prevented. The Arab world was growing more assertive and nationalist mobilization grew after the defeat of the Arabs in the first Arab-Israeli war, that of 1948–49.

Yemen, far from being isolated was therefore increasingly affected, as Nasserist nationalism swept the Arab world in the 1950s. Here history had its impact on Yemen in a paradoxical manner as seen in other countries that had not been dominated by formal colonialism and had therefore avoided the even limited, instrumental, modernization of colonial regimes. When change did come, as in Afghanistan and Ethiopia, two other conservative monarchies that had repulsed colonial encroachment, it was all the more sudden and impatient.

In 1962, after over four decades of isolation under the Imams, a radical revolutionary military coup occurred in San'a. Freedom and liberation, the end of all oppressive rulers and their associates, were proclaimed. Appeals were made to "the great Yemeni people" and to "progressive soldiers and merchants." The old regime was denounced in the language of Islamic repudiation as tyrannical ("dhalim" and "istibdadi"), but also as "clerical," "kahnuti"). Resistance from tribal forces and from the new Imam, who had just survived the coup, nonetheless developed, aided by Saudi Arabia and Iran. Up to 70,000 Egyptian troops came in on the side of the newly proclaimed republic. A civil war lasting eight years then ensued, with peace only being achieved in 1970 with a compromise, under which most republicans and royalists returned home, but a minority, including the Imam himself, and his close family, were not allowed to do so.

Meanwhile in the South, an area dominated by Britain from the port of Aden since 1839, an attempt to forge a new federation of the over two dozen Sultans and Sheikhs ruling the region met with growing resistance: once the revolution in the North had exploded, it was only a matter of time before armed resistance would begin in the South. This occurred in 1963 and four years later, faced with a widespread armed uprising across the South, now led by the radical National Liberation Front, the British departed in considerable disarray: a new state, the People's Republic of South Yemen, known from 1970 as the People's Democratic Republic of Yemen, was proclaimed.

Nor was this the end of the story. The two regimes of North and South, while both proclaiming their commitment to Yemeni unity, were in practice rivals and were in a state of greater or lesser armed conflict till 1982. For its part the new revolutionary Republic in the South declared its intention of promoting revolution in the rest of the Arabian Peninsula, starting with the southern Omani province of Dhofar, which bordered South Yemen. Guerrilla war continued there until 1975. Thereafter ideological confrontation, between oil-rich monarchies and impoverished revolutionary regimes, continued. The 1990 unification of North and South Yemen ended the formal division of the two countries, but even after that fighting revived: in 1994, after months of growing tensions between the still two distinct regimes of North and South, a civil war broke out, ending, after seventy days, with the conquest of the South by the stronger forces of the North. The entry of North Yemeni troops into Aden on July 7, 1994 marked the final point in a history of upheaval, conflict, and ideological dispute that had begun with the North Yemeni revolution of September 26, 1962. Only then did the thirty-two-year span of the South Arabian revolutions come finally to an end.

It is against this background over the past century and a half, of a self-induced fragmentation into the two states of North and South, then of British and Ottoman rule, popular mobilization, stagnating Imamate and, finally, civil war that this study by Dr. Gabriele vom Bruck is written. In one sense, her story is straightforward: Yemen, in common with many other Muslims states, has a group of notables, the sadah (singular sayyid), who are believed to be descended from the Prophet Muhammad. There are, in a Muslim world of one and a half billion, believed to be 10 million such people. Their number, social status, and economic activity vary from country to country, but wherever they are found the sadah are associated with a sense of authority and of being the legitimate interpreters of tradition within Islamic theology and culture. Whether it be in the revolutionary elite of Islamic Iran, the leadership of Hizbullah in Lebanon, the Afghan resistance, or in contemporary Pakistan, to be a Sayyid is to be someone. I was once examiner of a Ph.D. by a candidate who let me understand, in no uncertain terms, that he was descended from the Prophet.

The story in Yemen is complex, rich, and constantly changing. The sadah were not, in any simple sense, a class, though classes there certainly were, and are, in Yemen. They were in general associated with the Imamate system, but not in any direct sense either controlled by the Imam or able to control him. Some lived lives generally integrated into the broader economic system, some performed special functions. Mention of the sadah is to be found in many accounts by travelers of pre-1962 Yemen, as well as in the fine study by Abdullah Bujra of social stratification in the eastern part of South Yemen, Hadramaut, in the early 1960s, on the eve of the revolutionary time.

Dr. vom Bruck's study, methodologically rich in its use of interview, memory, and culture, is both a study of the sadah themselves and a rich and detailed analysis of modern Yemeni society as a whole. It analyzes the traditional place and legitimation of this group within Yemeni society, the impact of the Revolution of 1962, and the various forms of adaptation to the new republican and more secular circumstances that the sadah have explored. There are chapters on historical

background, lineage, childhood, family, marriage, and extensive discussion of the changing political and economic positions of this elite.

Dr. vom Bruck's study is based on extensive work in Yemen,
focusing on informants in a range of families. She has gained unique
access to the world of the sadah while, at the same time, maintaining
a critical, and comparative, perspective on her material. In its subject
matter, in its sources, both oral and written, and in its depiction of
Yemeni society as a whole, it is a unique and most valuable study.
Through the lens of this study of the sadah the author provides access
to the history of modern Yemeni society as a whole, and to the inter-
section of political power, wealth, lineage, and locality that define
much of that history. All of this material is enriched by telling quota-
tions, anecdotes, and insights that bring the sadah and indeed Yemen
as a whole very much to life. In sum, it can be said that in addition
to being an anthropological study of high quality, Dr. vom Bruck's
work makes a significant contribution to four broad areas of social science research: the history and anthropology of modern Yemen;
changes in forms of identity; the anthropology of memory; and the
social consequences of revolution.

In my own work on Yemen, North and South, which began with
research in Aden in 1970 and has focused on modern political and
social movements, I have not studied carefully, but have repeatedly
been aware of, and puzzled by, the phenomenon of the sadah. The
revolutionary regime in the South, lasting as it did from 1967 to
1994, was committed to a radical egalitarian programme, including
among its slogans "Workers, peasants, fishermen, Bedouin and nomads
unite!" as well as the call, probably unique in the whole annals of
twentieth century revolutions, "Arm the Women!" Yet older forms of
identity and loyalty persisted beneath this layer of radical rhetoric and
policy. Several of the leading members of the southern regime were
sadah, among them Presidents Haidar al-Attas and Ali Salim al-Bidh.
What this meant in practice, be it in terms of popular perceptions
of their authority, or the systems of patronage and loyalty they were
contained within, I cannot say. It may be that in some situations it
meant not a great deal, much less than, for example, membership of a
particular tribe.

Here there may be a contrast between North and South. My overall sense was that while the revolution in the South was in general
more socially radical, it was not so hostile to the sadah as that in the
North, if only for the reason that in Hadramaut the sadah were on
the side of change, against the ruling Sultans, while in the rest of the
South they were not of such great significance. In the North it was
another, often more violent, matter. Yet, as Dr. vom Bruck indicates,

with the peace of 1970 many sadah settled into life within the new republic. This was not, however, something that could be taken for granted and, as ever, the position of the sadah was affected by broader changes within Yemeni society. On my most recent visit to Yemen, in February 2004, I was given, as is conventional, an official guide by the presidential office. After a few days it emerged that my companion was himself a sayyid, indeed a member of one of the former ruling families. I noted that he was careful about whom he discussed this with. He said that over the past two decades things had improved a lot, but that there were now new clouds on the horizon, as a result of hostility to the sadah propagated by newly influential conservative Sunni, or Salafi, forces coming from Saudi Arabia. Whatever the doctrinal reasons for this, within Wahhabi thinking, one political factor is evident: the Saudi state had been established in the 1920s by driving out the Hashemite rulers of Hijaz or western Arabia, whose descendants now rule Jordan, and this anti-sayyid orientation was now having effect in Yemen. If nothing else, this illustrates two of the points that underlie Dr. vom Bruck's study, that Yemeni society itself is constantly changing, not least in response to the impact of external forces, and that the position of the sadah, far from being fixed by history or faith, is itself shaped by the broader social, political, and, not least, ideological context in which they are located. The story, of which Dr. vom Bruck has made such a fine analysis, therefore continues.

Fred Halliday
London School of Economics, July 22, 2004

Introduction

Locating Memory: The Dialectics of Incorporation and Differentiation

To know the Saiyids is to comprehend at least something of their great ancestor, the founder of Islam . . . The Saiyids . . . are families, clans, in which special qualities, virtues of a supernatural kind, and nobility, sharaf, are held to reside.

(Serjeant 1957)

With these words, the late Robert Serjeant introduced his subject *The Saiyids of Hadramawt* in his Inaugural lecture at the School of Oriental and African Studies in the summer of 1956.[1] Few of my Yemeni friends who identify themselves as *sadah* (sing. *sayyid*)—who claim direct descent from the Prophet Muhammad—would feel at ease with such flattery. They have learnt that to be called a *sayyid*, like being called a "woman" or "Jew" or "black," may be heard and interpreted in diverse and conflicting ways, as an endorsement or an insult.[2]

A few years after Professor Serjeant delivered his lecture, a revolution brought the Yemeni Imamate, which had been occupied by the descendants of the Prophet since the ninth century, to an end. The Imamate was sanctioned by the Zaydi-Shiʻi school of Islam which requires the spiritual and temporal supreme leader, the Imam, to be a recognized descendant of the Prophet and to possess profound knowledge of the religious sciences.[3] According to Zaydi teachings, as heir to the Prophet's legacy, the Imam becomes the embodiment of Prophetic authority by means of erudition and symbolic acts centering on prayer, generosity, and the banishment of evil. In 1962 a republic was announced which claimed legitimacy through the abolition of political authority based on birthright, and the power of the *sadah* declined.

The Yemeni Imamate endured for over a millennium, yet its history and elites have attracted little scholarly attention. This may be attributed to the common view that with the decline of the pre-Islamic Minaean and Sabaean states, the Yemen was marginal to the Islamic centers of Syria and Egypt. Within the Ottoman Empire until the end of World War One, it derived its significance mainly from being the closest point to the holy sites in Mecca. Furthermore, because unlike the South Yemeni sultanates the Imamate was one of the few states in the region which was never colonized by Western powers, it was generally assumed to have been isolated. It is probably fair to say that the cultural and political influence other Muslim states exerted on the Yemen outweighed the impulses they received in turn. However, the absorption and indigenization of these various currents, and the creation of local intellectual and political traditions would appear to merit further historical research. These processes have occurred in a different guise in the country's more recent history during which it has become a playground for superpower rivalry and Islamist agitation. As a nascent democracy the Yemen was vulnerable to the turbulence of the ending of the Cold War and the outbreak of the Second Gulf War in 1990, and was pilloried for its political stance, and denied aid. As Joffé (1997: 11) observed, the Yemen was one of the first states that faced the full force of the new global constellation of power. "Many of the problems that face the wider Middle East are being experienced in a more concentrated and immediate form in the Yemen and this, quite apart from its potential geo-political role in the new world of 'geo-economics' which now is emerging, make it of acute relevance today." Thus, it does not deserve its reputation as a geographically and culturally remote—and by implication irrelevant—corner of the Arab world. This point is worth brief development before the main theme of the book is taken up.

During the early Islamic period, Yemenis played an important role in the Muslim conquests of North Africa, Iraq, Iran, Syria, and Egypt. Within its borders, Yemen's contribution to the early development of Islamic culture was modest in comparison to its earlier achievements; this has been attributed to the diasporic activities of the descendants of the warriors who chose to settle in the newly conquered territories. Nevertheless, several towns became centers of Islamic Traditionalist science, attracting teachers from places such as Basra (Madelung 1987: 174). In the tenth century, the work of a local historian and geographer, Ahmad al-Hamdani, reflected the concerns of the classical Arabic geographical literature. Al-Hamdani located the Arabian Peninsula within the Ptolemaic earth and universe. A couple of centuries later the western Sunni town of Zabid boasted its first religious college, which was older than al-Azhar in Cairo (Daum 1987a: 26). At that time, religious schools based on religious trust properties (madrasas) were introduced. A certain style of biography of rulers, known outside Yemen only from occasional examples, developed in Yemen into a regular genre of historical literature. The earliest Yemeni biography, which became a model for successive ones, was that of the first Zaydi ruler, the founder of the Imamate (Madelung 1992: 129–30). The country was exposed to diverse trends such as the teachings of the Shi'i Imam Muhammad al-Baqir (d. 735) and of the founder of the Hanbali school of law, Ahmad b. Hanbal (d. 855), whose ideas were fundamental to the legal constitution of Saudi Arabia. There were also influences of the Andalusian mystic Ibn al-'Arabi (d. 1240) (Madelung 1987: 174–5). Scholarly activity in Yemen was such that it provoked the founder of one of the Sunni schools of law, Imam al-Shafi'i (d. 820), who lived in Cairo, to state that "one must go to [the Yemeni town of] San'a [to meet scholars] even if it's a long journey (*la budda min san'a wa- 'in tal al-safar*)" (M. Zabarah 1999, Vol. 1: 21).

Letters to and from other countries also refute the myth of Yemen's isolation. For example, Zaydi scholars exchanged views with their counterparts in Tabaristan in contemporary Iran. The Spanish-born Maimonides, who lived in Fez and Alexandria, communicated with Yemeni Jews, and letters from Yemen were found in the Geniza of Cairo.[4] In the late eighteenth century, Zaydi rulers exchanged letters with the Sharif of Mecca about Napoleon, discussing the French occupation of Egypt and the prospect that they might conquer Aden as well (Mahmud 1983: 6). In recent years books that elsewhere in the Arab world were deemed lost were found in Yemeni libraries and subsequently republished, and Yemen's neighbors have at last begun to acknowledge its intellectual tradition.

Viewed in the context of South Arabia's history, Yemen has a remarkable tradition of statehood from the pre-Islamic period onwards. Despite the fact that during the Islamic era it did not emerge as a separate political entity before the twelfth century, from the ninth century onwards there is evidence of Imamic state administration in the form of land taxation and rules concerning landownership (Madelung 1992: 189–90). After Yemen's adoption of Islam, the establishment of the Imamate in the ninth/tenth century has been hailed as the second most significant date in its Islamic history (Daum 1987a: 25). Ever since that time the Zaydis, who developed their own doctrinal school, remained a force on the Arabian Peninsula until the twentieth century. The Imamate, which often coexisted with other states, was ruled continuously by members of the Prophet's House. In comparison with other Shi'i (and most Sunni) states in the Middle East, it was unique because it was ruled by a dynasty which hardly ever became "dynastic" because doctrinal constraints discouraged hereditary succession from father to son. Moreover, unlike the monarchies of the Peninsula which were established in the nineteenth and twentieth centuries, the Yemeni Imamate (referred to as a kingdom since 1926) was not created through the process of European colonialism. Its ruling families, which are the subject of this book, contributed significantly to the creation of a religious and judicial tradition unparalleled on the Peninsula. Historical works, poetry, and legal and political treatises were produced in both city and countryside. The *sadah*, from whom the rulers of the Imamate were principally recruited, have attracted far less academic interest than their Sunni counterparts in the southern part of the country, above all the Hadramawt. Since the late sixteenth century, many Hadrami *sadah* left Yemen and enjoyed great religious prestige in India, Malaysia, Singapore, and Indonesia where they embarked on careers as teachers, judges, and rulers. It might surprise some readers that the present king of Malaysia belongs to one of those émigré families (Azra 1997: 249; Dale 1997; Ho 2002; Gilsenan 2003).

Those Hadrami immigrant families gained fame as leaders of the anti-colonial resistance around the Indian Ocean. By the time confrontation with various European powers had come to an end, the country the Hadrami pioneers had left was itself at the center of regional and global rivalry. When the deposition of the last Zaydi Imam in 1962 provoked a civil war between his followers and Egyptian-inspired republicans, Arab nationalist and monarchic forces competed with each other on Yemeni soil. Establishment of a socialist regime in the South after independence from Britain in 1967, implicated Yemen in the Cold War—the end of which coincided with the unification of Yemen in 1990. With the emergence of new political and religious elites, the country has become one of the theaters of Sunni reformist activity, with both a local and global impact. As members of the former elite of the Imamate, who have been the main proponents of Zaydi Islam, which is labeled "heretic" by some members of the reform movement, the *sadah* are at the center of the struggle of identification that has accompanied recent processes of state transformation.

Historically, the elite crowned by the Imamate derived its position from its noble descent, its tradition of scholarship and public service, and its reputation for piety.[5] Wealth served to enhance status but was not a significant status marker in its own right. The great majority held the status of "religious notability." As already noted, only descendants of the Prophet could occupy the highest political office; at the less elevated echelons of government, they shared power and lifestyles with the *qudah* (sing. *qadi*, judge) who derived their status from their (or their forebears') occupation as government-employed judges. The *qudah* were, in Mosca's terms, a secondary elite—those who allowed the rulers to rule (quoted in Eickelman 1985: 9). Members of both social categories held leading positions in the government and contributed to the promotion of scholarship. For reasons I will explain later, this book focuses primarily on the *sadah*.

The *sadah* (also referred to as Hashimites) have ruled in different parts of the Middle East for centuries, most notably in North Africa. They provide the ruling dynasties in Morocco and Jordan who even today claim legitimacy through descent from the Prophet.[6] In the early twentieth century, when the power and prestige of the old elites were already waning, the Sharif of Mecca and his son 'Abdullah (the first king of Jordan) had not yet abandoned the dream of a united Arab nation under Hashimite leadership. However, it remained an unattainable goal.[7] In 1958 King Faysal II of Iraq was deposed, followed in 1962 by the Yemeni Imam. In spite of the survival of the dynasties in Morocco and Jordan, the overthrow of the Yemeni Imamate attested to the decline of Hashimite power in the Middle East.[8]

This study analyzes the unique encounter of one group of *sadah*—those of the Yemen—with the post-revolutionary situation. It seeks to broaden our understanding of hereditary elites at a historical juncture—events that led to the removal of legal privileges and require radical reorientation and adaptation. In spite of the undiminished pride of birth displayed by these elites, studies of elites in decline as diverse as those of the French aristocracy (Magraw 1983; Wright 1995) or of the religious elite of Morocco (Eickelman 1976) have documented culturally specific modes of readjustment and interpretation of this experience. Rather than focusing on recent political history,[9] the book's goal is to examine the relationship of experience, social practice, and moral reasoning among this hereditary elite in the context of revolutionary change. A central feature of republican state formation has been the separation of political office and religious status. Mindful of the new political and social currents in Yemen and elsewhere in the Arab East and parts of North Africa, the *sadah* have reconstituted themselves in new professional and other vital social spheres. They pursue this exploratory project within the framework of their codified ancestral tradition in an effort to avoid the cultural dislocation elites have suffered elsewhere. The *sadah*'s historical and personal memory highlights the discrepancy between their experience of reverence and privilege on the one hand, and of persecution and stigmatization on the other. The book examines how these various features of remembrance translate into social practice, which only partly mitigates their predicament.

One of the central questions the book raises is how and in which contexts this elite invokes religious orthodoxy when the basis of their authority has to a great extent been eroded. The *sadah* are caught up in a dialectic of remembering and forgetting within a context in which descent-based authority with its inherent claims to legitimacy is no longer endorsed. The descent metaphor, which was the core of the Imamate's political culture and the *sadah*'s principal self-defining criterion, has been challenged by new constellations of power. The sense of pastness and self that emerges from this challenge is partially mediated through the recitation of religious orthodoxy, the sayings and deeds of great scholars from whom they have descended. The *sadah* consider these as an important component of their historical and genealogical consciousness because it was initiated by their founding ancestor, the Prophet. Quoting from the Qur'an, the Sunna, and the statements of Imams is a common practice among Shi'is; among the *sadah*, kinship endows this practice with even greater significance. Classical theological and historical studies portray the *sadah* as interpreters of religious texts. This study is concerned with the impact of specific historical circumstances on these interpretations, examining how men and women apply them to contemporary moral and social issues, which are significant to their own lives. The traumatic events of 1962 have added moral impetus for painstaking inquiries into domains of practice that were closely associated with state authority: profession, marriage, and the fundamentals of rule, which will be analyzed in the chapters to follow.

The recitational practice central to the investigation is anticipated through remembering historical knowledge as laid down in orthodox texts. As this knowledge awaits interpretation and reinterpretation, it is never freed from the dictates of either past or present. It is a future-oriented commentary on both past and present; while transitions are contemplated and certain practices repudiated, continuity with the past is maintained through the process of re-citation. It is thus a way of coming to terms with a transformed reality while continuing to cultivate modes of reasoning and to engage with forms of knowledge production which, for centuries, have provided a central raison d'être for the 'Alid self. ('Alids derive descent either from 'Ali and the Prophet's daughter Fatima, through either of their sons al-Hasan and al-Husayn, or from one of 'Ali's children by other wives.)[10]

The Old Elite and Republican Nationalism

In the twentieth century, the experience of the (North) Yemeni elite and the *sadah* in particular was both different and similar to that of other Arab elites. In most parts of the Middle East and North Africa, the twentieth century saw the decline of the elites. They had already been weakened by reforms in the nineteenth century, which annulled many of their privileges, and the collapse of the Ottoman Empire and the abolition of the sultanate (1922) and caliphate (1924) precipitated their decline (E. Burke 1993: 14).[11] In contrast, in the Yemeni Imamate, the longest lasting Shi'i power in the Middle East, the elite enjoyed a remarkable continuity until the mid-twentieth century. Like other parts of the Arab East, North Yemen had come under Ottoman suzerainty in the nineteenth century but escaped European 'colonial encroachment, thus achieving independence much earlier than most countries in the region.[12]

Therefore the social, economic, and political transformations which dramatically altered the lives of people of all social backgrounds in large parts of the Middle East from the late nineteenth century onwards, had far less impact on the inhabitants of North Yemen. Yet the absence of European colonialism does not imply stagnation. Although the elite was left unscathed, it was affected by state policies and turmoil. Post-independence rulers attempted to combine traditional styles of government with limited reforms. Some of these weakened the 'ulama (religious scholars), the most powerful professional class in the country who had always recruited the Imams and acted as "kingmakers" (and regicides) in their own right. Like their counterparts elsewhere, and indeed the Imam himself, they were "servants of that other power manifested in the Divine Message" (Gilsenan 1982: 52), but their role was not restricted to interpreting the law and bestowing authority upon the holders of power. In the 1940s intra-elite disputes on political and moral authority led to deep divisions, culminating in an uprising by 'ulama and men of letters who demanded a constitutional government. The position of the 'ulama in the twentieth-century Imamate was therefore substantially different from that of their Moroccan counterparts. According to Waterbury (1970: 83), their authority has not been sufficient to allow them to censure or challenge the Sultan.

Several Yemeni 'ulama took issue with the patrimonial character the Imamate had assumed and the curtailment of their autonomy. In joining and indeed spearheading the anti-regime movement, the *sadah* inadvertently played a central role in precipitating the internal erosion of the state which had officially represented them. The Imams' failure to implement profound reforms and the dissatisfaction and radicalization of army officers (many of them low ranking *sadah*) were root causes of the Egyptian-inspired revolution of 1962 which finally eliminated the political power of the elite. By virtue of their exclusive rights to the supreme leadership and their kinship with the former rulers, the *sadah* became the main targets of the revolution.

In the Imamate, 'Alid identity was unproblematic in so far as its law and institutions defined the *sadah* according to religious criteria and reserved a special place for them in the body politic. In the republic, which has committed itself to the elimination of religiously sanctioned hereditary distinctions, there has been a Tocquevillian shift from a society formally ordered by claimed descent to one in which all citizens are nominally equal. Republican legislation no longer recognizes the *sadah* as a distinct social category. This, however, does not sufficiently explain the precarious nature of 'Alid identity in present-day Yemen. 'Alid identity has remained politicized because the Imams were always *sadah* who, inter alia, have been ascribed a will to power and allegiance to the ancien régime by successive republican governments. The media demonize the Imams, portraying them as tyrants who monopolized the country's resources for their own good. These attacks have served to assign moral credibility to the republic and to delegitimize 'Alid claims to government positions which wield actual power. Like other nationalist-flavored rhetoric, this counterpoise to Imamic notions of righteousness reveals an obsessive sensibility over-attuned to its object of denunciation (Nairn 1997: 171). The *sadah* have experienced the defamation as a personal affront, which has contributed to highlighting and perpetuating their sense of distinctness even while many have attempted to play down this feature of their identity.

Nationalist projects, like colonial ones, are often predicated on a tension between contingent notions of incorporation and differentiation (Stoler and Cooper 1997: 10). By turning subjects into citizens, the republic has aspired to homogenize the society in the pursuit of creating a Yemeni national identity at the expense of other parochial identities.[13] The constitution denies legal and political recognition of cultural and religious difference.[14] Other interests notwithstanding, this urgency has been due to the fact that "North Yemeni society is more socially fragmented than that of any other country in the region with the exception of Afghanistan" (Halliday 2000: 68). However, in a paradoxical play on the imagery of ancient Arabian warriors and the empowered nation-state, there is official endorsement of and even identification with the *qabaliyyah* ("tribalism").[15] This is significant not least because for centuries, Zaydi rulers unsuccessfully attempted to create a new unified Muslim identity which would transcend particularist tribal identities. The creation of the notion of "citizen" (*muwatin*) and the fusion of disparate groups and factions in one single social category, the *qaba'il/qabilah* (tribespeople), enables the government to satisfy both the requirements of state building and the desire of its main constituents for recognition as the preeminent and most authentic component of Yemeni society. By invoking the *qabilah*, Yemeni nationalism defines for itself the parameters of inclusion and exclusion. In an interview with an Arab magazine, entitled "*Na'am, kullna qaba'il*" (yes, we are all tribespeople), President 'Ali 'Abdullah Salih replied to the question about the extent Yemen had succeeded to move from the stage of "tribalism" (*qabaliyyah*) to that of the state:

> The state is part of the tribes, and the Yemeni people are an ensemble (*majmu'ah*) of tribes. Our urban and rural areas all consist of tribes. All the official and popular state institutions are made up of the tribes/tribespeople (*qaba'il*). (*Al-Majallah*, October 1986: 17)

Here the President endorsed a masculinized ethos of nationhood based on the image of the *qabili* warrior; one which later came to haunt him when tribesmen began to abduct tourists and oil company personnel in the 1990s.[16] The "qabili" (sing. of *qaba'il*) has become a metaphor for the Yemeni nation, which is the most populous but economically weakest state on the Arabian Peninsula.

As head of one of largest tribal confederations, the influential Speaker of Parliament Shaykh 'Abdullah al-Ahmar considers himself to be the guardian of the state. He exalts the *qabilah* as the fundamental and unchanging nucleus of Arab society:

> The *qabilah* is the foundation (*al-asas*) of our Arab society, and to deny the *qabilah* is to deny our authenticity (*al-asalah*) and our ancestors . . . It is the foundation and it is natural. And God says [in the Qur'an] "we have created for you nations and tribes so that they would know each other." So why do we approve of one part of the verse and try to omit the second part, and why do we say this nation and that nation but not this tribe and that tribe? God clearly says "we created you nations and tribes" . . . I am the protector of the revolution, the republic, and unity, and now I add democracy; democracy which does not overshadow the *qabilah* because the *qabilah* is above everything except religion. (*Al-Thawrah*, May 3, 2000: 3)

This discourse alludes to issues of distant origins and ancestral descent which feature in conceptualizations of social categories and nationalist ideology, shaping moral judgments about people's capacities and dispositions. As Comaroff (1995: 244–5) remarked recently, "there seems little doubt that ethnic and nationalist struggles—in fact, identity politics *sui generis*—are (re-) making the history of our age with a vengeance. . . . The impact of the recent politics of identity has been notably ambiguous, having had both a liberating and a dark side." In the Yemen, the cultural construct *sayyid* has been an essentializing and mobilizing device useful both to apologists for Imamate rule and for hardcore republicans. The latter warn that the descendants and kin of the Imams constitute a threat to the infant republic. Debates over who is "really" Yemeni and who is not have continued unabated even after the reconciliation following the civil war between supporters of the Imam ("royalists") and of the newly installed republic in 1970, centering on the *sadah* as well as on other categories such as people disadvantaged by birth. The newly established elite, made up of army personnel of predominantly tribal background, merchants, and technocrats, represent themselves as either "more Yemeni" than the *sadah* or disclaim their Yemeni identity. Some depict the *sadah* as "foreigners" who, after immigrating to Yemen from different parts of the Middle East since the ninth century, unlawfully subjugated the local population to their rule.[17] Unlike the majority of the inhabitants of Yemen who trace descent to Qahtan, the putative eponym of the "southern Arabs," by reason of their northern Arabian origin the *sadah* consider themselves to be the offspring of 'Adnan, the ancestor of the "northern Arabs."[18] According to Detalle (2000: 53), "in genealogical treatises, which are very popular in that part of the world [the Arabian Peninsula], they [the descendants of Qahtan] are referred to as Arab Arabs, i.e. genuine Arabs. Descendants of Adnan are called Arabised Arabs." Distinctions between Qahtani and 'Adnani are endorsed by those who so define themselves. Essentializing self and "other" in post-revolutionary Yemen has buttressed a national collective identity with a fiction of pure origins in the ancient southern Arabian kingdoms.[19] Therefore, as Kugelmass (1995: 298) observed in another context, "constructions of otherness are part of a much broader process of mythologization, a process that is closely tied to resilient nationalism." A few years after the reconciliation (see figure 0.1),[20] the prominent politician Muhsin al-Ayni remarked:[21]

> Even if nowadays the Yemeni people are ashamed and unhappy about their underdeveloped standard of living, they will take it upon themselves to build a better life and a brighter future. Like *our* ancestors who established their civilization and left us the archaeological monuments in Marib and Sirwah, the Yemeni people will once again build a new life under the banner of justice, freedom and peace.

As Dirks (1992: 14) reminds us, "nationalism is . . . a system for organizing the past that depends upon certain narratives, assumptions, and voices, and that continues to have important stakes

Figure I.1 The National Pact (window sticker)

throughout the social and political order." In Yemen anti-'Alid discourse is an outgrowth of a nationalist movement that began in the late nineteenth century. The external "other," the Turks and the British on whom Yemeni nationalist ideology has centered until the 1960s, can be brought into relationship with the unwelcome historical "other within."[22] Zaydi history was marked by struggles against internal rivals and foreign invaders, struggles that were led by the Imams who were perceived as liberators by their indigenous supporters. Whilst aspiring to spread Zaydi ideas beyond the borders of the Imamate, the Imams saw themselves as guarantors of national unity against external threats. One of the last ones, Imam Yahya (d. 1948), saw his life's work in bestowing his subjects with the gift of an unpossessed future—to reverse a term coined by Khilnani (1997: 157). His biographer, the Qadi al-Shamahi (1937: 3–4), praised the Prophet's progeny for preserving the divine law

> at a time when many nations deviate from religion and observe [man made] laws [introduced by the colonial powers] which are more brittle than a spider's web . . . God has protected the Yemeni people from the force of this flood by a descendant of the Seal of the Prophets . . . our lord, the Leader of the Faithful al-Mutawakkil 'ala'llah Yahya ibn Muhammad, the descendant of the Prophet of God . . . May God help him and reward him for what he has done for Islam . . . he revived with his strength and knowledge what had been studied in religion, following the path of his forefathers who are related to his ancestor, the Prophet. He did all this with his pen and sword, making the Qur'an the judge of his people, his nation, and himself in private and public matters.

Since Yemeni nationalism shifted from the external to the internal other, the Imams have been paradoxically cast in the role of colonizers by opponents of the Imamate and some republican politicians and intellectuals.[23] The descendants of the former liberators, the Imams, are represented as potentially disloyal members of the nation that attempts to locate itself in a newly bounded space and time. This exteriorization of a certain group of people coincides with the nationalization of local descent lines. The Qahtani Yemenis, by extolling *their* ancestors, "transform [a certain period of] history into national history, legitimizing the existence of a nation-state in the present-day by teleologically reconstructing its reputed past" (Karakasidou 1997: 17). As under the Imams, descent lines are once again important, the fundamental difference being that the lines that are established back to the prophet Noah via Qahtan are genealogies of national descent.[24] Pasts which cannot be incorporated are particularized, consigned to the margins of the national and even denied recognition (Alonso 1994: 389).

In some respect contemporary nationalism is a mirror image of Imamic politics of exclusion based on differences of descent, but there is a notable difference. The model of genealogical history held by the

Zaydi *sadah*, who entered Yemen in the ninth century on a peace mission (see chapter 1), is one in which the sacred and the political merge. In the genealogies they constructed, with the Imam, the embodiment of Prophetic authority as the pivotal figure, other genealogies were subordinated. Kin ties were politicized such that only those related to the Prophet by blood qualified for the supreme leadership. The mythic and moral symbolism of blood was tied to the memory of Prophetic revelation. Power spoke *through* blood; blood was a reality with a symbolic function (Foucault 1978: 147). However, beyond the symbolism of blood there were also ways to incorporate the Qahtani Arabs (Qahtaniyyun) into a grand historical scheme whose focal point was the establishment of the Muslim *ummah* or "community." The glorification of primordial essences was secondary to claims to a distinct moral order.[25] By referring to themselves as *ahl al-bayt* ("People of the House") or *awlad al-nabi* (descendants of the Prophet), the Imams employed universalist notions which transcend parochial divisions. Furthermore, notions of supreme leadership formulated by the early Zaydis in the idiom of knowledge and preeminent heredity were commensurable with those held by their tribal supporters.

The new nationalist agenda in post-revolutionary Yemen has been designed to eliminate notions of descent-based social difference, but has not entirely succeeded in doing so. This is demonstrated by the assumption of power by the former General Secretary of the Socialist Party of the People's Democratic Republic of Yemen (the former South Yemen), 'Ali Salim al-Bid, after unification. Al-Bid became Vice President. In the northern part of the country, it was al-Bid's social origin rather than the ideology he represented which stirred people's emotions. Al-Bid is known to be a *sayyid*, and appearance on the political stage evoked memories of the ousted Imamic rulers.[26] He was thus stigmatized even before his political competence was put to the test. His former colleagues, who also joined the government of the unified Yemen, were bewildered by the stereotyping of the *sadah* they encountered. Meanwhile the *sadah* of the north hoped that the appointment of *sadah* to influential political posts would signal the end of their exclusion from high office. At men's afternoon gatherings (*maqyal, madka*), it was said by those who hold anti-'Alid sentiments that the *sadah* "had been thrown out through the door of San'a and returned through the window of Aden." They employed a type of speech that I had often encountered in the 1980s. By the time of unification, concepts of the *sayyid* as "other" had lost none of their injurious potency. But just how "other" was al-Bid? Appeals to the prerogatives of the Prophet's posterity had not been part of his political vocabulary, nor had he publicly referred to his 'Alid identity. His failure of public self-identification and his socialist background caused irritation among those who perceive the *sadah* as politically motivated defendants of the Muslim faith who proudly proclaim their descent credentials. Was this loosening of the link between al-Bid's social location and certain kinds of conduct a kind of "talking back" or was he even lowering himself? Some time after the power sharing agreement had been brokered by North and South, I discussed these dramatic developments with a young man who identified himself as a *qabili* from the northern part of the country (who will be referred to in detail in chapter 10). He commented on al-Bid's assumption of office "they say that 'Ali Salim al-Bid is a *sayyid* but I don't think he is. He is a socialist, and being a *sayyid* means nothing to him. He never talks about that."

This problem of representational incoherence of identity labels emerges from several chapters. The remarks made by the *qabili* raise questions as to whether the *sayyid*'s sense of self is molded by his or her genealogical history, and whether being born a *sayyid* constitutes an ineluctable reality such that forgetting means self-erasure. The book is an inquiry into why, whether, and how "being 'Alid" matters to those who carry the label. Al-Bid's example demonstrates that in specific contexts descent affiliation may be emphasized or obfuscated, and that people who are placed into certain identity categories are expected to conduct themselves in certain ways.

Historicized Memory

The events surrounding al-Bid's appearance on the political stage of San'a set the scene for the analysis of the cultural politics of memory which forms the core of the book. Centering on the post-revolutionary era, it explores how memory negotiates between religious orthodoxy and current social and political agendas. My main interest is in how cultural understandings and social location shape the ways in which subjects remember and then actively reconstruct memories in the process of making sense of their lives and of formulating responses to adversity. The focus on the situated, real-life memory of individual *sadah* has been motivated in part by anthropological studies of Muslim elites, which tend to center on collectivities.[27] Indeed, anthropologists concentrating on phenomena such as the displacement of the madrasas (traditional institutions of higher education) and revolutions have largely failed to engage with the nature of personal memory of elites. The way these events impact upon their subjectivity—the emotional and intellectual absorption of the experience by persons positioned in specific fields of relational power—has been sidelined. Experiences of this kind do not merely have the potential to transform established relationships and patterns of interaction, but they also accentuate lived moral values as well as contentious political agendas. By opening a window into "oral life worlds" (Fischer and Abedi 1990: xix), the book intends to complement such works as the famous biographical collections of Muhammad Zabarah (1940, 1979, 1984),[28] Ahmad al-Shami's autobiography (1984), and Messick's study (1993) of elite reproduction through textual analysis.

The *sadah*'s engagement with their world is shaped by both biographical memory and knowledge which Tulving would characterize as "semantic," part of which has been conveyed by the great 'ulama. Tulving (1972, 1983) distinguishes between different kinds of "memory systems" which foster two different kinds of consciousness. "Semantic memory" which structures transmitted knowledge, especially facts about the past, operates independently from autobiographical, "episodic memory" which affects our subjective sense of ourselves.[29] Bloch (1996: 217, 229) argues that an understanding of these types of memory requires establishing a link between the subject and the past which is the object of memory.[30] Beyond storing information, the socially induced practice of remembering concerns people's engagement in history, which is often a morally purposeful action. This point has been elaborated by Lambek (1996: 235) who, based on research among Malagasy Muslims, conceives of memory as "a form of moral practice." This practice, one might add, is one that allows the past to be "recognized by the present as one of its own concerns" (Benjamin quoted in Yoneyama 1999: 30). Lambek's notion is pertinent to the book's concerns in as far as fundamental moral issues are at stake in the memory processes that are described. As elaborated earlier, for many *sadah* one way of engaging with the past is to rationalize and sanction their practices with reference to the corpus of legal judgments and actions of their learned antecedents (*taqlid ahl al-bayt*, "Tradition of the Prophet's House") which is conceived as moral action.[31] It is this memory that ties 'Alid genealogical history to the founder of Islam. This ideal construction implies that in certain respects this memory is dehistoricized because it links the *sadah* to metaphysical truth, which exists beyond changing events. The past gets sucked up into the present and disconnected from time (see Bloch 1996: 222). As self-styled carriers of the Prophet's message, however, the *sadah* act as subjects and objects of history. They conceive of history as a narrative of the creation of 'Alid kinship that begins with the birth of the Prophet. History is embodied in the activities of their ancestors and thus in kinship relations.[32] For those who have suffered this history and consider it a source of pride, living with it is both inevitable and desirable, however painful it might be.

To a certain degree, the *taqlid* sustains historical consciousness; it helps to moralize and reproduce it. The stored knowledge that makes up the *taqlid* forms part of the "historicized memory" (Nora 1989: 14), which provides a primary reference point for those who engage with it on a daily basis. The Zaydi doctrine permits the believers—with the exception of those who have achieved the degree of *ijtihad* and thus capable of independent reasoning—to adhere to the judgments of the Imam recognized by the Zaydi community or to follow any other past or living *mujtahid* (a person who practices *ijtihad*). As a product of "reified" memory (Gruzinski 1990: 143), the *taqlid* is immutable, but it is continuously expanded and testifies to reoriented perspectives. These processes of reasoning, which are central to the book, lie at the core of the *taqlid* and do not only occur when history takes a detour from its expected tracks.[33]

Whereas much has been written about the formal relationship between *mujtahids* and their followers in countries dominated by the Twelver Shi'a, studies have in general not been undertaken about the motivations of ordinary men and women for allying themselves with particular religious leaders and for considering certain verdicts as meaningful (for an exception, see Torab 2002). Generally, explorations of Shi'i practices have privileged ritual over those moral commitments.[34] Moreover, we derive most of our knowledge about the contemporary Shi'a from the literature on Iran which has shaped our perception of the nature of Shi'i Islam. The unique features of the Arab Shi'is have been overlooked (Nakash 1994: 3)—particularly of those living on the Arabian Peninsula. For the most part, it has not only been ignored that Yemen had the most durable Hashimite rule in the Middle East, but little attention has been given to the subjective engagements with Zaydi-Shi'i principles after the downfall of the state which for centuries had invoked, interpreted, and trespassed them. In comparison to the demonstrative religious style of the Iraqi Shi'is, witnessed by millions of television viewers soon after the fall of the Ba'th regime in April 2003, Zaydi practices are characterized by austerity of ritual.[35] As one man explained, "I went to visit Najaf for the love of Imam 'Ali's beliefs, not because it was my duty to go there." By examining how and why members of the former ruling establishment and their descendants voluntarily submit to the authority of religious texts, the book is concerned to throw light on practices that remain under-researched.

People who most frequently draw on the *taqlid* are elite *sadah* who were brought up during the Imamate and enjoyed a predominantly religious education. The *taqlid* provides one of the key intellectual idioms through which the *sadah* make sense of what it means to be a *sayyid* in the republic, how one should marry, and whether one should indulge in conspicuous consumption; and what one's relation to and within the state should be. Those who have not undergone years of religious instruction, among them *sadah* who have merely enjoyed a basic education at the *maktab* (basic educational institutions attached to the mosques), are not necessarily unfamiliar with the *taqlid*. They gain this knowledge at the daily afternoon gatherings held at the houses of the 'ulama and other learned men or in other social contexts, for example during office conversations or ritual occasions. Most women possess no more than a rudimentary knowledge of religious sources, and cannot draw on as many as the men, but like them they refer to these sources to demonstrate their commitment to their faith and to safeguard their interests (see for example, chapter 7). As Dakhlia (1996) found in her research on Tunisia, the memory of even the illiterate is shaped by written words they never read; words which in their eyes possess greater truth value than the oral tradition. Since the challenge to 'Alid authority has intensified in the past decades, some non-*sadah* refuse to engage with *taqlid*-related arguments (chapter 7).

The diversity and flexibility of the *taqlid* is such that it bears many of the hallmarks of what has been conventionally described as the "Little tradition," thus demonstrating again the obsolete distinction between literate and nonliterate traditions. The Yemeni Imamate was rationalized in terms of the verdicts of eminent religious authorities who have commented upon every aspect of life. With respect to these verdicts, which make up a substantial part of the tradition I examine, it can be seen as a "local" tradition because the overwhelming majority of Zaydis live in the Yemen.[36] However, its leitmotiv is not merely "local" for it comprises sources familiar to all Muslims. Whenever I witnessed debates during which people relied on these verdicts to support their arguments, no one ever demanded that the actual texts be produced for authentication. This may be explained with reference to the authority of the speakers, but it also shows that the epistemological devaluation of memory as a source of knowledge which tends to go along with the spread of printing (Fentress and Wickham 1992: 14) has not yet occurred.

The *taqlid* concerns the forms and acts by which commitments are made rather than a rigid set of rules people submit to without further reasoning. In day-to-day contexts, it informs the practical judgments people make about how to live morally sound lives in the face of different kinds of constraints (Lambek 2000: 315). Elaborating on the function of memory as a vehicle for communicating social commitments and identifications, Lambek (1996: 248) notes that "remembering comprises contextually situated assertions of continuity on the

part of subjects and claims about the significance of past experience."
He treats memory not as a neutral representation of the past but as a
claim or set of claims about the past (1996: 240, 248).[37] As Bartlett
(1932: 256, 309) stressed long ago, memory proceeds by (conscious
or unconscious) selective omission; whatever is remembered inevitably
involves discriminating what is to be suppressed and obliterated. This
process, which involves interpretation, deliberation, and distortion, is
influenced by present-day concerns and predicaments.[38] Thus, when
religious orthodoxy and memory fuse in the idiosyncratic accessing of
knowledge, it is always accompanied by elements of suppression.
"Forgetting" operates as memory's twin in the work of scrutinizing
the past. Like other written texts, the corpus of verdicts, which make
up the *taqlid*, are detachable from their original contexts and can travel
from one period or social context to another. Recitational performance serves to put an authoritative spin on social and political practice.
It constitutes an interpretation of this practice and is therefore always
contested and political.

Among the *sadah* claims and counter-claims about morally appropriate behavior, which are taking place in politically fluctuating circumstances, can be seen as ways of maintaining respectability and of
envisaging alternative social realities. Adopting novel practices based
on reasoning within the idiom of the *taqlid* serves to reappropriate
what might otherwise be an alienating world. In the aftermath of the
revolution major shifts in the interpretation of the *taqlid* constitute a
moral rearmament. By reason of the contestability of the verdicts, an
indeterminacy is built into this identity construct which, as a matter of
fact, often gains stability through the commitment to this tradition. In
the current circumstances, this commitment speaks of a search for a
coherent self that is manifested in a link to the past; the meaning
derived from that past is an ethic workable for the future (Fischer
1986: 196). Situating themselves within this ideological framework
helps the *sadah* to transcend an ambivalent placement between a
scorned past and a future of uncertain fulfillment. It testifies to their
will to maintain integrity despite the categories used by others to pre-
define and constrain them (see chapter 10).

By employing the language of legitimation, the *sadah* seek to ensure
the bearing of the past on the present, often (though by no means
always) discounting contingent social, economic, and political imperatives. Utilizing elements of the *taqlid* is also a statement about what
has happened to them in the recent past and about what they wish to
become. At the crossroads of memory, where an anticipation of future
is realized through recourse to historical knowledge, this knowledge
is articulated and actualized in a present awaiting interpretation.

Memory becomes a form of historical imagination or, as Klein (2000: 128) has put it, memory reworks history's boundaries.

*   *   *

Before the themes discussed are developed in a number of core chapters, chapter 1 presents data on doctrinal issues and the House of the Prophet. Against the backdrop of conflicting interests pursued by monarchic and republican forces in the Middle East, chapter 2 traces recent historical events which led to the decline of the Yemeni religious elites. Chapter 3 rounds up part I, which provides a framework for understanding the living conditions of present-day *sadah*. This chapter deals with the internal organization of the Prophet's House in the Yemen, its genealogical divisions, and the significance of patronymics and social etiquette. Portraits of eminent families over time are followed by descriptions of the social milieu in which children's sentiments and sensibilities were shaped in pre-revolutionary Yemen (chapter 4).[39] Some people's life stories are traced through different chapters. Chapters 3 to 5 also speak to issues that have been raised in studies about elites outside the Middle East. For example, in her study of the French bourgeoisie, Le Wita (1994) notes the importance of genealogies, the remarkable social diversity of a group whose members nevertheless feel they belong to an immutable, hereditary body, and the notion that one is born a bourgeois but also learns to become one.

Part II concentrates on how the *sadah* acquire and recall memories through their membership of the learned elite, thus focusing on issues of transmission which were central to Halbwachs's (1992: 54–83) pioneering work on memory.[40] Chapter 4, as well as several subsequent chapters, highlights the ways in which cultural and historical imagery frames the memories of personal experience. Drawing on biographical data and the basic tenets of Zaydi theory, chapters 4 and 5 explore how knowledge considered to be authoritative is gained, validated, and legitimized. The intimate relationship between kinship and knowledge which these chapters aim to explain is fundamental to key Shi'i notions of authority. Through the performance of religious knowledge moral relationships, as well as persons, are created who are capable of mastering their environment. In the Imamate, this mastery was closely related to the "textual domination" so well described in Messick's *The Calligraphic State* (1993).

Because in the Imamate knowledge was a rationale of the state and a constituent element of moral and political authority, it must certainly be conceived as a material factor analogous to land or kinship connections (Lambek 1997a: 134). My analysis of the constituent features of traditional elite status focuses on the moral and political economy of knowledge rather than land.[41] Even though the old nobility used knowledge for material advantage, neither they nor low status individuals considered landownership (chapter 3) to be the prime criterion of their status.[42]

Several chapters (especially part III) examine how historical knowledge in the form of the *taqlid* informs the organization of the *sadah*'s everyday activities in a political context generally hostile to inherited status based on descent. The principles by which the marriages of 'Alid women are informed are dealt with in chapters 6 and 7. Among Zaydi 'ulama, these principles have been an issue of dispute for centuries. Scholarly debates over the management of reproduction which identified the Imam's subjects as bearers of certain substances and statuses, served to classify them as different kinds of persons in conformity with the criteria that determined legitimate rule. In the Imamate unions of 'Alid women and non-'Alid men were discouraged because in accordance with patrilineal rules, they neutralize the effects of descent and blood. Thus, not only was the ordering of sexual relations between these social categories fundamental to Imamate rule, but it was also structured in terms of gender. After 1962, the *sadah* have contracted an increasing number of marriages with the Qahtani elite. These unions are of symbolic significance for they are interpreted as a renunciation of claims to rule. They too expose tensions between what is considered to be Islamic and Yemeni tradition, and what is considered to be a humiliation and the pursuit of profit.

Like these chapters, chapter 8 demonstrates how legal judgments are used to explain behavior which the *sadah*'s forebears would have reckoned to be incompatible with their status. This chapter shows that during the Imamate, engagement in trade and commerce was considered an improper activity for religious scholars who were expected to embark on a career based on knowledge of the scriptures. Nowadays *sadah*, who are newcomers to the world of commerce, explain their decision to engage in economic activities with reference to those ancestors who were merchants. Chapter 9 draws attention to the moral dilemma faced by the affluent among the old San'ani elite who, like others who choose to adhere to the tenets of their faith, cannot reconcile their desire to adhere to religious principles of modesty with the requirements of a consumer society. Consumption, however, helps maintaining their elevated social position in post-revolutionary Yemen.

Part IV looks at social and intrapsychic engagement with vocabularies of difference. Chapter 10 analyzes depictions of the *sadah* as adherents of Imamate rule and challenges to the authenticity of their genealogies. Negative stereotypes, along with alternative interpretations of history and religious ideologies, offer a basis for both contesting 'Alid supremacy and for reproducing existing power relations. These investments in rhetorical dislocation pick up themes alluded to in Part III. Chapter 11 traces forms of linguistic injury which stem from stereotyping, exposing the *sadah*'s volatile place in the republic. Bloch (1996: 230) argues that in certain respects people react to major upheavals such as revolutions or conquests, in terms of the way they represent themselves to themselves in history. During my fieldwork references to the history of 'Alid persecution since the death of the Prophet, focusing on Imam al-Husayn's martyrdom as the quintessential emblem of 'Alid vulnerability, were not infrequently made. The memory of the revolution has become integrated with this history. The fusion of memories of the history of the *ahl al-bayt* and of persecution in the 1960s may point to a partial mental escape due to (self-)censoring mechanisms. Memories of this history, which of course were cultivated by the Imamic state, have acquired different meanings and functions in the republican era.

Other memories of the events of the 1960s (chapters 2 and 11) highlighted the misery suffered even by members of the 'Alid nobility, revealing that master narratives of 'Alid rule with their utopian rhetoric of a just social order did not prevent profound moral scrutiny and desire for regime change. In many cases, however, it was through their silence about these events that people were telling a story about their lives. The politics of memory in contemporary Yemen has inspired self-styled declarations about heroic commitment to the revolution, but has discouraged the victims from speaking out—even in the company of their children. This, as David Cohen argues, is not the consequence of forgetting, but of continuous acts of control in both public and private places.[43] They produce a self-awareness of surveillance whereby silence points to embodied and historically situated knowledge which remains tacit. Silence becomes a formative event in these uncertain and provisional projects of self-making; it also throws a new and different light on those known as principal producers and interpreters of knowledge.

With Tulving's theory of a two-tiers memory system in mind, chapter 11 suggests that autobiographical memory may recover "semantic" memory and vice versa. Autobiographical memory also allows us to infer that differences between the memory regimes established by the *sadah* and their adversaries are never absolute, and that on some occasions they interrogate the past in similar ways. It is demonstrated that interpretations of shared history are subject to criteria such as relative rank, but internal hierarchies are by no means the only factor that accounts for the idiosyncratic selection of memory. Moral concerns that may or may not be coupled with strategic interests also provide motivational force. Focusing on how individual actors operate along multiple axes of power and difference, chapter 11 documents the tensions between personal experience and normative/stereotypical categorizations of *sayyid*. They result in often contradictory processes of articulating knowledge about the past whereby an acquired sense of pride in one's ascendancy conflicts with the official discouragement of invoking a past unworthy of remembering in terms other than debasement.[44] These processes expose alternative moments of revisiting and departing from trauma. According to Caruth (1991: 10), "trauma is a repeated suffering of the event, but it is also a continual leaving of its site." Unsurprisingly, some *sadah* feel that they no longer ought to pronounce names the articulation of which enforces remembering that has inevitable political implications. It is argued that the erasure or suppression of historical memory aims at neutralizing their sense of marginality and non-recognition as full citizens of the republic. Endeavors at initiating self-displacement are conceived of as enabling survival in a new political setting where, in certain respects, the traditional mark of honor has become a stigma. The present circumstances, which render their lives precarious, encourage exploration of the ways in which power and morality operate in the formation of memory regimes. These considerations are vital to an understanding of memory work that focuses not only on *what* is remembered but *how* (Stoler and Strassler 2000: 9).

A core theme of the book is how through recourse to the *taqlid* as a form of practice the past is evaluated and interpreted. Chapter 12 revisits questions regarding the legitimacy of the Imamate by showing how contemporary 'Alid scholars place the memories of the rulers' instigation of violence and fraternal discord in their inextricable relation to traditions of Zaydi thought centering on social justice. This historical perspective enables personal memories of victimization during the revolution to fuse with those of their antecedents' sometimes ruthless pursuit of power. This chapter also contemplates how as a result of the upsurge of radical anti-Shi'i thought, Zaydi practices have become mnemonic sites of legitimation which nonetheless do not withhold allegiance to the republic.

* * *

My data derive mainly from high-ranking *sadah* most of whom held positions in the governments of the Imams. This was not a deliberate choice, but the coincidental outcome of my earliest encounters with Yemenis. I had already met the family of a former governor before I went to the field; soon after my arrival in San'a, where most data were collected, I became acquainted with men and women of the old, well-established families who introduced me to their friends and relations. Furthermore, my main interest was the link between descent, knowledge, and political authority.

In spite of the sensibilities deriving from these people's most recent historical experience and the social etiquette commonly pertaining to elites, I was able to share much of their everyday life. As Zur (1997: 75) reminds us, "private memories may be open secrets but they are anything but public facts." Kramer (1991: 8), writing about biographical data about Middle Eastern personalities, remarked that "the intimate sources for the lives of rulers and leaders remain even more inaccessible. There is nothing Middle Eastern about the desire of the powerful to erect defensive barriers between their personas and themselves, but the powerful of the Middle East sometimes have erected barriers so high as to be insurmountable, even decades later."[45]

People were certainly aware of the strategic advantages of withholding information about their personal lives, which others might interpret as weakness.[46] Yet I found the task of maintaining discretion while making these people's lives transparent (Lambek 1997b: 37) more daunting than dealing with the "defensive barriers." During the process of writing, I felt obliged to impose self-censorship and to conceal their identities more than I would have liked to; more concrete data would have added strength to their profiles and drawn out continuities between aspects of life stories taken up in different chapters. I have used pseudonyms whenever this seemed appropriate—a necessity which is itself an indication of the politics of memory in contemporary Yemen. Some biographical details have been changed in order to render identification more difficult. These changes have been made with careful consideration of the overall analysis.

Most members of the old nobility welcomed my interest in them as well as in Zaydi Islam—at an historical juncture where others sought to dismantle the authoritativeness of the rulings of Zaydi 'ulama, of their genealogies, and the history of the Imams. The value they attach to knowledge has almost always overridden considerations of gender. Doubtlessly others had reservations about a study that focused attention upon them during a difficult and painful period of accommodation. Despite reassuring me that "knowledge is an ocean without boundaries (*al-'ilm bahrun la sahil la*)," and that no knowledge is unworthy of investigation (*ma-fish 'ayb fi-'l-'ilm abadan*), some of those who so generously contributed to the study might receive a work which goes beyond the data contained in biographical dictionaries (sing. *tarjamah*) with some skepticism. I can only reiterate the plea recently made by the historian Zayd al-Wazir in his study of a Yemeni religious movement, where he asks for a dispassionate analysis of matters of academic interest without personalizing the issues involved (2000: 27–8). To those who would have preferred not to remember I ask forgiveness for opening old wounds.

Men figure more prominently than women in this study. Kristeva (1989) has rightly argued that collective identities are often represented as masculine. I am aware of this tendency and am prepared to accept the criticism. Although I was conscious of "differences within"— both in terms of relative wealth and gendered world views—women concealed their lives much more than men. It is likely that their sense of vulnerability as dependants (there were few professionals) and of female propriety served to censor memory, thus shaping narratives of self. Perhaps it is the untold narratives which partly define women.[47] Some women share their problems with their close friends or kin; others feel that they ought to cope on their own. While I was living in Falih where similar etiquette prevails, a newly married bride, unable to tolerate the demands made on her in her new household, collapsed in tears. All the women present pretended not to notice and none attempted to comfort her. A few minutes later the young woman collected herself and the matter was forgotten; nobody brought it up again. Few women expressed curiosity about historical and doctrinal matters. "You have to talk to the men about *'ilm* (knowledge)" they would say. However, many took a keen interest in daily events as they unfolded in Yemen and the Middle East at large since unification of the two Yemens and the Gulf crisis in the early 1990s. The availability of satellite television has certainly broadened the scope for discussion at their daily afternoon gatherings, the *tafritah*.

Many students of Yemeni society have written about the advantage of gaining information at the daily male and female gatherings. To some extent this is true because researchers have access to a considerable number of people without prior appointments. For example, Carapico (1998a: xi) maintains that "for research purposes a qat-chew [of the mildly stimulating leaves of a scrub] amounts to a 'focus group,' for people speak openly and the guest is entitled to introduce a discussion topic, listen in on other business, and to take discreet notes." However, at least among the urban higher status groups, conversations which involved all participants were highly selective and clearly took my presence into consideration.

A final word is needed about the omission of data on the *qudah* ("families of judges") who belonged to the ruling elite of the Imamate.

I could not have dealt adequately with them and the *sadah* within the already limited space. Let me stress that many of the cultural representations discussed here are shared by members of the old elite and indeed other Yemenis, but the *sadah* have a particular relation to them.

During the autumn of 2004 and spring of 2005, a dispute between a young Zaydi scholar and the government developed into hostilities leaving hundreds dead. Subsequent commentary on the events in the press and by ordinary Yemenis has touched upon several issues discussed in this book, and has rekindled debates about the place of the *sadah* in contemporary Yemeni society. As these occurred after this book went to press, I have not covered them. I do hope that by the time of its publication, a political solution to the problem has been found, and that the effort of a great number of *sadah* toward building a prosperous democracy will not be forgotten.

Part I

Framings

Chapter 1

The House of the Prophet

Question the people of Remembrance, if it should be that you do not know.
Surah 21:7

The Doctrine of the Zaydi Shi'a

The *sadah* who are the subject of this book belong to the Shi'a school of Islam. In the first and early second Islamic centuries, the Shi'a was characterized by allegiance to the Prophet's Family and repudiation of the claim put forward by the Umayyads to be the rightful rulers of the community (Daou 1996: 11). Shi'i teachings focus on 'Ali, the Prophet's paternal cousin who, by reason of his blood relationship with him and the outstanding services he rendered to the cause of Islam, is perceived as his legitimate successor. According to Shi'i interpretations of early Islamic history, 'Ali and many of his kin died as martyrs whilst defending their just cause against the overwhelming force of their opponents who unlawfully deprived the Prophet's descendants of their legitimate rights. Shi'a Islam centers on the personal allegiance to the Prophet, claiming that the Qur'an can only be fully understood by him and his Family (Schubel 1996: 187).

The term Shi'a, "the party of 'Ali" (*shi'at 'Ali*), represents the movement's claim of a privileged position for the Prophet's Family (*ahl al-bayt*)[1] as political and religious leaders of the Muslim community. "Apart from the fact that the term carried with it a great deal of honor, it was not the honorific side of it which brought about controversy, but rather the fact that it represented . . . a major component in the fight for power and leadership in Islam and a highly important element in the search for the legitimacy of rule" (Sharon 1986: 169). Some traditions declare support and love for the *ahl al-bayt* a religious duty and animosity toward them a sin (ibid.: 172). The *ahl al-bayt* are prohibited from receiving or handling alms (zakah), which is considered unclean. Instead they are entitled to portions of the *khums*, the fifth of war booty, unappropriated land, and mineral resources which are not privately owned (for example, gold, oil),[2] and of the *fay'*, property acquired by the Muslims without war effort. When upon the Prophet's death Abu Bakr, a man who belonged to his tribe (Quraysh) but not to his immediate kin, established the caliphate, the House of the Prophet lost its privileges and patrimony.[3]

Shi'i claims to 'Ali's rightful succession to the leadership are based on Qur'anic verses and statements attributed to the Prophet (hadith). One focuses on the event of Ghadir Khumm, a place where the Prophet stopped on his return from his last pilgrimage to Mecca and made an announcement to the pilgrims. There he uttered "He of whom I am the lord, of him 'Ali is also the lord (*man kuntu mawlahu fa-'Ali mawlahu*)." This hadith is interpreted as 'Ali's investiture as the Prophet's successor.[4] Other factors referred to by the Shi'a are 'Ali's closeness to Muhammad, his marriage to the Prophet's daughter Fatima, his profound knowledge of Islam, and his exceptional merits in its cause. 'Ali was the second person to acknowledge Muhammad's divine mission after his wife Khadija. He declared that the *ahl al-bayt* had the right to the leadership of the Muslim community as long as there remained a single one of them who knew the Qur'an and Sunna and was a true believer (Madelung 1996: 420). Claims to the leadership were thus based on heredity, knowledge, and piety.

The Shi'a emerged during the first civil war which broke out upon the death of the Caliph Uthman, the Prophet's third successor. According to Madelung (1997: 1), "no event in history has divided Islam more profoundly and durably than the succession to Muhammad." 'Ali claimed the caliphate but was defeated by Mu'awiyah b. Abi Sufyan, the founder of the Umayyad dynasty in Syria (661–750). Thereafter his followers recognized his eldest son al-Hasan as his successor. Al-Hasan, the second Imam after his father, abdicated in favor of Mu'awiyah, but the Shi'is continued to regard him as the rightful leader. After al-Hasan's death, they encouraged his younger brother al-Husayn (d. 680) to oppose Mu'awiyah and to reinstate the *ahl al-bayt* to their rightful office. They considered Mu'awiyah to be an oppressor. Al-Husayn, the third Imam, declined to rise against Mu'awiyah, remaining loyal to his brother's treaty with him according to which he would not rebel against him. When however Mu'awiyah was succeeded by his son Yazid, al-Husayn and other Shi'i notables refused to pay allegiance to him. After offering their support to al-Husayn, the tribal leaders of Kufa (southern Iraq) abandoned him after intimidation by Yazid's governor. Al-Husayn neverthe-less led the revolt against Yazid's army. He and over twenty relatives were killed at Karbala on 10 Muharram 61/10 October 680. Since then the images of martyrdom and betrayal, and of the triumph of oppression and injustice over legitimate leadership have figured prominently in Shi'i thought and ritual. To Shi'i believers, muharram rituals—which are not performed in the Yemen—convey the message that Shi'is *must* remember.

Following al-Husayn's death in battle, the Shi'a had no leadership from among the progeny of 'Ali and Fatima. Al-Husayn's surviving son, 'Ali Zayn al-'Abidin, became the fourth Imam. The first schism within the Shi'a occurred after his death. One branch recognized al-Husayn's grandson Muhammad b. 'Ali (known as al-Baqir), the fifth Imam who systematized Shi'i law; the other his half-brother Zayd. Zayd's rebellion failed and he was killed in Kufa in 740 (Ende 1984: 88; Madelung 1996: 423). In spite of his efforts on behalf of the *ahl al-bayt*, those who later became known as the Twelver-Shi'is did not back him because he was not designated as Imam by either his father or brother and because he refused to declare illegitimate the rule of Abu Bakr and 'Umar (Haykel 1997: 193, 204). Al-Baqir's son Ja'far al-Sadiq (d. 765) then became the sixth Imam. The Imami Shi'a, which later developed into the Twelver-Shi'a, began to emerge as a distinct Shi'i group in the early 'Abbasid period during Ja'far's life-time, but later they became synonymous (Daou 1996: 12; Kohlberg 1976a). Ja'far elaborated the teachings of his father. Henceforth the Shi'i school of jurisprudence has been referred to as the Ja'fariyyah. Ja'far prohibited uprisings against the authorities; his preaching cen-tered on the arrival of the mahdi ("the rightly guided one," redeemer) in the distant future. Political quietism replaced the earlier phase of the Imams' revolutionary activities. The succession of Imams continued until the twelfth, Muhammad al-Mahdi, who disappeared in 874. According to Twelver-Shi'i legend he went into occultation (*ghaybah*) and is expected to restore just rule (Arjomand 1988: 2–3).

The dissent which occurred during Zayd's rebellion led to the gene-sis of a separate Shi'i branch known as the Zaydiyyah, which gained most of its following at the periphery of the Muslim heartlands, notably at the Caspian Sea and the Yemen. Nonetheless, Zaydi and Imami Shi'is share central ideas about the nature of legitimate rule. Notions of kin-ship with the Prophet, coupled with the idea of the Imams' divine right to rule and their supreme knowledge (*'ilm*), became the hallmark of the Shi'i theory of the Imamate. One of the most important features of Shi'i notions of leadership is its legitimation through *'ilm*. Regarding Imami ideas about the twelve Imams, Arjomand (1988: 3) writes that "this partial derivation of the authority of the Imams from *'ilm* had no counterpart in the Sunni theory of Caliphate and constituted another distinctive feature of the Shi'ite theory of authority."

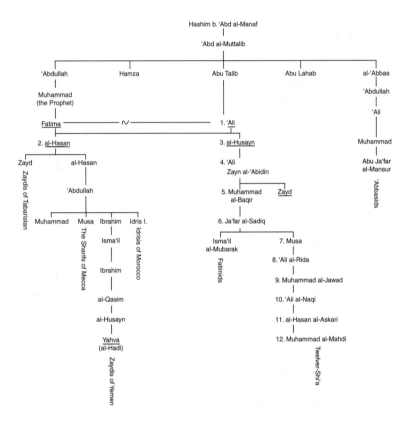

Figure 1.1   Genealogy of the Hashimites (The numbers refer to the Imams of the Twelver-Shi'a)

Unlike the Imamis, the Zaydis do not recognize a hereditary line of Imams but are prepared to pay allegiance to any qualified descendant of al-Husayn and al-Hasan.[5] In their eyes, the first Imam after al-Husayn was Zayd b. 'Ali.[6] After al-Husayn's death, the legality of the Imamate was contingent upon a formal summons (*da'wah*) and armed rising (*khuruj*) against oppressive rulers. The Imam must be recognized by the 'ulama and other notables who pay homage (*bay'ah*) to him (Madelung 1996: 423). Like God's unity (*tawhid*) and justice (*'adl*), the Imamate is a religious principle (*asl din*) in Zaydi thinking (Haykel 1997: 192). Although in theory there has to be at all times at least one 'Alid who fulfills the qualifications for the Imamate, the Zaydis are not obliged to support him unless he proclaims the *da'wah* and calls for "rising" against unrighteous rulers. The period of an interregnum is not fixed. "Thus there could be long periods without a legitimate Zaydi imam" (Madelung 1988b: 86; personal communication).

The condition that any pretender to the Imamate must be a recognized descendant of the Prophet through 'Ali and Fatima is a distinguishing doctrinal feature of classical Zaydi Islam. Al-Qasim b. Ibrahim al-Rassi (d. 860), who systematized the Zaydi teachings, argued that an Imam becomes recognizable through ordaining the good and prohibiting the reprehensible (*al-amr bi-'l ma'ruf wa-'l-nahy 'an al-munkar*). He must be the most excellent of the Prophet's descendants (Madelung 1965: 142). Zaydi doctrine became rooted in the Yemen through al-Qasim's grandson Yahya b. al-Husayn (d. 911), who took the title al-Hadi ila'l-haqq (Guide to Truth), which was carried by the 'Abbasid caliph Musa (Madelung 1965: 4, 86–94; Sourdel 1971).[7] During the history of the Imamate his teachings, known as the Hadawiyyah, remained predominant. According to the Hadawiyyah, the eligibility of claimants to the Imamate is limited to the descendants of 'Ali and Fatima.[8]

Zaydi doctrine lists fourteen qualities a candidate for the Imamate must meet, of which the following are the most important: (1) membership of the Prophet's House by descent through 'Ali and Fatima; (2) a high proficiency in scholarship and the ability to pronounce an independent judgment in legal cases (*ijtihad*); (3) the ability to resort to the sword; (4) being male and of free descent; (5) being just and generous; (6) being willing to oppose tyranny and injustice (Sharh al-azhar Vol. 4: 518–24). The Imam's prime task is the implementation of Islamic law (i.e., leading the communal prayer, collecting taxes, waging war, and carrying out legal punishments). Specific legal principles (*ahkam makhsusah*) are issued by the ruling Imam and come to characterize his Imamate. (These are also referred to as the Imam's *ikhtiyarat*, debatable legal pronouncements, which were not necessarily followed by judges and ordinary citizens.) The Imam may also appoint his governors (Sharh al-azhar Vol. 4: 518, 529). The Imam's office as the head of state was not separated from his role as the supreme guardian of the faithful.

Although theoretically "ordaining the good and forbidding the evil" is incumbent upon every believer, these duties were considered to fall particularly heavily on the Imam and the scholars from among whom he was chosen.[9] As one Zaydi scholar put it to me, "in the Zaydiyyah the most important issues are *al-amr bi-'l ma'ruf* and the *khuruj*. The one closest to God practices *al-amr bi-'l ma'ruf* and rises against the oppressor (*zalim*)." As noted by Madelung (1988b: 87), with their strong revolutionary motivation, the Zaydis "posed a more immediate threat to the established caliphate and the peace of the

community than the quietist Imamiyya." With respect to the principle
of *khuruj*, which was not mentioned by al-Qasim b. Ibrahim, it is crucial that in theory the Zaydis must rise against anyone who trespasses
the law, including the ruling Imam. Although the Imam possesses
God's mandate to rule, his authority is subject to critical scrutiny by
both the 'ulama and believers. Such a notion is predicated on the
assumption that with the exception of 'Ali, Fatima, and their sons, the
Imams are not infallible and sinless (*ma'sum*) (Kohlberg 1976b: 98).
The Imamate can be relinquished if the Imam lacks any of the
required qualifications. During the Imamate, these were often questioned by the Imams' rivals, and their offences against the law provided fertile ground for challenges.

The Rise of a Muslim Nobility: The Banu Hashim

According to Madelung (1997: 16), the Qur'an accorded the *ahl
al-bayt* an elevated position above the rest of the faithful. Their special
status derived from their kinship with the Prophet, from which
resulted the attribute of purity which according to Shi'i interpretations is highlighted in the holy scriptures and the regulations of
the Shi'i and most Sunni schools of law (Madelung 1992: 24–5). In
Arabic speaking cultures, the broadest generic label for the Prophet's
kin is Banu Hashim ("clan" of Hashim). Named after the Prophet's
great-grandfather, Hashim b. 'Abd Manaf, the Hashim clan belonged
to the Quraysh, the Prophet's tribe. The head of the Hashim clan, Abi
Talib, took care of his orphaned nephew Muhammad, the future
Prophet (Noth 1987: 12).

In subsequent centuries, the name Hashim was adopted by several
movements and ruling powers. The movement against the Umayyads,
which gained momentum about 700, was called al-Hashimiyyah.
In the eighth/ninth century, the members and supporters of the
'Abbasid house which overthrew the Umayyads were referred to as
Hashimiyyah. The 'Abbasids based their claim to power on their
blood tie with the Prophet through al-'Abbas, the Prophet's uncle.
One of the slogans of the Hashimiyyah movement was that those
scions of the Prophet's Family upon whom the Muslim community
had agreed should rule (Nagel 1987: 103). In his famous poetry
"al-Hashimiyyat," the Shi'i poet al-Kumayt (680–743) expressed
reverence toward all descendants of Hashim rather than just 'Ali and
his descendants.[10] The Prophet's kin were thus included into the
category of the Banu Hashim (Madelung 1992: 6–7).

In several Muslim cultures, most notably Morocco and Egypt,
the descendants of the Prophet are referred to as *ashraf* or *shurafa'*

(sing. *sharif*, noble, honorable).[11] Both before and after the rise of Islam, persons from highly esteemed tribes or families were referred to as *ashraf*. The *sharif*, a person of noble blood, was contrasted with the *da'if* (pl. *du'afa'*) or "weakling," that is, of humble descent.[12] In famous genealogical works that appeared in the ninth/tenth century, the term *sharif* was standardized as denoting persons of eminent status. These were either descendants of the Prophet or of ancestors of the southern and northern Arabs, Qahtan and 'Adnan. Later the title *sharif* became reserved for the 'Alids. In the eastern provinces of the Yemen the descendants of the Prophet are referred to as *ashraf*.[13]

The etymology of the term *sayyid* (lord), which was already used in ancient South Arabian, is similar to that of *sharif*. *Sayyid* means the master in contrast to the slave, and the husband in relation to his wife (Qur'an 12:25).[14] In the Qur'an (for example, 33:67) the heads of tribes or the Prophet Yahya (John) (3:39) were also referred to by this term. It is likely that the term *sayyid* was used as a title for the 'Alids at about the same time as *sharif*. In the hadith collections of al-Bukhari the Prophet referred to his grandson al-Hasan as a *sayyid*. Other sources report that the Prophet called 'Ali "the sayyid of the Arabs and the sayyid of the Muslims." From the fourteenth century onwards, the term was used by slaves to address their masters, and it was applied to a variety of holy men, Sufi masters and leaders of shrines (van Arendonk 1996: 329–33).

In tenth-century Yemeni historiography a *sayyid* or *sharif* was someone who would nowadays be called a shaykh (tribal leader) (Dresch 1989: 169; 191 n. 10). In the Yemen the title *sayyid* has been used exclusively for a descendant of the Prophet after the Zaydi state became more stable a couple of centuries later. The Prophet's female descendants are referred to as *shara'if* (sing. *sharifah*, honorable).[15] In contemporary (northern) Yemen the terms *sayyid* and *hashimi* (a descendant of Hashim) are used interchangeably, but those 'ulama who assume that the descendants of 'Ali and Fatima ('Alawiyyun-Fatimiyyun) take precedence over other members of the House reserve the title *sayyid* for them and refer to the others as *hashimiyyun* (Hashimites).[16] The majority of contemporary Yemeni *sadah* consider themselves to be descendants of al-Hasan (Hasaniyyun). Few trace their descent from al-Husayn.

The Dispersion of Muhammad's Posterity and the Origins of the Zaydi Imamate

When asked about 'Alid immigration in the Yemen, Sayyid Muhammad b. Muhammad al-Mansur, one of the leading Zaydi scholars, began by recapitulating the harassment the Prophet's kin suffered in the early caliphates.

> When the Umayyads came to power [in the seventh century], they persecuted and oppressed the Hashimites. During the Umayyad and 'Abbasid period they were imprisoned and killed; many were forced to flee. The 'Abbasids were Hashimites but there was competition between them and the descendants of 'Ali. They dispersed all over the Islamic world. Idris b. 'Abdullah b. al-Hasan b. 'Ali went to Morocco where he married. All Moroccan Hashimites are descended from him. They went to Turkey, Yemen, Azerbaijan, Iran and Afghanistan. When the period of oppression was over, in all those places they were well known as 'Ali's descendants (*awlad 'Ali*). God promised Muhammad that his line would continue. 'Ali used to say: "Whoever survived the sword will have enough children to multiply and survive."

The Prophet sent 'Ali to the Yemen as his representative in 631 (Madelung 1997: 18).[17] Until the arrival of the Zaydis in the Yemen two centuries later, Islam was sustained by the Umayyad and Abbasid dynasties.[18] In eighth-century Arabia, fortune was not on the side of the 'Alids. A descendant of al-Hasan, Muhammad al-Nafs al-Zakiyyah, met his death at the hands of the 'Abbasid caliph al-Mansur in 762. In the Yemen, however, the 'Alid cause was promoted by a grandson of Ja'far al-Sadiq, Ibrahim, and the 'Alids were supported by al-Shafi'i, founder of legal school named after him, which is still predominant in the southern and western parts of the country.

Prior to the Zaydi penetration of the Yemen, there were brief episodes of Zaydi rebellious activities in Khurasan (Iran) in 834. In Iran Zaydi ideas were spread by supporters of al-Qasim b. Ibrahim. The first Zaydi state was established in 864 in Tabaristan on the southern coast of the Caspian Sea where Yahya b. al-Husayn was considered the chief religious authority even while other 'Alids held the leadership. He was forced to leave Tabaristan after having been addressed as Imam by his followers (Madelung 1988b: 88). A few years later Yemeni tribes invited him to the Sa'dah region during a period of disarray, requesting his mediation.

Sayyid al-Mansur went on to explain:

> In the ninth century, the governors of the Yemen could no longer control the country. The heads of tribes decided that an external mediator be called to settle their disputes. There were many problems among the tribes which the 'Abbasid rulers could not solve. Yahya b. al-Husayn was asked to come to the Yemen to impose order. In the beginning, the people cooperated, but later he encountered problems. He disapproved of his soldiers stealing peaches from people's fields; the farmers were

complaining about that and when he heard this he decided that he did not want to be the head of such troops. He only stayed for two to three years and returned to Madina [where he was brought up]. When the situation in the Yemen deteriorated, he was promised support if he were to return. However, he was not accepted by all and fighting broke out between his soldiers and those who opposed him.[19] He also asked the leaders of other regions to support his *jihad*. He thought if he had 3,000 supporters he might even conquer Baghdad. However, few people were willing to fight for him because he had set certain conditions for the soldiers. They must not take booty and should fight only in order to spread the word of God and to abolish injustice. All those who fought for him did so without expecting a reward. In the end he was poisoned. Thereafter his son Muhammad became the leader but like his father, he encountered many problems. He was succeeded by his brother Ahmad.

The Hashimites' fortunes went up and down. This is what it was like for them. Before al-Hadi's arrival there were Hashimites in the Yemen. They were descendants of 'Abbas b. 'Ali. They did not come forward because al-Hadi was in a much better position than them for he had been asked to intervene; people had chosen him, he was more knowledgeable, and he had a better reputation. It was due to al-Hadi's efforts that the knowledge of Islam became more widely distributed. Al-Hadi had followers from Tabaristan, Egypt, and the Hijaz who were not all Hashimites. Seventy came from Tabaristan, they were called al-Tabariyyin. There were about two hundred and fifty people from elsewhere.

Yahya b. al-Husayn laid the doctrinal, legal and military foundations for the Imamate. Alluding to the Shi'i notion that in times of evil God will send someone of the *ahl al-bayt* to restore justice, al-Hadi was quoted as saying of his endeavors in the Yemen that "Islam was dead and revived through the *ahl al-bayt*" (Landau-Tasseron 1990: 256). He extended his rule from Sa'dah to Najran in the north and, temporarily, to San'a in the south, but he faced difficulties in winning over the important tribes (Madelung 1987: 176). The Imamate consolidated itself to a greater degree by the late twelfth century, a development which was aided by the immigration of a great number of Zaydis from the Hijaz, Iraq, and Iran (Gochenour 1984: 149, 209). The *ashraf* of Mecca, who had connections with the Zaydi Imams and controlled most of the Hijaz until 1925, were Zaydis until the late fourteenth/early fifteenth century (Mortel 1987).

Notes on the Hijrah and the City of San'a

During the formative period of the Zaydi state in the Yemen, the followers of the first Imam al-Hadi entered Yemeni society through the institution of the hijrah, a formal relationship of protection granted by a tribe. The term hijrah refers to a formal agreement between the tribes and certain individuals or social groups. In the case of the *sadah* and other social categories such as the *qudah*, the granting of a status of protection by the tribes is an expression of respect for their descent and the revealed knowledge they represent. The meaning attached to this type of agreement is different from that between the tribes and the lower service categories (*mazayinah* or *du'afa'*, see later). The hijrahs were established either by invitation of a particular tribe or by request of the ruling Imam. Some of these agreements still have at least formal validity. In accordance with the hijrah agreement, those protected are exempted from the payment of collective bloodmoney and the obligation to fight. In theory, any infringement of their integrity (including their families and estates) has to be defended by the hijrah-granting tribe.

For the rural *sadah*, the hijrah provided the principal material base. Initially, the *sadah* were given arable land and surrounding watershed lands (Gochenour 1984: 166). According to Shaykh 'Abdullah al-Ahmar, the *sadah* were hijrah because of their descent (*nasab*), piety (*din*), and knowledge (*'ilm*).[20] They were given gifts (*sadaqah*) such as land because the *qaba'il* wanted them to live in dignity and expected baraka ("blessing") in return.[21] The Shaykh's grandfather had given his best land to the *sadah* seeking to ensure his children's welfare and the prosperity of his crops. The *sadah* were also rewarded for their mediation between conflicting parties, the provision of literary services such as teaching, reading and composing documents, and the writing of amulets. Some were employed by the Imams as governors, tax collectors, and judges.

The northern town of Sa'dah, where the Imamate began, was the first Zaydi hijrah in the Yemen. In conversations, Zaydi 'ulama explained the origin of the hijrah in terms of their need to safeguard their religion when the state was weak and unable to enforce shari'ah law. From the medieval period onwards the course of history was not determined by the urban centers but by those enclaves in the midst of tribal territory (Dresch 1989: 172; 1991: 257). Contenders for the Imamate often emerged from the hijrahs which were either hotbeds of dissent or centers of state power. Indeed, their power correlated with the strength of the Imamate: when the Imamate was weak or there was no Imam at all, the *sadah* had hardly any influence beyond the boundaries of the hijrah.

The nobility was by no means confined to the cities. Their situation differed from that of other elites such as the Ottoman urban notables who resided mainly in the cities and were influential in the rural hinterland because of their position in the city (Hourani 1968: 45). Urban and rural nobles were linked through kinship and shared political interests. There was a degree of cultural homogeneity between the towns and the hijrahs, but mobility was restricted because of the arduousness of the journeys. Urban nobles had the highest regard for the scholarship and humble lifestyles of their rural peers, and they took shelter there during periods of conflict with the government. The hijrahs were important centers of learning. The sons of the urban learned families would spend several years in one or two of the reputable enclaves such as Kawkaban and Shaharah. Once the Imamate was overthrown, Zaydi learning declined in the hijrahs and new, Sunni-oriented institutions have been established (see chapter 12). Not since the civil war (1962–70), when some San'ani families which had been linked with the former regime escaped persecution by fleeing to their hijrahs, have they had political significance (see chapter 2).

The *sadah* experienced the recent political upheaval most tangibly in San'a, from where the 1962 revolution began. The bulk of the research on which this book is based was carried out in San'a. Today the city is the capital of the unified Republic of Yemen with 1.7 million inhabitants of an overall population of 19 million. Pre-revolutionary twentieth-century San'a was not rigidly divided according to status differences. High court judges and barbers often inhabited the same districts. In contrast to the patterns prevailing in parts of the rural areas (see, for instance, Gingrich and Heiss 1986), social distance became apparent in crucial domains such as political authority and marriage, and had few spatial correlates. Another reason for such status cohabitation may be sought in the fact that wealth and status did not necessarily coincide. Some of those classed as people "without roots" who pursued despised occupations, among them the butchers, were prosperous whereas some of the nobles were very poor. Divisions of labor and occupational specializations, based to an extent on cultural outlook, were crosscut by patterns of social, economic, and ritual exchanges. Families who had traditionally been associated with learning were not concentrated in one particular area, but were keen to reside in the vicinity of mosques in the quarters of al-Talhah, al-Fulayhi, al-Madrasah, and the Great Mosque. The quarter which was more markedly set apart was Bir al-Azab which was wealthy. Its fountains and gardens gave it a suburban flair. Built in the sixteenth century, it was the seat of the Turkish Governor (*vali*) and inhabited by some Yemeni senior officials and merchants.[22] Having left San'a in the early seventeenth century after fierce fighting with the Zaydis, the Turks re-occupied it in 1872 (figure 1.2).

Before the disintegration of the (recently rebuilt) wall which physically and symbolically maintained the city's distance from the rural

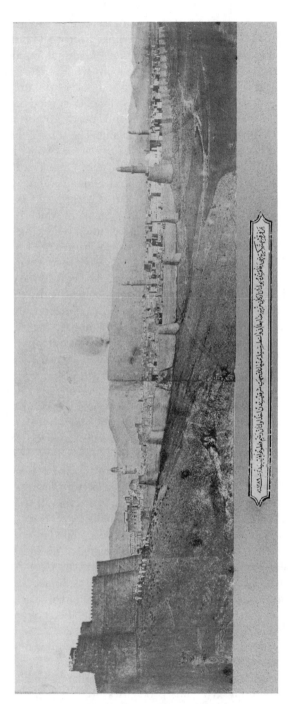

Figure 1.2 San'a, 1872

hinterland, travelers who did not reach it by nine p.m. had to stay in nearby villages as the six gates of the city were closed every night (Serjeant and al-'Amri 1983: 147). With the exceptions of the hijrahs, the hinterland was associated with impious and unruly tribes. Some inhabitants of the city (referred to as San'anis) still hold negative stereotypes of the tribesmen, many of whom have now settled in the city. Some of these stereotypes are fostered by memories of the 1948 revolt and the kidnappings during the 1990s. Following Imam Yahya Hamid al-Din's assassination by a rival in 1948, his eldest son Ahmad regained power after allowing loyal tribes to sack the city.

Since 1962, when the Egyptian forces entered San'a in support of the republicans, many changes have occurred.[23] They constructed a great public square called Maydan al-Tahrir, located between the old city and Bir al-'Azab, where the tank that attacked the last Imam's palace used to be located. Two new shopping streets were joined to the square, and have become the main commercial focus of the new part of the city, rivaling the old suq. The town expanded to the west where the university was founded in the 1970s. On the northern and southern sides of the walls of Bir al-'Azab new residential suburbs sprang up. The city has been spreading west, south, and north and even eastwards toward the slopes of Jabal Nuqum. The style of two-storey houses used by Turks during their occupation has been adopted. A slightly modified version has become especially fashionable: residential quarters are on the upper level and shops and garages are on the ground floor (Lewcock 1986: 53–4).

Since the 1970s, the composition of the population of San'a has altered considerably. There has been a steady influx of immigrants from all parts of the republic including the former PDRY which has resulted in greater social heterogeneity.[24] Some of the southern migrants are prominent professionals and businessmen; many work as shopkeepers and taxi drivers. By the mid-1980s the Sunni merchants, who are largely responsible for the development of the private economic sector, had given San'a a much more mercantile face (see chapter 8). Previously uncommon names such as Ghanim, Shamsan, and Faraj are heard in the streets of the city. The shop labels ("Talk of the Town," "Mademoiselle") in the new parts of the city demonstrate its cosmopolitan aspirations.

Many people who have been resident in the old town for centuries prefer to build new houses rather than to convert the old. The less affluent move into the old city (Piepenberg 1987: 101) (figure 1.3). The newly constructed houses—some of them bungalow-style—have fewer multi-purpose rooms than the old ones. A considerable number of these houses are inhabited by nuclear families who consider the

Figure 1.3   San'a, the old city 1986

Figure 1.4   San'a, suburb of al-Haddah 2005

older dwellings, which used to accommodate several generations, to be inappropriate. Where extended families do live together, they have chosen to abandon the former joint residence for a cluster of two to three houses where about three to five nuclear families live in close proximity to each other. While some features of traditional San'ani buildings, particularly stained-glass windows (sing. *qamariyyah*) have been preserved, the new style—is remarkably different (see figure 1.4).

As a result of the emergence of suburbs for the wealthy, the distance between different status groups has taken on a geographical dimension that was previously far less pronounced.[25] Nowadays San'a is undergoing unprecedented construction activity. Wealthy Yemeni expatriates, merchants, and government employees, who are not merely a tiny elite, invest heavily in the development of new neighborhoods and are interested in political stability. The new neighborhoods indicate that economic achievement has become a status indicator no less (or perhaps even more) significant than heredity. Observers are confronted with a bewildering reality where this building boom would appear to contradict the often cited economic decline since the Gulf War of 1991. Before these issues are discussed in chapter 9 in relation to new consumption regimes, chapter 2 goes back in time and examines the social order and decline of the Hamid al-Din Imamate.

Chapter 2

The Zaydi Elite during the Twentieth-Century Imamate

With respect to Imamate rule, distinctions between religious functionaries who relied on their noble descent and notables who held state-derived authority cannot meaningfully be drawn.[1] Senior military personnel who were often drawn from those families also enjoyed a good reputation whilst less prestige was accrued to the merchants. The elite was referred to as *al-buyut al-kubar* (great houses), *'aliyat al-nas* (people of distinction) or simply *al-nas* (people of esteem). Those who were prosperous but not connected with religious learning such as Bayt al-Sunaydar, a great merchant house, were referred to as *al-wujaha'* (those who have "social visibility").[2] The court nobility and the distinguished 'ulama, some of whom belonged to the royal court (*al-diwan al-malaki*), claimed the highest rank. The elite monopolized senior administrative posts, the courts, and the higher teaching and endowment institutions. Some held high positions in the army. The core of the learned *buyut al-kubar* were also referred to as *al-buyut al-'ilm* ("houses of learning"). This term is used predominantly by people with a traditional education to describe people of their own kind. Their descendants who have been to university do not often use the term to specify their social position, nor do the opponents of the old learned elite.

Wealth was derived from land and political positions. Although most of those whose ancestors had been senior government employees possessed land and real estate, by no means all who belonged to the nobility by birth were also privileged in matters of wealth. The *sadah* constituted an amorphous and varied category, which included people of almost all professions and levels of income. The *sadah* of San'a carried out all kinds of occupation, but did not work as barbers, butchers, vegetable sellers, sweepers, and musicians (*al-akhdam*) at the very bottom of the social pyramid.[3] In the countryside landless *sadah* who were not employed by the government lived off Qur'an recitation and as tenant farmers. Some could barely survive without recourse to charity in the form of small portions of the harvest. When the botanist Hugh Scott traveled in the Yemen during the late 1930s, he noted that

> some of these nobles are busy men, holding responsible posts, for instance those of the Imam's sons who are ministers, and the Amirs of provinces whom we had met. But noble birth is not a necessary qualification for high office, as we have seen in the case of the prime minister, and many of the blue-blooded are unemployed. Such men have little or no occupation save to stroll about the streets or gather together in the *mifrajes* or reception-rooms of friends, where for hours at a time they converse, blow smoke through the water-pipes, and chew *qat*. (Scott 1942: 131)[4]

Zaydi dogma prohibited the Imam and senior officials from pursuing economic activities in the areas they governed lest they be tempted to misuse their offices for their own ends. In practice this meant that those who claimed membership of the high-ranking section within the elite had to remain aloof of the market place. There were very few big Zaydi merchants (see chapter 8). They succeeded in claiming elite status but, like those *sadah* who were craftsmen or traders, were accorded less prestige than the scholars. In line with doctrinal requirements, the *sadah* had exclusive access to the supreme leadership, and had opportunities to recruit their kin into the state service. However, they had few privileges in the sense of vested legal rights. Certain privileges granted to the *sadah* by the majority of legal schools, such as their entitlement to the *khums* and the *fay'*, were often disregarded in practice. The Imam was entitled to distribute these funds among the Prophet's descendants at his discretion. During the reign of the Hamid al-Din, the treasury (*bayt al-mal*) had no provision for the *khums*.[5] Because in theory the *sadah* were prohibited from accepting alms, the salary of those who worked as tax collectors had to be derived from sources such as the *jizyah*, the tax paid by the Jews rather than the zakah.[6]

Unlike the Prophet's descendants who lived in countries such as Egypt, Morocco, and Syria, Yemeni Zaydi *sadah* were never subject to the supervision of a *naqib al-ashraf*, a registrar whose duty it was to administer genealogies, enter births and deaths in the records, evaluate claims to 'Alid heredity, and punish pretenders. Elsewhere the *naqib* had moral authority over the *sadah* and even judicial powers, ensuring that their conduct did not conflict with the norms of etiquette applying to their station. He supported their claims on the treasury and urged 'Alid women to refrain from marrying men of lesser status.[7] In Yemen, because political authority rested with the *sadah* who naturally had a keen interest in maintaining their genealogical records, there was no requirement for officials such as the *naqib al-ashraf*. 'Alid houses kept their own records, and the government was solicitous to thwart attempts of non-'Alids to make illicit claims to membership of the House. In the 1790s, an imposter, a *muzayyin* (a member of the low status service stratum), was sentenced to flogging at a San'ani court. A drum was strapped to his back, the man was led through the streets and anyone could beat the drum (al-Shawkani n.d., Vol. I: 235–6).[8] On occasion the Imams provided *sadah* (and non-*sadah*) with documents confirming their descent status. The recipients of these documents were mostly civil servants who were assigned to areas outside the capital (Meissner 1987: 354–5).

Since the establishment of the Imamate, which privileged the *sadah* in matters of rule, there was a continuous upward flow into the ranks of the nobility through the appointment of judges (*qudah*) whose descendants have continued to carry their title.[9] It was wealth or proximity to centers of religious learning that enabled men to move up the social ladder and become *qudah* or *fuqaha'* (sing. *faqih*, learned men who did not hold official posts as judges).[10] During the earlier centuries, they owed their governmental posts to competence in legal matters rather than their pedigrees; those who assumed power and influence began documenting their genealogies and became founders of houses.[11] According to Qadi Muhammad b. Isma'il al-'Amrani who holds a post at the Presidential Office and teaches at the Great Mosque, in the Yemen the term *qadi* came to designate a *tabaqah* (pl. *tabaqat*; stratum, social category) at the time of al-Mahdi 'Abdullah (r. 1816–35). He referred to Imam 'Ali as the ancestor of the *qudah* (*jaddhum*) because he had been sent to the Yemen by the Prophet in order to mediate between tribes and to spread the teachings of Islam (see chapter 1, p. 36). The title *qadi* was originally attributed only to government-appointed judges, but it was also given to particularly loyal retainers by the Imams and eventually became hereditary (compare Douglas 1987: 4).

The *qudah* held prestigious portfolios in the civil service and greatly contributed to Zaydi scholarship. They influenced the tribes on behalf of Imams they favored and some, among them most prominently Muhammad al-Shawkani (d. 1834) who served as Chief Justice (*qadi al-qudah*) under the Qasimi Imams, lent legitimacy to practices which contravened Zaydi doctrine.[12] In spite of their significant social prestige, the *qudah* considered themselves disadvantaged because they were excluded from competition for the supreme leadership. By being so strongly implicated in the control of the national and local government and the judiciary of the Imamic state without ever exercising authority as rulers, they were ideologically polarized. Both contemporary *sadah* and *qudah* hold that the Imams granted the *qudah* prestigious positions because they feared those *sadah* who might have competed with them for the highest office. According to a Yemeni saying, the *qudah* are a substitute for the *sadah* (*al-qudah badil al-sadah*). Whether they actually "remained learned commoners at heart" (Mundy 1995: 173) remains an open question.

Like others who were not engaged in religious learning, the paramount shaykhs also fell into the category of the *wujaha'*. Some held leadership roles for several generations, were big landowners and had urban aspirations. Several, most prominently those who had affinal ties with the ruling elite (for example, Bayt Abu Ras and Bayt Shirhan) took up residence in San'a and Ta'izz. The destiny of pretenders to the Imamate and indeed the Imams often depended on the shaykhs.

One of the crucial attributes of elite status was the possession of genealogies of varying degree, which distinguished the elite from the *menu peuple* (*al-'ammat al-sha'ab*), a category comprised of artisans, shopkeepers, carpenters, and such like. Among them were those of lowly birth who were said to lack social worth (*qalil 'asl*) in contrast to those labeled as *sharif al-asl* (of pure descent).[13] A plethora of labels was and still is occasionally used for members of the lower service classes.[14] In San'a, circumcisers and barbers, and people who serve at houses during ritual occasions were called *mazayinah* (sing. *muzayyin*). The status of the butchers was slightly higher because some of them were fairly prosperous. The occupations pursued by the *mazayinah* were degrading also because they provided salaried services in personalized exchange relationships.[15] People whose social status was similar to that of the *mazayinah* were the vegetable growers or *qashshamiyyun* (growers of garlic, radishes, and onions), bath attendants, brickmakers, sweetmakers, musicians, bloodletters, saddlers, tanners, dyers, blacksmiths, innkeepers, and shoemakers (vom Bruck 1996).

In the republic, which commits itself to the abolition of hereditary social distinctions, some of those deemed to be of lowly birth have obtained an education and high political office.[16] However, discrimination, which derives partly from assumptions about their foreign origin, persists. They are presumed remnants of the Persians who ruled the Yemen in the sixth century. According to this view, toward the end of their rule their descendants were permitted to settle in the country on condition that they would accept professions despised by Yemenis. Others hold that their ancestors were thieves, adulterers, or deserters who took refuge among other tribes who were forced to engage in degrading service.

Intra-Elite Conflict

As pointed out in the introduction, in the first half of the twentieth century the trajectories of the lives of the elite were shaped by intra-elite disputes about the nature of knowledge and political authority rather than European colonial rule. Of all the Arab countries, only parts of the Arabian Peninsula remained free of European control. Whereas South Yemen was under British rule until 1967, the North became an independent state after Ottoman withdrawal in 1918. Thereafter Yahya Hamid al-Din, who had been formally declared Imam in 1904, aspired to rebuild the Imamate after periods of anarchy and foreign occupation. As Haykel (1997) has cogently shown, he "reinvented" the Imamate as a patrimonial state, which it had already been under his predecessors of the House of al-Qasim. (The Qasimiyyun ruled, almost continuously, from 1597 until 1962.) After the "period of disorder," as the troubled nineteenth century is referred to by Yemenis, and Ottoman control, the Imam's policies clearly demonstrate his desire to establish a centralized and autocratic nation-state with the long-term goal of liberating the South from British rule. Insisting on his sovereignty over the Protectorates, he wanted to unify the Yemen under his authority (Lackner 1985: 16, 19). He was willing to implement reforms which would further his goals as long as they did not threaten either the independence or the religious character of the state.

Imam Yahya's adoption of the title "king" in 1926 facilitated dealings with the European powers and served his interests to institute a patrimonial state, but he represented himself first and foremost as the *amir al-mu'minin* (leader of the faithful). He wanted the Yemen to be part of the Islamic *ummah* rather than a state among several independent nation-states. The emblem of the state was a red flag with five stars symbolizing the five pillars of Islam, the five *ahl al-bayt* who are mentioned in the "Mantle hadith," and the kingdoms' five provinces (*wilayat*).[17] The nationalization of the memory of the *ahl al-bayt*, however, always maintained its universal referentiality.

It is difficult to assess the impact of Ottoman occupation on Yemeni society and the elite in particular.[18] Two factors testify to the unshaken power of the local elite. First, according to the Treaty of Da"an in 1911 the Ottomans granted the Imam jurisdiction over the Zaydi areas; second, Ottoman officials could not advance their claims to hegemonic power by taking Yemeni elite women as wives. On the other hand, Yemeni nobles of both *qadi* and 'Alid descent, among them men of the royal house, married Ottoman women.[19] Imam Yahya's appointment of a Turk, Raghib Bey, as foreign secretary, and one of his son's marriage to his daughter demonstrate both the confidence with which the Yemeni elite incorporated members of the former occupation force and their desire to reshape the administrative apparatus in anticipation of cooperation with the Western powers. Ghalib Bey knew several foreign languages and had considerable diplomatic experience.

Several features of Ottoman administration were maintained by Imam Yahya. He sought to establish a regular army and to intensify diplomatic activity, but trade and commerce remained low on the political agenda. During the reign of the Hamid al-Din no powerful mercantile class emerged which might have posed a threat to the religious elite. The Imam was deeply influenced by Zaydi values of asceticism, and his anxiety of renewed occupation provided an obstacle to extensive economic transactions with foreign powers. Thus the official newspaper, *al-Iman* ("Faith"), which was published since 1926, wrote "the East, and especially the Muslims, must unite and produce their own goods, and run their own commerce. If you lose economic independence you lose political independence" (Obermeyer 1981: 188). The Imam was convinced that the "flag follows trade"; nonetheless, in the late 1920s he signed commercial treaties with Italy and Russia, and later his son 'Abdullah went to the United States to establish trade relations (ibid.: 185, 191). His moderate investment in "development" has been much criticized in the following years during which conflicting demands were made on him. Those who espoused more radical reforms were opposed by conservative 'ulama who even objected to Prince 'Abdullah's plans to invite an American consortium to tap the country's oil wealth. Ultimately, the Imam himself took a conservative view.

Early on, the Imam extended his power from the center to the periphery by pacifying the tribes in the east, and he took control of the Shafi'i areas which had cooperated with the Ottomans. Some Sunni shaykhs lost the positions they had held under the Turks (Douglas 1987: 41). In the 1930s and 1940s the Imam weakened and alienated powerful men of reputable houses by transferring their offices to his sons.[20] His diplomatic activities and his investment in the army attest to his desire to increase Yemen's internal and external security. The military academy continued to be run by Turks. In 1947 the Yemen joined the United Nations and the Arab League in spite of the latter's anti-Hashimite Egyptian domination (Susser 1995: 9), presumably in order to secure support against the British if it should be needed (M.A. Zabarah 1982: 37–8). The Imam arranged a treaty of friendship with Iraq, and in 1934 he signed a border agreement with the British and Saudi governments.

His cultural policies marked a break with tradition. He maintained several key features of the Ottoman-introduced reforms focusing on greater systematization of education. The early years of his rule were marked by important changes in the organization of knowledge that had begun in the late nineteenth century. In 1877 the Ottomans introduced the printing press in the Yemen which precipitated bureaucratization and changes in attitudes toward knowledge. Imam Yahya ordered the printing of a number of books. He appointed a history committee and oversaw the production of a "unified" Yemeni history, which signaled a break with previous historiography. In 1925 he opened a new library at the Great Mosque of San'a and overhauled all its operations from cataloguing to lending (Messick 1993: 115, 120–2). A decade thereafter he introduced as new code of judiciary procedure according to which judges were obliged to give reasons for their judgments (Würth 2000: 160, n. 21). In the late 1920s and mid-1930s, students were sent for training as pilots to Rome and to the military college in Baghdad. In 1946 five students, all members of learned families, were sent to the Ma'had 'Ali, a diplomatic school in Cairo. A year later forty students of different backgrounds selected from schools in San'a, Dhamar, and Ta'izz were sent to Lebanon at the expense of the Lebanese government.[21]

The learned elite had access to the BBC World News and translations of European literature that was brought to North Yemen from Cairo via Aden. The men who cherished ideas of consultative government (*shurah*) and of Muslim reformers such as Jamal al-Din al-Afghani and Muhammad 'Abduh advocated wider reforms than the Imam was prepared to implement. Poetry had always been an integral part of cultural life, but the Imam was less inclined toward the "new knowledge" (*al-'ilm al-jadid*), a term used to describe science, geography, and *belles lettres*. He associated the "new knowledge" with the culture embodied by the European colonial powers and feared for the moral contamination of his people and their alienation from Islam. He anticipated that the influx of literature from Egypt would be accompanied by the introduction of cinemas and cabarets, of which he had received vivid descriptions from Yemenis who had gone abroad.[22] This type of education was of course also a threat to his rule.

Historians' appraisal of Imam Yahya as an anti-reformer who subscribed to "repressive traditionalism" (Carapico 1998a: 30) have failed to appreciate the consequences of his educational policy which constituted a significant component of his endeavor to establish a centralized administration. Restrictions on the independent administration of the teaching institutions led to a bureaucratization and curtailment of authority of the most influential professional class in the country, the 'ulama (vom Bruck 1998b: 165–7).[23] In the Imamate the mosques were the principal teaching institutions. Madrasas, teaching institutions based on endowments, were only established in the sixteenth century in the Zaydi centers of Kawkaban, San'a, Thula, and Dhamar.[24] In 1926 Imam Yahya founded the *Madrasah al-'ilmiyyah*, a college comparable to the Jami' al-Qarawiyyin in Fez (Morocco) and the Farangi Mahall in India (Metcalf 1978: 112). During Imam Yahya's rule seven hundred and fifty students were enrolled in the college. The institution was designed to train the future administrative elite. The style of teaching differed greatly from that practiced at the mosques, and students were given a choice of disciplines to study. Those students who aspired to superior posts in the government had to spend about twelve years at the *Madrasah al-'ilmiyyah*. It was a college that was to produce legal administrators rather than jurists entitled to exercise independent judgment (*mujtahids*). According to Sayyid Muhammad al-Mansur, with regard to the teaching of the religious sciences and the Arabic language at institutions such as the *Madrasah al-'ilmiyyah*, the Yemen was among the best nations in the Arab world (see also chapter 12).

Under Imam Yahya's predecessors renowned scholarly families had been entitled to administer the income from mosque endowments. Later on government control was enforced through a dual policy of standardization of teaching methods and the transfer of pious endowments from local teaching institutions to the treasury which sponsored the *Madrasah al-'ilmiyyah*.[25] Although the 'ulama represented the state bureaucracy, they had previously enjoyed relative autonomy in law and education. These policies clearly weakened their position. The discontent of the 'ulama also focused upon the patrimonial state the Imam had created. Several policies were conceived as violations of Zaydi tenets: the Imam's approval of the nomination of his eldest son Ahmad as crown prince and the delegation of governorships to his sons, his intolerance of criticism leveled at him by those of his own rank, and his refusal to open his grain stores during the famine of 1943. But it was not personal grievances and frustrated ambition alone that nurtured the growing resentment of nobles, men of letters, and shaykhs who aspired to profound social, economic, and political reforms. The dissatisfaction of the Sunni reform-minded scholars also focused on their lack of representation in the government and the

Imam's fiscal policies in the south which they regarded as unjust. The reformers, several of whom had studied and traveled abroad,[26] aspired to a government which was guided by the principles of " 'consultation and constitution' (*shuriyyan dusturiyyan*)" (Dresch 1989: 238), laid down in a Sacred National Charter (*al-mithaq al-watani al-muqaddas*).[27] They asked for a parliament, freedom of speech, infrastructural expansion, and modern schools. As noted, in 1948 the Constitutional Movement culminated in an uprising and the subsequent assassination of Imam Yahya, which was crushed by his son Ahmad a few weeks later. The Constitutional Movement was a corrective movement that aimed at retaining the Imamate. It is unlikely that a government conceding wider political representation would have weakened the ruling elite.

Imam Ahmad followed his father's policy of moderate reforms. Hospitals were built, and foreign engineers were asked to install generators to provide electricity for the major towns. Palestinians were hired to launch a new newspaper (*al-Nasr*, "The Victorious") and to run a school for girls; they and Egyptian teachers also instructed students at the secondary schools where English and geography were taught. The Imam established links with Eastern European countries and diplomatic relations with the Soviet Union and China, and signed treaties with several other countries (M.A. Zabarah 1982: 40; Douglas 1987: 171–2). Like his father, Imam Ahmad was criticized by conservatives, for example for legalizing radio music (on this subject, see chapter 9). The greatest challenge, however, came from foreign anti-monarchic forces who supported reform-minded officers, students, and an estranged tribal leadership. The *buyut al-ʿilm*, on their part, had played a significant role in weakening state authority. Demands for reforms accompanied by political violence in the late 1940s contributed to destabilizing the Imamate.

Toward Revolution

In the 1950s, as nationalist movements in the Arab world and internal and external opposition to his rule grew stronger, Imam Ahmad became increasingly more isolated. In 1955 fraternal dispute resulted in the execution of two of his brothers, a decision that was greatly resented by many *sadah*. The religious establishment was also divided over the question of succession when Imam Ahmad declared his son Muhammad as the heir apparent (*wali al-ahd*). Internal tensions were heightened by events elsewhere in the Arab world. Surely the Imam was keenly aware of the significance of the deposal of King Farouk of Egypt in 1952 and the assassination of King Faysal II of Iraq in 1958, though he denounced the latter as a "servant of the Christians" and supported 'Abd al-Karim Qasim, the republican prime minister. In the Hijaz, the power of the Sharif Husayn had already been broken in the 1920s, and the reinstallation of the sharifian Sultanate in post-colonial Morocco can hardly be perceived as a revival of Hashimite power in Arab politics. Another important factor was the Irshad reform movement in Hadramawt and many Hadrami diasporas which challenged traditional Hashimite privileges from the mid-nineteenth century onwards.[28]

From 1953 onwards Cairo radio broadcasted a program called *Sawt al-'Arab* (Voice of the Arabs) through which President Nasir spread the idea of the Egyptian revolution to other Arab countries. *Sawt al-'Arab* became a propaganda instrument directed against "despotic" Arab monarchs and Western imperialists. The leader of the Yemeni Union in Cairo which opposed Imam Ahmad, Qadi Muhammad al-Zubayri, transmitted programs focusing on the "occupied South" (the Aden Protectorate), the democratic rights of the individual, and "Muslim Unity" (Douglas 1987: 176–8). In the 1940s, al-Zubayri had asked Prince Ahmad in his eulogies to "fly us wherever you want, we are a group . . . following you in glory" (Taminian 1999: 216), but later he became disillusioned. In a book (n.d.) published about three years before the revolution he attacked colonialism "from without" and despotism "from within," challenging political authority based on birthright (Serjeant 1979: 100). It is noteworthy that al-Zubayri did not conceive of clear-cut divisions between Sunnis and Zaydis, arguing that the latter, rather than considering themselves to be ruling over the Sunnis, "feel a bitter vehement resentment that it is a particular stratum of Hashimite families which enjoys the divine right to govern, that is specially favored with it and passes it, turn about, among the ambitious male members of it, generation after generation, feeling a consciousness of superiority and distinction over the rest of the sons of the populace" (quoted in Serjeant 1979: 107).

Criticisms of the *sadah*'s prerogative of power which now became prominent had hardly been publicly voiced since the days of Nashwan b. Sa'id al-Himyari (d. 1117), a poet and politician whose son-in-law al-Ju'ayd al-Wadi'i portrayed Imam al-Husayn as a "heretic." Nashwan's poems which present a scathing attack on the Prophet's descendants have appealed to those reformers striving to revive the Qahtani-'Adnani conflict after the 1962 revolution. In reply to a poem by a member of the Qasimi dynasty attacking him, Nashwan wrote:

Quraysh has died
And all creatures are mortal
. . . . .

I tell you: the inheritance of the prophets is not worthy of us
And you want to tell me that prophets have eternal life?
The Prophet came from you, and he died too,
And now we are supposed to believe a prophet from you to be divine!
Stop all this nonsense
To which no-one can give his approval,
There is greater merit in turning your back on folly!

. . . . .

You have extolled [Imam] al-Husayn, that heretic,
And I was the first to tear up his paper.

. . . .

How enraged you became when you heard it said
That your Imam was dead.
An Imam is not immortal;
There is no shame if an Imam is killed,
Killing for a good deed is like a fountain to drink.
With Muhammad the era of prophets was gone,
Even Muhammad himself did die.

. . . . .

Be proud, then, of Qahtan more than
of any other mortal.
For men are like shells
and Qahtan's sons are like pearls!

. . . . .

The one most worthy of power
Is the one most pious among them, and they should confide in one
Who does not neglect those
That do right and follow the law.
It does not depend on his skin,
Be it white or be it black,
Whether his ear is pierced, or his nose.
Wake up at last, Shi'ite! Come here!
Too long have you been slumbering.[29]

Qadi Isma'il al-Akwa' who dedicated some of his work to Nashwan after the September revolution was among the reformers who were imprisoned in 1948. In a work published in 1963 Muhammad Nu'man, the eldest son of Ahmad Nu'man who was one of the few Sunni thinkers who took part in the uprising of 1948, recapitulates speeches made by some of the detained scholars who had gathered to discuss Yemen's future. The author quotes Qadi Muhammad al-Sayaghi, who acted as Imam 'Abdullah al-Wazir's secretary in 1948, as saying that the construction of a new society required the elimination of two things (*quwwatayn*): (1) the ruling elite (*al-asyad*; masters) [that is, the *sadah*] who impose their influence (*al-siyadah*) on the nation in the name of religion, dogma, and school (*madhhab*); (2) ignorance and false beliefs. Ahmad al-Marwani, himself a *sayyid*, proposed that most people of the southern parts of North Yemen (*al-yaman al-asfal*) had already understood the aims of the Free Yemeni Movement (*al-Ahrar*) and were more willing to accept a constitutional monarch than the people of upper Yemen (*al-yaman al-'ala*). Because the people of those areas were more prepared to accept the descendants of the Prophet, the reformers had to concentrate their efforts there. These people [the *qaba'il*] were opposed to the regime but would readily bury their grievances once the sons and daughters of the Prophet would call upon them to support their cause. They did so because they wanted to plunder the shops and houses of San'a rather than to protect their honor and dignity (Nu'man 1963: 54–5, 112–13).

In a letter addressed to Muhammad Nu'man by 'Abd al-Rahman al-Iryani in the 1950s he reasons that it would take a long time to convince people in the north that no other than a *sayyid* should exercise power. (Al-Iryani became the second republican president in 1967.) The tone of the letter is pessimistic, the writer asserting that (non-'Alid) Yemenis had lost their confidence and dignity and took it for granted that a *sayyid* should rule.

> Every Yemeni thinks he should be ruled rather than rule himself. You come from one of the greatest families in the Yemen, but have you ever thought to be a ruler, or has any of your Qahtani friends? Your answer will be no. This means that we accept our inferiority while on the other side, *every child of the Banu Hashim is thinking about ruling the Yemen*. This shows you that there is a big difference between us and them and how dangerous it is. (quoted in Nu'man 1963: 33–4; emphasis added)

By the mid-1950s, men like al-Zubayri and Ahmad Nu'man wanted the Imam to be more "progressive" and to establish a constitutional monarchy, but the idea of a republic was not yet born. Students like 'Abdullah Juzaylan, who was trained at the Military College in Cairo, and Muhsin al-'Ayni, who studied in Lebanon and France, asked for more radical change.[30] The harshest attack on the Imamate and the *sadah* came from al-Baydani, one of the leaders of the Cairo-based Yemeni opposition who later became deputy to the first republican president. He had been attached to the Yemeni educational mission and later employed in the Imam's diplomatic service. He labeled the Imam a "despot" who kept his subjects ignorant so as to render them compliant. The *sadah* were collectively held responsible for the country's misery and accused of bigotry, nepotism, and oppression. Advocating a Qahtani republic (*al-jumhuriyyah al-qahtaniyyah*), al-Baydani called for their annihilation or eviction from the country.[31] A primordial and national category, Qahtani, was thus defined in opposition to *sayyid/* 'Adnani.

Whilst this agitation intensified, Imam Ahmad could hardly win the support of any faction because he had alienated many important forces. He had killed and imprisoned members of his family, leading scholars, and shaykhs. He neither fulfilled the expectations of those 'ulama who urged the establishment of a consultative Imamate (*imamah shuriyyah*) nor of the students who had completed their higher education abroad. The Imam disapproved of the students' nationalist ideologies, which were inspired by Egypt, and refused to assign them to senior posts which were reserved for the 'ulama.[32] Many took part in the 1962 revolution and later held prominent positions in the new republic.

In the late 1950s the political climate was marked by anti-'Alid sentiments which the *sadah* describe as *'unsuriyyah* (racism; here: discrimination). Several *sadah* expressed the opinion that the Imam encouraged the *'unsuriyyah* in order to increase their feeling of insecurity and to cause them to rally around him. He wanted them to believe that the 'Adnaniyyun might be expelled from the country, and was quoted as saying "You don't understand that I am sitting on a nest of snakes and scorpions, and you will see what happens once I am gone." The *qudah* who held political authority had benefited from the *'unsuriyyah* because the opposition targeted the *sadah* rather than the ruling elite to which they belonged. Some of the *sadah* who had been critical of Imam Ahmad renewed their loyalty to him after finding themselves confronted with al-Baydani's attacks. Others doubted al-Baydani's influence on the population.

After his death on September 19, 1962 Imam Ahmad was succeeded by his eldest son Muhammad al-Badr. Al-Badr announced the establishment of a consultative council (*al-majlis al-shurah*), the release of all child hostages, and that the zakah be based on self-estimation (*amana*) all over the Yemen.[33] He was deposed by revolutionary officers loyal to the Egyptian President 'Abd al-Nasir seven days after he had assumed power.[34] In the North, during the incipient stages the revolution had limited tribal support. It was welcomed by the Sunni population of the south and the Aden Protectorate. Thousands of men left for the north to defend the revolution against its enemies, among them tribesmen from the hinterland, northerners who worked in Aden, students, and tribal leaders from border regions (Lackner 1985: 37).

Al-Badr's rule has been subject to diverse evaluations by the *sadah*. Some maintain that his erudition was wanting, that his political alignments were volatile, and that he could not provide leadership at a critical moment of the Imamate's history. Others, among them opponents of the Hamid al-Din, hold in the Imam's favor that he announced reforms during his short period of rule which fulfilled most of the demands made by the Ahrar. Indeed, some argue that Nasir had planned the military coup in Yemen long before al-Badr came to power because his troops arrived within forty-eight hours.[35]

Resistance to the revolution, supported by Saudi Arabia, centered mainly on the countryside with Zaydi majorities, the main cities being controlled by the republicans. The conflict gained a new dimension through the confrontation between Arab nationalists and monarchists. For President Nasir the engagement in the Yemen provided an opportunity to move out of the isolation which had resulted from the break-up of the union between Egypt and Syria, and he sought control of the oil fields on the Peninsula. King Sa'ud, who regarded the establishment of the United Arab Republic as a threat to the Arab monarchies, was alarmed by the prospect of the establishment of a Yemeni republic, under Nasir's influence, at Saudi Arabia's southern border. Against the backdrop of the historical rivalry between the Hijazi Hashimites and the Saudis, the latter were likely to have little sympathy for the Yemeni Hashimites who, moreover, belong to the Shi'a which has no official Saudi recognition as an Islamic doctrine. Yet, when a new era in Arab politics was ushered in and several monarchies were overthrown, these considerations were of minor importance. After 1958, the Saudis were no longer concerned about a Hashimite alliance between Iraq and Jordan (Maddy-Weitzman 1990). When the new Iraqi regime withdrew from the Arab Union with Jordan and the latter was left isolated, the kinship link between King Husayn and Sharif Husayn b. 'Ali was made light of and Jordan and Saudi Arabia moved closer together (Podeh 1995: 103). The conflict between different Yemeni factions drew to a close only after Egypt's defeat in its war with Israel.[36] The Saudis abandoned their efforts to reinstall the Yemeni monarchy as soon as the threat of Arab nationalism on the Peninsula under Egyptian leadership had been lifted. In 1970 reconciliation took place between Yemeni royalists and republicans.

The majority of the elite of the Imamate who suffered atrocities in the revolution were *sadah*. According to Haykel (1997: 280), a number of *sadah* were summarily executed on the grounds of being the historical oppressors of the Qahtani population. The *sadah*, on their part, saw the revolution as directed against them despite the fact that most republican leaders have publicly attacked only the Hamid al-Din. From among the *qudah* who had held the majority of seats in Imam Ahmad's Cabinet of 1955 and had been loyal civil servants (Douglas 1987: 267), only two were executed.[37] The *qudah* attributed responsibility for the Imam's policy to the *sadah* alone and placed emphasis on their tribal (Qahtani) origin.[38] No one seems to have counted those executed; apparently they numbered less than fifty, three of them members of the royal House.[39] The house of Imam Ahmad's deputy in San'a became the headquarters of the Egyptian army; other houses were occupied by the republican army. Those of the Imam's officials who were prepared to negotiate with the new regime were silenced because their influence on the population was feared.

When the Palace was attacked my brother's friends wanted to take him to his hijrah in the countryside, but he refused to leave his family. He and my father were arrested. They were both on death row. The soldier who was asked to execute them refused to kill my father because of his age. They [the republicans] were determined to kill those people who could have got support in the countryside had they managed to escape. My brother was killed at the age of forty-one. Many of our family were ready for change. In my childhood everybody was talking about this subject. Even my brother was willing to cooperate with the new regime, he wanted change. He did not run away even while he could. He refused to give his daughter in marriage to the Imam's brother because he thought they [the Hamid al-Din] had lost their legitimacy and were an obstacle toward opening the country. (The brother of one of Imam Ahmad's senior civil servants)

My father was stationed in one of the southern provinces. The governor of a neighboring province, a friend of my father who himself fled after the revolution was announced, advised him to leave. However, my father said he had done nothing wrong and stayed. He was executed on the second day of the revolution. My mother was eighteen and had three children. We had to leave our house and were moved to another official's house that had also been confiscated. One of the new presidents, who had often been a visitor at our house, did nothing to help us. At last al-Hamdi [president 1974–77] returned our property, but we lost most of our books. My mother often told us that our father was a scholar and that he would have wanted us to study hard. It was a strong motivation for us. (The daughter of one of the Imam's deputies)

Many *sadah* who had been linked with the Imamate fled the capital because they had been outlawed and their property confiscated; some rejected the Egyptian military presence in their country. Scholars who had criticized Imam Ahmad and were imprisoned by him were not pardoned by the republicans.

I had protested against the Imam's nomination of his son as crown prince. After 1962 I was not released from prison because I had only opposed the monarchization of the Imamate. I was labeled a counter-revolutionary by the republican officer who visited me in prison. He accused me of seeking to correct the Imam's policy rather than to overthrow him. I replied to him that most revolutionaries had been with the Hamid al-Din because they had served al-Badr. (The son of one of Imam Ahmad's governors)

Like this scholar, the *sadah* had always been a potential for subversion because during the Imamate the supreme ruler was drawn from their midst. However, they had disputed personalities rather than the state. In 1962 a great number of the officers who participated in the revolution were *sadah*.[40] The day after al-Badr's palace was shelled, Yemenis woke up to a radio speech about the glorious revolution made by 'Ali Qasim al-Mu'ayyad, a member of a renowned house, who in 1962 became a member of the Free Officers' Association (*Tanzim al-dubbat al-ahrar*). He introduced his speech by saying "this is the radio of the Yemen Arab Republic." Like other revolutionary officers of 'Alid and *qadi* background, he had not benefited from the rule of the Imam. Even today some of the formerly disadvantaged *sadah* express their resentment of the privileged ones in no uncertain terms. Tensions between them crystallized during the revolution, and families were harassed by both the new government and their kin.

My father was abroad when the revolution happened. The government threatened my mother that they would take her house and her land or her young son as hostage in order to force my father to return home and to stand trial. She had already sent my older brothers to Aden. My youngest brother was only three years old and too young to follow them. We dressed him in girl's clothes so that he would not be kidnapped. Many people treated my mother badly, especially those of our family who were less well off. They told her they would come and occupy her house. They put up the radio loudly to annoy her, deliberately showing indifference to her state of anxiety and vulnerability. (A governor's daughter)

Sadah who played important roles in the revolution claimed that they were ready to sacrifice their lives and those of the rulers to whom they were related. As revolutionaries, they lost the trust of the *sadah* who suffered in 1962, and they were suspect in the eyes of others.

I was brought up in a mountain village where my father worked as a government employee. Our life was as bad as everyone else's. The women had to go down to the valley to fetch wood. It took three hours down and five hours up. My mother and my older brother left at 4 p.m. and spent the night in a house in the valley. At 4 a.m. they started their journey back. Those who had no house in the valley had to make the trip in one day.

Once my brother accompanied my father to the Imam's residence in Ta'izz. My brother was astonished to see the difference between his village and Ta'izz which was a modern city. He was keen to become educated. Two years after the trip he attended one of the high schools in San'a without my father's consent. Later he came to visit us and convinced me and my other brother to join him. I came to study in San'a in 1956 where I was taught by Egyptian teachers. They told us about Nasir and the Egyptian revolution, and that the Yemenis were primitive. I listened to *Sawt al-'Arab* programs. I became pro-Nasir and resented the rule of the Hamid al-Din. There were no roads, no universities. It was hopeless. Later I entered the military academy in San'a. In the army there was a movement against Imam Ahmad. Some officers had connections with the Ahrar and al-Zubayri. Nasir wanted us to be the pioneers of the revolution. When Imam Ahmad returned from Rome [where he had had hospital treatment], he sent the Egyptian teachers back after the uprising of Shaykh al-Ahmar.[41] For six months the students had no support. Al Badr convinced the Imam to bring Russian instructors, but they did not train pilots. I graduated at the end of 1961 when the Free Officers' Association was established. By that time the Imam had almost lost control.

The young cadets wanted to have a new regime. In 1948 the problem was that one Imam was replaced by another. The Imam sent several *sadah* to study at the military colleges. He thought they would be happy to receive an education. Then they turned against him. Seventy per cent of the revolutionary officers were *sadah*. We wanted to be nothing but Yemeni citizens and to see social differences eliminated as they had been in Iraq. We resented the distinctions between *sayyid* and Qahtani which were based on the old enmity between the Bedouin of the desert and the people of the mountains. The *sadah* were ruling but my family was living as poorly as everyone else. Once an Imam was opposed, all *sadah* had to suffer the consequences. It was time to end their rule.

In 1962 we thought that the resentment against the *sadah* was only a temporary phenomenon which would soon disappear. When the anti-'Alid sentiment was strong in 1962, we realized that we would have to suffer for a while for the national cause. I heard my ancestors cursed all the time. Neither Hashimi nor Qahtani trusted me. Some of my relatives owned land among the tribe where they were hijrah. The sharecroppers were encouraged to refuse them their share. Some were forced to sell their land cheaply. If they refused to sell the land was taken. My father-in-law was forced to sell his land for very little. Even today [1998], there is land that has not been returned to their owners, but some *qaba'il* give them a share of the harvest. I regret that the discrimination against the *sadah* has continued. It was not what I had intended. Under al-Iryani [president 1967–74], the discrimination against the *sadah* in the government became institutionalized. It was his philosophy. He thought that the *qudah* had rightfully replaced the *sadah* and that the latter should be kept out of government. After the June 1974 Reform Movement, there was no racism against the *sadah* in the government. However, many people in the government want to keep the *sadah* out of good positions. They look at them as a danger to the state and they benefit from that. They say that if they [the *sadah*] occupy leading positions they will become too powerful because they have respect among people. (Yahya al-Mutawakkil, a revolutionary officer)

Although official revolutionary rhetoric targeted the deposed dynasty, in the eyes of ordinary people the revolution was directed against the *sadah* collectively. According to the son of a former primary school teacher, "the revolution occurred in order to finish the *sadah*" (*qamat al-thawrah 'ala shan al-qada' 'ala-'l-sadah*). They were exposed to ridicule and abused in the street. Their headgear, the *'imamah* (see chapter 9), was thrown on the ground, and boys would shout in San'ani dialect: "*Sidi bidi ma nashtish, hubb al-rukbah ma nashtish*" (*Sidi* [my lord] we do not want, nor do we want to kiss his knee).

I was already in my fifties when the revolution happened and I was too old to change my outfit. I went into the street wearing my *'imamah*. Some people told me to take it off, but I refused. They threw stones at me and chanted "*ma 'ad fi-sh sadah*" (there are no more *sadah*). (One of Imam Yahya's secretaries)

In the 1960s, the honorific "sayyid" became a swearword. Prior to a conference in 1965 between the various factions which were involved in the civil war, a high ranking *sayyid* stayed at a shaykh's house north of the capital. Even though the shaykh sided with the republicans, he treated the *sayyid* with utmost respect. He personally brought him warm water to his room for his refreshment. Downstairs the shaykh's son was shouting at the houseboy (*duwaydar*) who had annoyed him: "*Raj'i* (reactionary), *kalb* (dog), *mal'un* (cursed), *sayyid*!" The shaykh was ashamed of his son's behavior and did not come under the *sayyid*'s eyes all day long.

Republican poetry which flourished during the civil war also challenged the status of the *sadah*.

Royalist poem (*zamil*)
Al-Tiyal mountain announced
And all the high mountains in Yemen responded
We would not turn Republicans even if we perished
From the face of the earth

If yesterday became today
Or the sun rose from the south
If the earth caught fire
And clouds in the sky rained bullets.

The republicans would chant:

Thousands of greetings to you, oh you the republic
We redeem you with skulls and blood
Today we join you, willing to sacrifice. We bring
down the *sayyid* from the zenith of the sky.[42]

It was several decades before the notion of the republic was fully understood in the northern region, and some people may not have desired quite such a radical transformation. There can be no doubt however that many Yemenis of diverse social background—among them senior officials—readily subscribed to substantial changes. Since the republic has consolidated and a parliament installed, there has been little enthusiasm for 'Alid rule. Explaining the durability of the Moroccan monarchy, Combs-Schilling (1989) argues that there is an indissoluble bond between its stability and cultural notions which are enacted in common rituals such as the Prophet's birthday (*'id al-mawlid*) and the Great Sacrifice (*'id al-kabir*).[43] "The rituals help to construct an experience in which the individual comes to see his or her self-definition as intertwined with the ruling power" (Combs-Schilling 1991: 663). Certainly these rituals create important links between potent symbols and the authority of religiously sanctioned leaders (vom Bruck 1999a: 186). Both Moroccans and Yemenis were ruled by Muhammad's descendants for centuries; yet central legitimizing notions which are invoked in rituals and manifest themselves in the doctrine of the Imamate have not inscribed themselves into the collective memory such that its abolition would constitute a threat to peoples' notion of self. The Yemeni Imamate had drawn on these symbols and lasted for over a millennium, but there is no evidence to suggest that "the position of king [the sharifian ruler] is written on every man's definition of self . . . and to lose the form of rule would be in part to lose himself . . . to move away from monarchic rule . . . would be a form of apostasy, an act of self-destruction" (Combs-Schilling 1989: 305–6; see 218). The case of the Yemen calls into question the author's conclusion that people's "self-definition as well as hope" (ibid.: 253) depends on the presence of a Hashimite ruler.[44] Combs-Schilling is however not entirely wrong in suggesting that the notion of 'Alid descent has an impact upon people's self-perceptions. It will emerge from subsequent chapters that it is not just 'Alid descent but blood descent in more general terms which continues to pervade perceptions of self and its relations with others. Yet rather than treating this association as a straightforward one, it is suggested that it renders the self-definition of both *sadah* and non-*sadah* problematic and paradoxical.

Examining the structural features of the House of the Prophet and other characteristics relevant to its persistence over time, chapter 3 provides further background data for the understanding of the *sadah*'s current situation.

Chapter 3

The Anatomy of Houses

Genealogy, as an analysis of descent, is . . . situated within the articulation of the body and history. Its task is to expose a body totally imprinted by history.

Michel Foucault

The House of the Prophet is made up of patronymic descent categories of varied size, which are referred to as *buyut* (sing. *bayt*, house). The term *bayt* designates the physical structure of the house and its inhabitants as well as patronymic units ranging from single households to large descent categories.[1] A descent category includes all people who share a patronymic.[2] From birth onwards, both men and women are members of such "houses." The larger ones include up to 30,000 people. All members are equally related to the founder of the house they belong to and whose patronymic they carry. Women maintain membership of their natal houses throughout their lives, but on contracting a marriage and particularly through producing children, a woman becomes closely affiliated with the house of her spouse.

Households, the smallest social and residential units, typically comprise two to three generations (a man and his wife [wives], his parents, his unmarried and married sons and their wives and offspring, and unmarried, divorced, or widowed daughters and sisters). Kitchens are markers of individual households. A man's wives are said to be living in separate households only when they have their own kitchens. Since the late 1970s, nuclear and neolocal households have become more common. Often houses are built for married sons next to the main house or in its neighborhood.

In the past there was one house for all. Now only one of my brothers lives in my father's house. He built houses for the others around his.

That way they are close to him but problems in one family no longer affect everybody. (A scholar's son)

Yemeni kinship terminology is idiosyncratic. The term *ahl* or *al* (kin group, descent category) is used interchangeably with *bayt* but the latter is used most frequently. On occasion, a married woman who is going to visit her natal family says she is going to see her *ahl*, but she would identify herself as a member of Bayt so-and-so. The term *ahl* is also applied to "imagined communities" such as the inhabitants of San'a (*ahl San'a*). In San'a, the term *'a'ilah*, a common Arabic word to denote the "family," is used to refer to a man's wife or his wife and children.[3]

The term *usrah* also has connotations of "family" or "lineage." Several *sadah* described the unity of the House of the Prophet by saying "*nahna usrah wahidah*" (we are one); others subsumed all descendants of Imam al-Hadi (*awlad al-Hadi*) under that category. When asked about the people who belonged to his *usrah*, a *sayyid* mentioned the fifty-four male members of his house. Another listed his parents, himself and his wife, and their offspring.

Houses are internally ranked according to the criteria of wealth, education and profession, and moral authority. Descent from great ancestors (Imams, 'ulama) junior to Imam 'Ali is also a rank criterion.[4] Houses are involved in various material and symbolic exchanges both "within" and "between" (Mundy 1995). Great emphasis is placed on the house as a moral entity whose name must not be tarnished. As such it is reminiscent of Lévi-Strauss's depiction of the house as "a moral person which possesses a domain that is perpetuated by the transmission of its name, its fortune and titles, along a real or fictive line, held as legitimate on the sole condition that this continuity can be expressed in the language of kinship or of alliance, and more frequently of the two together" (quoted in Lea 1995: 206).

Affiliation to the Prophet's House is automatic by birth. Non-agnates are not normally assimilated into houses, and membership cannot be obtained through affinity.[5] I was unable to establish whether "adopted" children were given their foster father's patronymic. One of the few cases I came to know about was a boy in the southern city of Ta'izz who had been left at the doorstep of a pious *sayyid*. He was treated like a child born to the family, though the older children referred to him as *afriki* (African) because of his dark complexion. Their kin in San'a maintained that children like him were not given the name of their foster father and therefore were not eligible to inherit or to claim 'Alid status.[6] There was agreement among the *sadah* concerning an "imposter" *sayyid* whose father had worked for

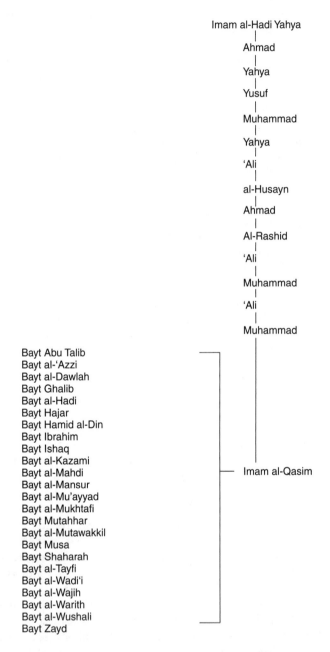

Imam al-Hadi Yahya
|
Ahmad
|
Yahya
|
Yusuf
|
Muhammad
|
Yahya
|
'Ali
|
al-Husayn
|
Ahmad
|
Al-Rashid
|
'Ali
|
Muhammad
|
'Ali
|
Muhammad

Bayt Abu Talib
Bayt al-'Azzi
Bayt al-Dawlah
Bayt Ghalib
Bayt al-Hadi
Bayt Hajar
Bayt Hamid al-Din
Bayt Ibrahim
Bayt Ishaq
Bayt al-Kazami
Bayt al-Mahdi — Imam al-Qasim
Bayt al-Mansur
Bayt al-Mu'ayyad
Bayt al-Mukhtafi
Bayt Mutahhar
Bayt al-Mutawakkil
Bayt Musa
Bayt Shaharah
Bayt al-Tayfi
Bayt al-Wadi'i
Bayt al-Wajih
Bayt al-Warith
Bayt al-Wushali
Bayt Zayd

Figure 3.1 The Qasimiyyun: Houses that trace descent to Imam al-Qasim b. Muhammad

one of the great houses in Shaharah. He had adopted their name and later obtained asylum in a European country after claiming persecution. Most *sadah* commented on the case with no more than a smile. Most of the various patronymic sections are branches of larger descent categories. A great number of *sadah* who carry different patronymics trace descent from Imam al-Mansur al-Qasim b. Muhammad (r. 1598–1620), the founder of the Qasimi dynasty. His descendants (al-Qasimiyyun) number about 40,000–50,000 (figure 3.1). The different ancestors of houses are distinguished according to genealogical depth. For example, Bayt Sharaf al-Din refer to Imam Sharaf al-Din (1473–1558) as *al-jadd al-kabir* (great ancestor), Imam al-Hadi as *akbar jadd* (a more senior ancestor), and Imam 'Ali as *jadd* (ancestor, eponym).[7] References to specific ancestors depend on their prestige as well as on the contexts in which they are mentioned. Apart from the Prophet Muhammad and Imam 'Ali, those referred to are usually men of recognized high status. If a person's mother or wife also belongs to the House of the Prophet and looks back to illustrious ancestors, they are also recalled.

The *sadah* explain the differentiation of sections of the House along patronymic descent lines by two factors.

(I) Natural growth of the original core group of *sadah* has led to the confusion of people who carried the same names (for example, Muhammad b. 'Abdullah). Individuals were given a patronymic in order to be distinguished from their namesakes. The patronymic passed to the next generation which constituted yet another branch of the House. At certain stages in history the various existing houses branched out of others when specific persons, the ancestors of the contemporary houses, became the founders of new patronymic sections. The people who belong to these sections also claim membership of the original house. Some extended families which form part of large houses (for example, Bayt 'Ali Husayn, Bayt Muhammad Hasan, etc.) are referred to by the professions of their most well-known members. Thus the house of the former governor (*'amil*) of San'a, Husayn 'Abd al-Qadir, is still referred to as Bayt 'amil San'a, a term by which this house is easily distinguishable from other branches of Bayt 'Abd al-Qadir who are one of the patronymic sections which split from Bayt Sharaf al-Din.

(II) The migration and subsequent dispersal of *sadah* all over the country also accounts for the establishment of branches. Those who left the places where their forebears had settled to take up administrative posts or as a result of political conflict, or for other reasons, often adopted new patronymics. The various sections of 'Alid houses are usually dispersed rather than localized. Most houses are represented in about five to six rural and urban places. The majority of members of a specific house may be concentrated in a specific area or town but none is entirely localized.[8]

Houses do not have formal leadership. In some cases, the position of the *ra'is al-'a'ilah* (leader or head of a house or extended family) is occupied by either the oldest man of a house or someone with a high reputation and influence. In the past, those who held influential positions in the government automatically acted as leaders. As one man of Bayt al-Iryani, a *qadi* house, told me, "we did not have a family head, but when Qadi 'Abd al-Rahman became president [in 1967], he was treated like one." These leaders' advice is sought and their views are respected, but they rarely have the authority to enforce decisions or to impose sanctions. Some determine the amount of *mahr* (indirect dowry) to be paid among members of the house and their potential affines.

> If someone wanted to marry a woman he [my uncle] did not approve of, he would argue. Members of his house and even outsiders came to ask for his advice in inheritance matters. During Ramadan he taught the children at his house and discussed religious works with his friends and kin at night. Once a boy played with a gun and shot his cousin, a young girl. Her mother asked for a large amount of blood money. The boy's family lived off a small plot of land; had they sold it they would have been ruined. My uncle ordered that the boy be hidden until the problem was solved. He agreed on a sum that satisfied the girl's mother. He himself contributed a sum and collected money among other members of the house. Three quarters of the sum was paid by other members of the house. (The nephew of a *ra'is al-'a'ilah*, a religious scholar)[9]

Genealogical Knowledge

The *ra'is al-'a'ilah* has excellent knowledge of the ties which bind his house to previous generations which might have carried another patronymic. Sebti (1986: 435), writing about the Moroccan *ashraf*, refers to their genealogies as an "*historiographie familiale.*" The meanings the *sadah* attribute to kinship relations are profoundly rooted in their historical experience. History itself is a genealogical map on which the living can trace their relationships to those people who are referred to in the documented histories. For the majority of the *sadah*, history is a form of kinship reckoning, and their moral construction of the person is tied to the memory of their ancestors. "Family history" and political history collapse into each other because the subjects of these histories are often identical.

Eickelman (1985: 118–19) notes that men of learning distinguish themselves from others through possessing a sense of history that encompasses knowledge of genealogies. People's use of extensive genealogies is often correlated to wealth and social class. By assigning each individual his place on the genealogical chart, "genealogy makes a title of social function, thereby transforming it into privilege" (Le Wita 1994: 5; compare Bouquet 1996: 47). Irrespective of their relative rank, Yemeni 'Alid houses possess written genealogies which demonstrate their descent from the Prophet.[10] Among contemporary *sadah*, these genealogical links range well over forty generations. In the Imamate, aptly characterized by its "*politique généalogique*" (Sebti 1986: 445), genealogies were a vital resource both for legitimizing claims to status and proving eligibility for the highest political office.[11] In the imagination of the collective Zaydi *ummah*, arboreal metaphors were mixed with metaphors of Prophetic substance, and constituted part of the moral discourse which accounted for the religious character of the social and political order.[12] In the aftermath of the revolution, the symbols of 'Alid identity represented by genealogies were no longer displayed in the *mafraj* or *diwan* ("reception-room"), and even today there are few. Everyone is conscious of their former political function, and non-*sadah* do not look at them merely as personal recollections.

Genealogical memory may be triggered by mundane, banal situations such as teasing among relatives. On several occasions, I had offered a girl help with her English homework. On realizing that she would be too shy to ask, her father told me in her presence that she was from Bayt Mutahhar, descendants of the founder of one of the branches of this large house to which both her parents belong. Mutahhar was a descendant of Imam al-Hadi in the twenty-seventh generation. There are seven generations between him and the girl who traces descent from him through her mother. The girl's father implied that Mutahhar's descendants were less socially competent because he had not been a scholar. Her mother was sensitive to this kind of joking, and it was avoided in her presence.

For most *sadah* genealogical knowledge is the prime criterion of self-identification. First encounters between *sadah*—irrespective of their nationality—are often occasions for evoking the genealogical imagination. When *sadah* are introduced to each other, they identify themselves by explaining their location on the genealogical grid. Younger *sadah* explain their relationship to the most renowned antecedent of their house. Thus, relationships are forged through genealogical framing.[13] One *sayyid* argued that the possession of genealogies had a moral impact on people. "If someone behaves badly, you first ask about his family. This is why genealogies are so important." In the countryside, characteristic signs of the presence of

Figure 3.2   Sayyid Yahya al-Sa'di, former governor of Sa'dah and Hubaysh, in front of one of his ancestor's tomb, assisting the author. Sa'dah province, 1985

learned houses are gravestones telling the genealogical histories of the deceased (figure 3.2).

The charts possessed by 'Alid houses include either all or a number of those men who carry the same patronymic. Women's names (with the exception of Fatima) are rarely entered because they bear children for other houses and in those charts, which record women's names, a termination sign is added to the name. Occasionally, however, when I undertook to update people's genealogies, some men insisted that the names of their beloved daughters be recorded.

Collective genealogical works of Yemeni 'Alid houses are incomplete, and often focus on major branches, depending on the author's affinity.[14] Since the 1970s, several 'ulama have been involved in producing new collective charts. For some, this exercise in identification and remembrance has become a matter of urgency since the Imams, who in the absence of the *naqib al-ashraf* validated genealogies, no longer pursue this task. "In uprooted memory, genealogy is not an abstract diagram but an essential structure of remembrance" (Bahloul 1996: 115). 'Ulama like the Mufti Zabarah (*al-mufti al-'amm*) understood the spirit of their time: before his demise in July 2000 he spent several years updating his father's famous genealogical works on the Yemeni nobility from the beginning of Islam until his time (he died in 1960). Making inquiries among people across the status spectrum, he added their data, including those of the circumciser living in his neighborhood, Ahmad Barquq (see n. 17).[15] Among the 'Alid *buyut al-'ilm*, genealogies form part of recorded family histories. Over the generations, assignments to official posts, migration to other parts of the country or abroad (with reasons), and the composition of literary works, have all been documented. Another historical record produced by these houses is referred to as *safinah* ("boat"; a reference to Noah's Ark). For example, one talks about *Safinat Bayt al-Mansur*. A *sayyid* described the *safinah* as a notebook filled with poems and ideas which one has heard or read elsewhere and likes to keep for posterity. It is through the *safinah* that future generations learn about their forebears' intellectual proclivities. They also gain knowledge about them through poems which are recorded in other peoples' *safinat*. Some of these *safinat* are well-known records of the sayings of famous people. It is likely that these works become objects of nostalgia among new generations for they memorialize both destroyed hijrahs and the ancestors' intellectual worlds.

Patronymics

Prestigious ancestry encoded in a patronymic has often been referred to as an element of high status.[16] Less well-placed people do not possess patronymics, and their names tend to change with each generation.[17] H. Geertz (1979: 346) pointed out that for people like the *sadah*, the patronymic is a kind of condensed pedigree. Like other learned people, the *sadah* chose their names by drawing on the repertoire of religious knowledge that had accumulated over the centuries, but they attached different meanings to this practice. Rather than merely enunciating their identity as Muslims, the names and titles carried by the nobility highlighted their kinship with prominent religious authorities and their special relation to religious knowledge that they transmitted, interpreted, and legislated. According to Bourdieu (1990: 170), "categorems" such as names and terms of address are "first and foremost kinship *categories*, in the etymological sense of collective, public imputations . . . As such they contain the magical power to institute frontiers and constitute groups, by performative declarations . . . that are invested with all the strength of the group that they help to make."

The patronymic is the most general identifier of both men and women, regardless of personal status (for example, *ibn* or *bint* al-Shami; a descendant of Bayt al-Shami). The patronymics of the elite that are rooted in the Muslim tradition (for example, al-Mutawakkil, al-Mu'ayyad, al-Mahdi) were institutions in their own right. According to Endress (1982: 175), in the Muslim Middle East hereditary names borne by all descendants of a particular family have appeared only recently. In the Yemen such names have been held at least since medieval times. Renowned houses such as Bayt al-Shami and Bayt al-Wazir were founded respectively in the twelfth and thirteenth centuries.

Patronymics of distinguished houses do not necessarily remain constant. Some houses are referred to by different names depending on the area. Bayt Zayd, one of the Qasimi houses which split from Bayt al-Mutawakkil, is known as Bayt al-Dawlah ("the governor's house") in Bani Hushaysh and as Bayt al-Mam (Imam) in Wadi Dahr where their ancestor Muhammad b. al-Hasan b. al-Qasim was an Imam. In the capital they are referred to as Bayt Zayd al-Mufrih. The father of the present household head who was called Zayd was known as a particularly serene and cheerful man. He was nicknamed *al-mufrih* ("the delightful").[18] As in Morocco (H. Geertz 1979: 354), it is common for nicknames based on certain characteristic features to become patronyms. Some testify to the personalized histories and self-images held by the old elite. The names often allude to the theme of knowledge. According to Bayt al-Mukhtafi ("the one who hides") of Sa'dah, their ancestor was never seen because he was always studying. Bayt Abu Talib trace descent to Ahmad b. al-Qasim who was called Abu Talib ("the father of the student") because he was dedicated to knowledge. Some patronymics are *nisab* (pl. of *nisbah*) which express a relationship to a person's place of origin (for example, al-Shami [from al-Sham, a local term relating to the northern part of the Yemen rather than Syria]; Zabarah [from hijrat Zabar]).[19]

The patronymics and certain first names and honorifics held by the Yemeni elite are components of a specific lifestyle cultivated by them. The demeanor expected of the bearers of certain patronymics is largely confined to the private domain. Some people insist that their name be used in conjunction with the article (*al-*) which elevates the carrier of the name (for example, al-Husayn, al-'Abbas). Honorifics centering on notions of truth, honor, beauty, light, and purity are used in place of personal names with distinguished religious connotations (for example, Jamal al-Din ["the beauty of religion"] for 'Ali; Sharaf al-Din ["the honor/ dignity of religion"] for al-Hasan or al-Husayn).[20] Unlike other personal names associated with the Islamic tradition which were popular among the wider population, those female names equivalent to male two-part, theophoric names (for example, Amat al-Rahman, "Servant of the merciful") were used exclusively by women of 'Alid and *qadi* background.[21] Prior to 1962, a woman was addressed as *sitti* (my lady) by women of lower status and as *ukhti* (my sister) by her siblings and same status women. Her brothers would also call her *karimati* ("what is precious to me").[22] The use of first names without prenoms is discourteous. Men address their elder brothers as either *akhi* (my brother) or *sidi* (my lord).

Property and Rights to waqf

As collectivities, houses do not normally engage in corporate activities— economic, political, or ritual.[23] Houses neither have a common territory nor are they property-owning bodies. Property is held by individuals or extended families. In northern Yemen land is rarely held by large patronymic groups. The majority of the *buyut al-ʿilm* of Sanʿa were and still are landed families.[24] Government employment and land were the main sources of wealth. Some affluent elite houses held a monopoly of a certain office or function such as the administration of public endowments (*waqf ʿamm*).[25] Unlike other religious categories in Sunni-dominated countries, the Zaydi *sadah* had no access to resources through administering shrines which are discouraged by Zaydi Islam.[26]

Today the majority of the *buyut al-ʿilm* are among the well-to-do, but few are as wealthy as the shaykhs and senior military personnel. They have maintained and increased their property by pursuing careers in newly established economic domains (see chapter 8). The revolution left few of the ruling houses impoverished. Most of the confiscated property of members who were connected with the previous regime, with the exception of that owned by the Hamid al-Din, has been returned. Some of the estates have been transformed into hotels.[27] Since they no longer testify to Imamic power, they have offered shelter to itinerant strangers and the new holders of power. The government has offered compensation or rent for some of the confiscated property owned by non-royals. There is dispute over land which was sold by the government; in some cases lawsuits are still in progress.[28]

Wealthy *buyut al-ʿilm* used to initiate different kinds of endowments. In accordance with Islamic family law, a person may alienate a third of his or her property from the heirs specified in the Qur'an and declare it as waqf. Some people determined that the beneficiaries be students, the poor of Sanʿa, and needy kin.[29] Certain endowments (*al-waqf al-daris*) obliged the beneficiary to read the Qur'an for the soul of the deceased donor. The donors specified the place and time where certain verses should be read—for example, in a mosque during the month of Ramadan. Students also benefited from the *waqf al-daris*. It was important for the wealthy to demonstrate their concern for the poor and needy. Most of the august *buyut al-ʿilm* possess *waqf dhurriyyah* (waqf for one's descendants), inalienable estates owned by a house or extended family which must not be divided or sold. This kind of waqf is the only joint property held by wealthy houses. The distribution of this waqf is supervised by a relative of the initiator or a scholar who is called *al-mutawalli*. For example, two men of Bayt Sharaf al-Din who live in San'a and Kawkaban are responsible for the distribution of income from privately owned and waqf land and the supervision of tenant farmers. These trustees are entitled to 10 percent of the income. This institution provides jobs for those who did not receive an advanced education.

Some houses benefit from income derived from family endowments which were made by several forebears. Thus, Bayt al-Mahdi possess waqf which was initiated by Imam al-Mahdi 'Abbas who lived in the eighteenth century, his son al-Mansur 'Ali, and Imam al-Mansur Husayn. One of the branches of this house enjoy an additional waqf made by a man who died in the mid-twentieth century. He stipulated that those women of his house who found themselves in difficult circumstances should read the Qur'an for the deceased and be given a 100 percent share like the men rather than 50 percent.[30] The waqf is looked after by one of the respectable scholars of San'a.

One man from Bayt al-Mutawakkil declared his library as waqf, allowing all members of his house to consult it in order to prevent partition. In order to avoid friction among their descendants and to display piety, some houses made waqf for the poorer branches and those who could not afford an education. For example, since the days of Imam Abu 'l-Fatah (d. 1052), the poor of Bayt al-Daylami have benefited from the waqf designated for them. Waqf is also established in the memory of those who died prematurely. When Muhammad b. Muhammad al-Daylami's plane was shot down by royalist forces in 1962, his kin made a waqf for a mosque and the poor in the area where he was killed.

With respect to "family endowments," genealogical memory may serve to articulate as a set of claims. Gilsenan (1982: 65) pointed out that genealogies are produced and maintained as "a social record of relations of power, marriage, and property." Genealogies serve as an *aide-mémoire* for claims to shares of endowments. Where there are perceptions of misuse of *waqf dhurriyyah*, those who have legal claims may choose to forget. In one instance, a branch of a large house decided to forfeit its rights, arguing that the conditions laid down by its ancestor had not been fulfilled by the trustees. The members of the branch reasoned that if they were to accept their share, they would be punished in the afterlife. One of the trustees had divided the waqf land and built houses for his sons on those parts he claimed for himself and his descendants. His angered kin argued that they could accept his action if the rent from the houses was to be used for sponsoring someone who was to read the Qur'an on behalf of their ancestor, but this was not what the initiator of the waqf had had in mind. Contemporary Yemeni law discourages the initiation of "family waqf" because they can be used to disadvantage certain members of the house.[31] Legal experts argue that by specifying for example—that the waqf is to benefit the advanced education of his or her descendants, a donor succeeds in excluding those who choose careers as craftsmen.[32] Although women are usually entitled to their share, their descendants have no legal claims to the property because they belong to their father's house. By contrast, the descendants of women who married within their houses are entitled to their share from the day of their birth. Some women waive their rights to their share in order to maintain an obligation on the part of their brothers to support them in times of crisis.

The Life Cycle of Houses: Bayt al-Wazir and Bayt al-Farisi

Houses have a tangible reality as physical manifestations of genealogical history and temporary power. Memory is rarely objectified in the anatomy of houses which were inhabited by generations long gone by; they become foci of sentimental attachment only when their loss is experienced as deracination (see later). Early on in my fieldwork, I was tempted to perceive the hijrahs as historicized landscapes saturated with "genealogical time" (Bahloul 1996: 115) and evoking a strong sense of belonging. However, no strong coherence between landscape, history, genealogy, and contemporary identities exists. Visits to these places with San'ani friends whose antecedents had lived there were not sentimental journeys. There is no sense in which the ancestors have become reified and embodied in houses, land, and tombs to which might be attributed mystical efficacy.[33] The houses people own in the hijrah do not carry the prestige which is attached to the fetishized family seats of the European middle classes (Le Wita 1994: 33–8).[34] However, those who possess houses in the vicinity of San'a delight in the quality of ease which they associate with them. They spend a few relaxed hours in these houses on Fridays, enjoying the view, fruit from the fields, and the company of relatives and neighbors.[35] Some have sold these estates without much hesitation. They do not have the comforts of a townhouse; some have simply been locked and abandoned, or turned into tourist hotels. Since the economy began to decline in the late 1980s, they have provided badly needed income.

Bayt al-Wazir

The history of Bayt al-Wazir demonstrates that processes of erecting, maintaining, and re-erecting houses often coincide with crucial events and "processes in the lives of their occupants and are thought of in terms of them" (Carsten and Hugh-Jones 1995: 39). Memories of the activities and positions held by men and women of these houses, and movements between houses through marriage, have been immortalized by those members who have acted as historiographers and biographers. These histories attest to the validation of memory as continuous practice which plays a significant role in the constitution of the self. The production of memory as well as its activation in the process of reading the "family histories" serves this end.[36] Beyond these discrete personal aspects of remembrance, these histories put forward truth claims which are public in nature and which may become collective memory.

Bayt al-Wazir has produced a large number of 'ulama and several Imams. Its history is intimately linked with the hijrahs they established and which grew proportionally with the political authority they wielded. This narrative tells the story of the dispersion of members of the house over time, the establishment of hijrahs from the tenth century onwards, the adoption of new patronymics, and their commitment to learning, a theme which will be elaborated in subsequent chapters. During the Imamate, Bayt al-Wazir members were appointed governors, judges, and teachers in different areas, and were involved in opposition movements and uprisings against the rulers.

Bayt al-Wazir trace descent to Imam al-Hadi, the first Zaydi Imam, through al-Da'i li-'llah Yusuf b. Yahya b. Ahmad b. al-Hadi. The first Imam to emerge from this patriline was Imam al-Mansur Yahya, a grandson of Imam al-Hadi Yahya. Members of the house stayed in Sa'dah for about one hundred years. Yusuf died in Sa'dah in 916. His son al-Qasim moved to Hijrat al-Jubjub where he taught.

Due to rivalry and fighting among Imam al-Hadi's kin, 'Abdullah al-Hajjaj b. 'Ali, a descendant of Imam al-Hadi in the eighth generation, left Sa'dah and settled in hijrat Waqash in Bani Matar.[37] His father 'Ali b. al-Qasim died in 1013. Later on 'Abdullah's son al-Mufadhdhal migrated to Bani Jabr near Marib in eastern Yemen where he established a hijrah called al-Furay' in Khawlan. He moved between Bani Jabr and Waqash and taught in both. His son Muhammad al-Afif b. al-Mufadhdhal (d. 1217) stayed in Waqash. He fought against the Ayyubids (1173–1228) and supported the rise to power of Imam alMansur 'Abdullah b. al-Hamzah (d. 1215). Before al-Hamzah came to power, Muhammad al-Afif had refused to become Imam; eventually he became one of al-Hamzah's ministers (*wazir*). Since that time, the descendants of Muhammad al-Afif have carried the patronymic al-Wazir. His living descendants refer to him as one of their main ancestors (*jadd bayt al-Wazir*). Al-Afif was a sympathizer of the Mutarrifiyyah, a pietist movement which flourished in Waqash in the eleventh century.[38] After al-Afif's death, Imam ʿAbdullah b. al-Hamzah declared it heretical and persecuted its members. Waqash was destroyed in 1214/15 but later rebuilt. It maintained its reputation of scholarship for another 200 years.

Al-Afif's nephews Yahya and Muhammad b. Mansur rose against Imam ʿAbdullah b. al-Hamzah after his suppression of the Mutarrifiyyah movement. Muhammad b. Mansur drove the Imam out of the area and claimed the Imamate himself. He assumed power under the title al-Mahdi and lived in Waqash and al-Furayʿ. His brother Yahya—one of the ancestors of Bayt Sharaf al-Din—established Hijrat Samr near Waqash after its demolition at the time of Imam al-Mansur in the fourteenth century.

One of al-Afif's grandsons, al-Afif b. Mansur, went to live in Hijrat Shazab in al-Sudah (Hashid). After Waqash had been destroyed, Shazab rose in prominence. Al-Murtada b. al-Mufadhdhal b. Mansur b. al-Afif (d. 1331/32), a scholar and poet, also moved from Waqash to Shazab. He had three daughters, Fatima, Umm Fadl, and Safiyyah who were famous for their knowledge; Safiyyah reached the degree of *ijtihad* (on her rulings, see chapter 6). (Imam Muhammad b. ʿAli al-Mahdi said of her that if she were a man, he would choose her as Imam.) When al-Murtada was visiting his uncle al-Afif he was asked to remain and eventually got married to al-Afif's daughter. He was referred to as *ʿalim ahl al-bayt* (the supreme scholar of the *ahl al-bayt*) because he was considered to be one of the greatest ʿAlid scholars of his time. His presence in the hijrah added to its fame. His son ʿAli (d. 1382) was one of the great ʿulama. ʿAli's son Ibrahim died before his father in 1380. His sons al-Hadi I, Muhammad, and Salah were the first to go to Sanʿa to study and teach, and to work in the service of the Imam. They were taught in Saʿdah, Sanʿa, and Taʿizz. They acquired a house in Sanʿa but retained close links with their relatives in Shazab, Waqash, and Saʿdah. They owned land in Saʿdah and eventually obtained more in al-Haddah near Sanʿa. At that time, Bayt al-Wazir lived in Saʿdah, Waqash, al-Furayʿ, Shazab, and Sanʿa.

Ibrahim's grandson ʿAbdullah b. al-Hadi I lived in Saʿdah and died in 1437. His mother was the daughter of the great Zaydi scholar ʿAbdullah b. al-Hasan al-Dawwari. ʿAbdullah's uncle Muhammad b. Ibrahim and his son ʿAbdullah b. Muhammad died from the plague. Muhammad b. ʿAbdullah b. al-Hadi (Ibrahim's great grandson) died in

San'a in 1492. He had a son called Sarim al-Din Ibrahim who was famous for his works on jurisprudence and poetry (d. October 1508).[39] He is referred to as one of the great ancestors (*al-jadd al-kabir*) of the different branches of the house of Muhammad al-Afif, notably Bayt al-Wazir, Bayt Uthman, and Bayt Mufadhdhal al-saghir. Sarim al-Din had three sons, Ahmad (*jadd* Bayt al-Wazir and Bayt Uthman), al-Hadi II, and Muhammad who was killed and left only a daughter. Ahmad had two sons Yahya, who continued the line of Bayt al-Wazir, and 'Abdullah, to whom Bayt Uthman trace descent. Uthman b. 'Ali (d. 1815), a descendant of 'Abdullah in the sixth generation, is the *jadd* of Bayt Mufadhdhal.

In 1504 Sultan 'Amir b. 'Abd al-Wahhab, the ruler of the Tahirid dynasty of "lower Yemen" (*al-yaman al-asfal*), conquered San'a. Several Zaydi 'ulama with a considerable influence on the population, among them the sons of Sarim al-Din, were exiled. Al-Hadi was taken to Rada' and Ahmad to Ta'izz. They were prohibited from leaving the city. Their brother Muhammad had died while defending San'a against the Sultan's forces. Sarim al-Din who had a special affection for his son Ahmad, the youngest after Muhammad's death, wrote poetry "of the kind that cures the heart" for him. Ahmad had one son Yahya; Yahya's son Muhammad, another great scholar, lived in San'a where he died in 1526.

A descendant of Sarim al-Din, 'Abd al-Ilah b. Ahmad b. 'Abdullah, went to the Bakili tribe in the region of Bani Hushaysh north-west of San'a as a mediator. He was the first of Bayt al-Wazir who was given the status of hijrah in Bayt al-Sayyid in the district of al-Sir. Others, among them several 'ulama of Bayt al-Wazir, left San'a and settled in the hijrah where they acquired land. One of 'Abd al-Ilah's great grandsons Uthman b. 'Ali b. Muhammad who was a *mujtahid* and *hakim* (judge) in eastern Yemen (al-Mashriq), lived in Hijrat Bayt al-Sayyid. Uthman became the founder of a branch which split from Bayt al-Wazir which is referred to as Bayt Uthman. After Uthman b. 'Ali (d. 1231), Bayt Uthman produced only two *mujtahids*, Yahya b. 'Abdullah b. Zayd b. Uthman and his grandson Yahya b. 'Abdullah b. Yahya, referred to as Yahya *al-saghir* (junior).[40] He was the teacher of Husayn b. 'Ali al-'Amri who became Shaykh al-Islam. Today there are very few 'ulama in Bayt al-Sayyid, and none in Waqash.

Hijrat Bayt al-Sayyid was the last one founded by Bayt al-Wazir. By establishing new hijrahs and remaining in most of the others, members of this house extended their power and influence in the country. Salah b. Ibrahim, one of the first who went to this hijrah, was blind but a good reciter of the Qur'an. He studied the sciences (*'ulum*) but did not reach the level of *ijtihad*. Salah lived for about eighty years and died in 1758. Salah's son al-Hadi IV (d. 1769) was one of the scholars who occasionally spent time in Hijrat al-Sir. He followed his father in studying at the mosques of San'a to recite and study the Qur'an. He failed to reach the level of *ijtihad*. Al-Hadi IV had a son Muhammad but little is said about him except that he married the daughter of Ahmad b. Isma'il al-Daylami who was the mother of his son 'Abdillah. Muhammad b. al-Hadi used to go to al-Sir in autumn, the time of the grape harvest. He built a house in al-Sir near the main mosque (Masjid al-Awsat). His son 'Abdillah extended the house by building a facade where the *diwan* facing east had stood.[41] He was the first who lived in al-Sir permanently and initiated a *waqf qira'ah*, an endowment for those who would read the Qur'an on behalf of his father's soul. They would receive the income from grapes which were grown on the waqf land. 'Abdillah's son Muhammad who became Imam in 1854, adopting the title al-Mansur (the victorious), was born in that house. He was educated in Bayt al-Sayyid and San'a and became a *mujtahid*. His rule was limited to the Wadi al-Sir; there were several pretenders at the time who claimed to be Imams. The Imam's sister practiced Arabic medicine. All ancestors from Imam al-Hadi Yahya down to Imam Muhammad b. 'Abdillah b. Muhammad (d. 1890) were *mujtahids* with the exception of two who did not reach this level of scholarship.

One of Imam Muhammad's grandsons, 'Ali, became an influential governor during the time of Imam Yahya.[42] He studied in San'a and al-Rawdah and was appointed *hakim* in Bani Matar north-west of San'a in 1912. In 1918 Imam Yahya appointed him *qa'id* (leader, governor) in al-Rawdah just outside the capital. A year later he became the governor (*amir*) of Ta'izz. One of his first tasks was to bring the local shaykhs who had gained considerable power during Ottoman rule (1849–1918) under his control.

As noted, in 1938 Imam Yahya precipitated the transformation of the Imamate into a patrimonial state by transferring several important offices to his sons. 'Ali al-Wazir, who was opposed to this policy, was replaced by the Imam's eldest son Ahmad and took up office in the northern province of Mahwit from 1939–48. During his governorship, his main residence was in San'a and al-Sir. He stayed in Mahwit until the beginning of the uprising of 1948 in which Bayt al-Wazir played a leading role.[43] In the newly established government headed by his relative Imam 'Abdullah al-Wazir, he was appointed prime minister, but left the office to Imam Yahya's son Sayf al-Haqq Ibrahim who had joined the opposition movement against his father. 'Ali al-Wazir became president of the parliament. On regaining control of San'a, Imam Ahmad ordered him and four other members of the house to be executed in June 1948.[44] In the same year three houses of Bayt al-Wazir in al-Sir and two houses in San'a were destroyed and their property confiscated. The demolition of

'Ali al-Wazir's house was distressing for his family for it constituted the material metaphor of genealogical time. The size and facade of the building with its unique blue stones demonstrated the symbolic capital represented by the "house." The remains of the house were taken to San'a and placed in front of one of the palaces, objectifying the Imam's victory.[45] The Imam annihilated certain manifestations of ancestral memory, houses and personalities, but the library of Bayt al-Wazir was taken to the Great Mosque of San'a. He saw their books, another repository of ancestral knowledge, as the heritage of all Muslims. They possessed an authority of their own which he respected.[46]

The history of Bayt al-Wazir demonstrates the processual nature of the house and its movements across time and space: the creation and decline of hijrahs, migration, and the establishment of new patronymic groups; the importance of knowledge which served as the basis of rank and hierarchy within houses (see chapter 5). The other example, that of Bayt al-Farisi, centers on "domestic time" (Bahloul 1996: 102). This case shows that houses may "grow" when households expand and buildings are added to old ones; men take additional wives, their sons marry and raise families. There is constant movement in and out of houses—women leave in order to join their husband's households, but return following divorce or widowhood; young men go abroad to study or take up posts outside the city where they were raised. The rooms of the house rarely "belong" to anyone for long. As H. Geertz (1979: 336) remarks, "the association of persons with rooms is by no means simple." Nowadays some previously overcrowded houses are abandoned spaces, speaking of a by-gone era; after the revolution, most of their former inhabitants rebuilt their lives in other parts of town. Bayt al-Farisi is one such example.

. . . and Bayt al-Farisi

Ahmad al-Farisi, who was born in the late 1880s, held a post in Imam Yahya's government. Like many of his colleagues, he had no office outside his house. (Often these men pursued their professional careers within their houses or in buildings in the courtyard.) As the number of his wives and descendants increased, Sayyid al-Farisi built a second house next to the old where he could act in a consultative capacity without disturbance. He had eleven children by four wives (figure 3.3).

The old house had four floors.

Ground floor: storage place for wood and grain, and a grinding stone.
First floor: kitchen, pantry, bathroom, maid's room, family dining room (women and children ate after the men).
Second floor: Ahmad al-Farisi's room; Amatullah's room (Amatullah was Ahmad's first wife), taken over by Safiyyah, the fourth wife,

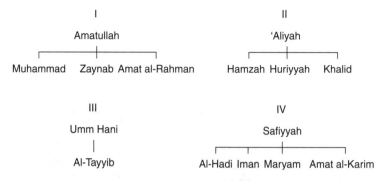

Figure 3.3 Sayyid Ahmad al-Farisi's wives and children

after Ahmad married her (by that time, his second wife, 'Aliyah, had died and the third, Umm Hani, was divorced); bathroom; a *diwan* where Ahmad kept some of his books and had meetings with clients and colleagues.

Third floor: bathroom; *makan al-wasat*, a multi-purpose room which was used by members of the family throughout the day and after dinner; two of the elder unmarried daughters slept there. One room was occupied by Amatullah's eldest daughter Amat al-Rahman. The other rooms were occupied by Safiyyah who moved to this floor after she started to have children. (Meanwhile Amatullah, who had lived on this floor, moved back to her previous room, her children having already grown up.)

Fourth floor: two rooms belonged to Hamzah and his wife and their two children; the single room to Khalid and his wife, who had no children.

The smaller two-storey house was built in the early 1950s. It had storage space and a bathroom on the ground floor; the *diwan*, where most of Ahmad's books were stored, was used for occasions such as weddings. Ahmad received visitors throughout the day in the *diwan* on the first floor; social gatherings were held there in the afternoon. Guests would take seats in the room next to the *diwan* when it became over-crowded. (Safiyyah's son used the room after he reached the age of ten and was no longer allowed to sleep in his mother's room.) Ahmad's oldest son Muhammad, a scholar, gave Qur'an lessons to the children of his brothers and sisters in the *diwan*. The second floor was occupied by Muhammad and his family. His oldest son Husayn lived in the small room next to the bathroom. The room next to Muhammad's belonged to his wife, Fatima. There were bathrooms on all floors (figure 3.4).

Figure 3.4 Layout of Ahmad al-Farisi's house

Ahmad's eldest daughter Amat al-Rahman married in the 1940s. She was divorced shortly thereafter and returned to her father's house. She was older than her father's second wife and had a strong personality which resembled her father's. He took delight in her assertiveness and strength of character. The younger children were afraid of her. It was said of her that "she ruled the house." As one of her sisters put it to me, "she embodied the house (*hiya kanat al-bayt*); her father's house was her kingdom. She took responsibility for everyone in the house, and she knew how to deal with people. She was an authority figure like my father. One could not really talk to her." Because she was strict, she was called *'ammah* (paternal aunt) by her younger sisters and their children who would normally address her as *khalah* (maternal aunt). When Safiyyah's children asked her for permission to go out, she advised them to consult *'ammah*.

After her husband's death in the late 1950s, Zaynab, Ahmad's second eldest daughter, divided her time between her husband's and her father's house. When she came to live with her family, she occupied her father's *diwan* when it was available or stayed with her sisters who lived upstairs. She later moved to the small house. Before her marriage in the 1950s, Huriyyah lived in the small room on the third floor opposite the *makan al-wasat*. After her divorce, she returned to the small house where she lived with her daughter in the *diwan* on the ground floor. After her former husband died, she moved to his house which was inherited by her daughter.

Sayyid al-Farisi was dedicated to his children's education but he sent only the boys abroad to advance their studies. He arranged marriages for his sons when they were about sixteen years old. Muhammad's first wife was his cousin's daughter, who died young. He then married a woman of the House of the Imam. During the time of Imam Ahmad, he worked as a judge in a southern city, spending much of his time away from San'a. In 1962 he was arrested and spent over a year in prison. On his release he left the Yemen and settled in Cairo where his younger brothers Khalid, Hamzah, and al-Hadi had completed their high school education. Their brother al-Tayyib, who was born to Ahmad's Lebanese wife, Umm Hani, lived with his mother in Beirut and only came to San'a for short periods. During those visits, he lived in Khalid's room while the latter was abroad. After finishing high school, Hamzah studied civil engineering in Russia and Czechoslovakia. There he took a second wife whom he never brought to San'a for fear of his father's disapproval. He was arrested during the revolution and died soon after his release from prison.

Al-Tayyib, who attended the Military College in Baghdad, was abroad when the revolution started. He returned to the Yemen several years later and began his career as a diplomat. Khalid studied political sciences and returned to the Yemen two years after the revolution. He married his brother's widow and wanted to settle down in San'a, but his family advised him to leave again because he might not be safe. He completed his doctorate abroad and also became a diplomat. He died in San'a in 2002. According to his sister Iman, during the revolution even poor Farisis "who knew nothing about politics were harassed. Anybody who carried that name had problems because they were related to 'Ali al-Farisi (who occupied a prominent position in Imam Ahmad's government) and my brother Muhammad."

Al-Hadi went abroad to study and spent eleven years in Cairo and Budapest. He took a degree in tropical medicine. On returning to the Yemen in the mid-1970s, he married and took up employment in the Ministry of Health. Before he built himself a house in the suburbs of San'a, he occupied the entire third floor of the old house. Several of his relations live nearby, among them his nephew 'Abd al-Qadir, the son of Muhammad and his third wife, Amat al-Khaliq. Muhammad's second wife has remained in exile. After her husband's death in 1960 Safiyyah, Ahmad's fourth wife, went to live with her brother in Dhamar. Their older brother, who had been a senior official in the Imam's government, was executed in 1962. Safiyyah spent some years in exile in Saudi Arabia, but later returned to San'a where she lives in her son's house.

Between 1962 and 1967, the house was inhabited solely by women. Their male cousins and friends of the family did the shopping for them.[47] After her son was killed and the women of the house were evacuated by the Red Cross, Safiyyah's mother insisted that her daughter move to her house, but none of the women was harmed. Some of their female kin whose husbands were killed or imprisoned stayed with their retainers while the house was occupied.

Until they got married, Safiyyah's daughters lived with their mother on the third floor. Safiyyah's eldest daughter Iman left the house in the late 1950s when she married a man of a renowned 'Alid house who is a close relative of her maternal uncle who was killed. In the early 1980s, her sister Maryam married a shaykh's son who works for the Security Forces. The couple have their own house. In 1985 Ahmad's youngest daughter Amat al-Karim got married to a Foreign Office employee of Turkish origin. Like many other women of 'Alid houses in the last two decades, the women of Bayt al-Farisi have begun to enter conjugal relationships with non-'Alids. During the years following the revolution, the life of the houses where they grew up has changed as significantly as theirs. Muhammad died in 2000, a few months after his sister Amat al-Rahman. What is left of Bayt al-Farisi are small, well-to-do families in the suburbs few of whom embody the domestic memory of the old house, which is let to rural migrants.

Part II

Growing to be 'Alid

Chapter 4

Snapshots of Childhood

This chapter draws biographical sketches of children whose families belonged to the Zaydi elite of the Imamate. Their fathers served in the governments of Imam Yahya and Ahmad. A substantial number of 'ulama have emerged from their houses who held positions as rulers, governors, judges, and teachers all over the country. The stories demonstrate how the children learnt about their social location, the diversity of religious affiliations, and notions of righteousness.

Ibrahim

One of the stories which is told by Ibrahim, one of 'Ali al-Wazir's sons (see chapter 3), describes how the assumption of religious and political leadership becomes a natural disposition; and how it motivates people and shapes their experience of what is right, good, just, dignified, and worthwhile (Lambek 1997a: 134). Ibrahim's story, set in the 1940s and 1950s, draws out the fundamental connection between the making of 'Alid identity and the appropriation of the world of the ancestors as landscapes of personal memory worlds—to borrow a term coined by Munn (1995: 87). Some passages maintain direct speech to show how this man's self-understanding and memorizing is realized within a certain narrative frame which is contingent on culturally specific coordinates and social location.[1] For Ibrahim, the names of Imams and key religious terms provoked a mnemonic passage from the historical *mise en scène* into his personal experience.[2] Ibrahim represents—or indeed remembers—one event in terms of another (P. Burke 1989: 102). He interprets his experience of the failed uprising against Imam Yahya, in which close relatives took part, in analogy to the revolts led by righteously guided leaders against unjust rulers during the early period of Islam. The theme of the denounced Umayyad rulers is also taken up in his account of the ritual which was performed when his brother accomplished reading the Qur'an. The pastness of the memory shared among Shi'i Muslims is actualized as his own past. He lives through the traumas of his youth by revisiting the mental worlds of his ancestors who became martyrs, configuring them into his personally lived identity space. Imam al-Husayn's tragedy in Karbala has become a narrative of trauma for generations of Shi'is, an historical event which almost transcends history, but the violence of the events experienced by the young Ibrahim in the late 1940s gives this event a remarkable actuality. For the boy, there was no escape from the memory worlds that are inhabited by his ancestors and which gave meaning to his personal traumas. In other words, biographical memory here is activated by a recourse to events in the lives of his ancestors, the remembrance of which has shaped Ibrahim's childhood memory. The memory of the failed uprising of 1948 is partially assimilated to the historical narrative of martyrs.

Autobiographical and historical memory merge and jointly shape Ibrahim's sense of self and the trajectory of his life. I pointed out earlier that some *sadah* interpret the 1962 revolution within the parameters of the history of harassment suffered by the *ahl al-bayt* since the days of Umayyad rule. In Ibrahim's story, personal trauma relating to the 1948 uprising is most explicitly situated in the collective memory of persecution. In other words, "the historical power of the trauma" (Caruth 1991: 7) which manifests itself in its inherent latency extends beyond autobiographical memory into a distant shared one. It remains an open question whether this manner of assimilating the traumatic event contributes to minimizing or intensifying its effects. Historical consciousness provides a pre-established framework for a repeated suffering of the event at the same time as it facilitates a temporal departure from it.

In 1939 Ibrahim's father took up office in the northern province of Mahwit. One of Ibrahim's cherished memories of Mahwit is a green hill called *masna'ah* which faced the family residence on the west side. It overlooked both the town of Mahwit and the terraced fields, some 100 meters away. The mountains were planted from their peaks down to the valley with wheat, corn, fruit, coffee, and flowers. The cluster of houses around the governor's residence was itself a veritable little town. The mosque where Friday prayers took place was at the foot of that hill. Between the market and the town there was a large square with the school that he and his brothers and sisters attended. Ibrahim's education was to prepare him to become a scholar—he was to distribute knowledge (*'ilm*) and to interpret it, to judge, and to govern. He lived up to these expectations not merely through his educational achievements. As a child he sent letters to his younger brother

Qasim addressing him as *Mawlana Amir al-mu'minin al-Mutawakkil 'ala -'llah al-Qasim b. 'Ali al-Wazir* (My lord, the leader of the faithful Qasim b. 'Ali al-Wazir). The exercise of government became the stuff of childhood phantasies. In his mind, supreme political leadership was intimately linked with knowledge. His social location demanded that scholarship be a matter of filial piety. From an early age, the children were instructed by a private teacher and they only went to school for a couple of years until they were about ten years old. They were also taught by their father. From the beginning, they learnt reading and writing and producing the sounds encoded by the Arabic characters. Within six months they had memorized whole extracts from Qur'anic verses and works of poetry, which they also had to write down. If they made mistakes, they were beaten. They learnt about the lives of the Prophet and Imam 'Ali, about Imam Husayn's martyrdom, and other formative events in the history of Islam. They were also taught about the body's ritual cleanliness. Emphasis was placed on both recitation and writing. The status of a memoriser was higher than that of a person who merely knew how to write. Recitation was said to promote the understanding of Arabic grammar and the correct pronunciation of words. The children were told that there were two reasons they had to know grammar well and to learn it by heart. Knowledge of grammar was required in order to understand the complex language of the Qur'an, and to assist the boys in their future careers. When they were a bit older, the boys were also trained to be eloquent. They had to give speeches and to recite in front of the class. Even while they had not yet reached puberty, they spoke at public rituals that were performed to honor children who had successfully completed reading the Qur'an (*khatm al-qur'an*).

It was explained to Ibrahim that the knowledge he had learnt by rote would facilitate reasoning and the finding of new meanings and answers. Learning would help the development of the mind, and of reasoning (*'aql*). In order to be fully recognized as adults it was important to demonstrate *'aql*. Some boys were treated as "men" even before they reached puberty. Ideally the boys would in future be able to practice independent judgment (*ijtihad*), but in order to do so they had to memorize the sources without having to consult them. Despite the emphasis on rote learning, they never memorized the entire Qur'an. They read it and then memorized a few verses. The girls studied with their brothers until they finished reading the Qur'an. The ritual marking the occasion was performed for them, too. When the boys started to attend school, their father hired a private male teacher for the girls.

Ibrahim's father had four wives, each with her own residence. Two of those, among them Ibrahim's mother, were Shafi'is. Ibrahim was never resentful toward the other wives and had cordial relations with them. They treated him like one of their own sons. It was through them that Ibrahim first came to know about different religious practices when he was five years old. When his father entered Ibrahim's room he found him laughing, and asked him why. The boy explained that he was laughing because of the way his *khalah*, one of his father's wives, and the *duwaydar*, the houseboy who ran errands, prayed. His father explained to him that they were Shafi'is.[3] Ibrahim then decided that he too was a Shafi'i because he loved his *khalah*. At that time he also came to understand that he was descended from the Prophet Muhammad. His eldest half-brother's wife told him that his nephews were descendants of the Prophet on both their father's and mother's side whereas Ibrahim was a descendant only on his father's side. Ibrahim was sad and went to his mother and told her about what he had learnt. She told him that his parents were descended from two prophets, one being Muhammad and the other Hud.[4] This news made him very happy. There was an expectation that the children would learn to recite the names of their ancestors back to the Prophet. The boys and one of their sisters who was of similar age would compete with each other to see who could memorize the furthest back, with Ibrahim's younger brother Zayd being able to go back to Imam al-Hadi, a descendant of Imam 'Ali in the ninth generation.

Ibrahim's mother Fatima, who had never learnt how to read and write, was resolute and socially aware. She was asked for advice by other women who felt they had been wronged by their husbands or brothers. For example, she would be approached by women whose husbands had claimed that they had not been virgins and asked for their *mahr* to be returned. When she felt that women had been treated unjustly, she transferred their cases to her husband. She offered hospitality to the poor and occasionally prepared food for the prisoners of the town.

Ibrahim was the son closest to his father. He had permission to sleep in his room. The governor used to get up early and recite the dawn prayer (*fajr*). He woke his son gently only after the sun had risen. He allowed Ibrahim slightly more time to sleep than he had allowed himself. The father used to have breakfast on his own and he would then act as an arbitrator among people who brought their cases to him. After lunch he issued legal judgment on the queries that he had received, and later in the day, he would teach students in his house for two hours. Sometimes Ibrahim attended the classes. After dinner the governor played with his children and would ask them what they had learnt that day. He wanted to know what they had understood from the texts rather than how much they could remember.

He explained to them the meaning of hadith and Qur'anic verses which they had written in their notebooks in the morning. Ibrahim and his older brother tried to learn by rote the *Matn al-azhar*, the central Zaydi work of jurisprudence. They disappointed their father. When their younger brother, Qasim, started to memorize it at the tender age of ten his father discouraged him. "You will only forget it all like your brothers before you." The boy's teacher was astonished to hear that his student was not going to continue memorizing the *Matn*, but because his father saw little point in it he did not insist. However, he succeeded in memorizing the *Matn al-hajib*, a grammar book. Ibrahim made sure that his younger brothers memorized it without making mistakes; if they did he would spank them.

When they had finished school, Ibrahim and his brothers would ride out, climb the hills, or swim in the small ponds. In Mahwit, only the boys were allowed use of the pond in their garden, but in San'a and Hijrat Bayt al-Sayyid however, their sisters could swim in garden ponds, which were used for the irrigation of grapes. Because the gardens were secluded, the girls did not have to give up swimming after reaching puberty. The children also played with their local peers. Horse riding was conceived as an appropriate activity for the boys, but they were not allowed to ride a bicycle—an activity considered to be both vulgar and one which displays a lack of self-restraint. Bicycles had first appeared in the capital San'a where the family had a house. Anxious to see bicycles, Ibrahim's brother Zayd once went to the center of the city. He was not supposed to walk about in the streets and, above all, he had to avoid the market place. While his eyes followed the boys riding their bicycles, his guard suddenly appeared next to him. He felt ashamed of himself and was afraid that his guard would tell his father where he had found the boy. In Mahwit, the children were allowed to play outdoors, but in San'a they had to confine themselves to the yard of the house. The city was said to be polluted (*wasikh*). They were not supposed to be out and about lest they might learn bad words from the market people (*awlad al-suq*) or listen to the lute which was *'ayb* (improper, shameful). In the company of a guard, the children would go to school, the mosque, the Imam's residence, or to visit relatives. When visiting their aunt, they could only avoid the suq by leaving the city through one gate and moving back in again through another.

The children were looked after by a Turkish nanny of whom they have fond memories. On returning home from their lessons, the children competed with each other to place their heads on her lap while listening to her stories. At night she continued telling them Yemeni stories and stories about the Ottoman sultans. Their father contemplated for a while employing a French or English nanny so that they would learn another language. But, being warned that she might not pray, he abandoned the idea.

The boys remember being addressed by the title *sidi* since they were able to understand the meaning of words. In spite of their elevated social position, their father did not want them to feel superior to other children by virtue of their noble birth. He taught them to judge themselves on the basis of their moral conduct. When Qasim was angry with a servant and told him that he was only a servant (*khadim*), his father asked him "What is the difference between you and him?" During a procession, when the boys were on horseback, they passed some of their fatherless schoolmates who greeted them with a deferential salute. The governor expressed his displeasure at the offhand greeting his sons gave in return by withholding their horses temporarily. At home, he shared his meals with his guards and gardeners.

The boys' older sister Huriyyah read *A Thousand and One Nights* to them when they were between eight and eleven years old. She simplified the classical Arabic into a more vernacular form so that they could understand it. These stories conveyed the idea of love to the children. Ibrahim's brother Zayd had at the age of six developed an affection toward their neighbor's daughter. He was eager to see her and play with her. He asked Huriyyah to let him put on some *'udah* (perfume oil) and his best clothes, and to take him to the girl's house. She told him that *'udah* was only used by women and refused to take him.

The family had a library, which contained about 3,000 books. Some were 800 years old and had come from Iran and Tabaristan. Ibrahim spent much of his time in the library. The father wanted his sons to be familiar with the religious sciences, history, and *belles lettres*. He gave Ibrahim books about the Russian revolution and an Egyptian translation of Goethe's Faust. There were always books and papers lying between heaps of qat in the room where men's afternoon gatherings took place. Literature brought by visitors from different places was discussed at these daily meetings. Illiterate men who attended learnt from the conversations being held. In the 1940s the governor's reform-minded friends and colleagues had occasional meetings at his house. The governor did not mind if his children entered the room when he had guests and they sometimes felt free to make their own comments. From the age of ten the boys took a keen interest in people's problems and tried to offer their advice, an endeavor which found their mother's approval.

The governor discussed technological innovations with his children, explaining that in America there were lifts in big buildings. He listened to the BBC Arabic service three times daily. He never traveled without his radio; among his entourage was a retainer who rode the camel which carried the radio and the batteries. Ibrahim was often asked by his father to follow the news on the radio and later to provide him with a digest, which was sometimes read by his father and his guests at their meetings. In this way the boy learnt much about what was happening in the world. As he was listening to the news about the rising power in the heart of Europe, he began dreaming about the wide streets of Berlin. Many Yemenis favored a German victory over Britain, anticipating that this would free the Arabs from British colonialism.

One Friday Ahmad al-Wazir, a remote cousin of the governor, a well-known scholar who taught in Bayt al-Sayyid (d. 2003), delivered the sermon (*khutbah*) in the mosque of Mahwit. Ibrahim still recalls that sermon.

He said that the holy Qur'an had created a rational revolutionary thinking in the world of mankind and that it had brought mankind from darkness to light. His sermon had an impact on me and I insisted on traveling to al-Sir (the regional "capital" of the area where Bayt al-Sayyid is located). When I informed my father that I wanted to go to al-Sir he told me "your ancestor Imam 'Ali appeared in a dream and told me that I would be with him soon, so I would rather keep you close to me." I was sad to hear this, but later I was given permission to go. During the summer months I was taught in al-Sir. Upon returning from Mahwit, my father arranged for me to meet the Imam and some of his officials so that they would know me. I recited poetry in front of the Imam which had been written by someone else, but I pretended that I had composed it. I emphasized the Imam's good qualities and asked him to absolve my father from his position in Mahwit. I wanted my father to live in al-Sir so that I could spend more time studying there without being deprived of his presence in my life. The Imam was impressed and sent a telegram to my father telling him that his son was one of the country's best young poets.

My new teacher in al-Sir was open to new kinds of knowledge and represented the reforms which the "Seal of the Prophets" had called for, namely that men should live in accordance with the spirit of their epochs, in the framework of exemplary morals, and constant progress. Imam 'Ali had said that one should prepare one's children for the future, for a life which is not your own. Through my new teacher and my father—may Allah bless him—I learnt about Jamal al-Din al-Afghani, Muhammad 'Abdu, Muhammad Rashid Rida, and 'Abd al-Rahman al-Kawakibi. I was also influenced by the writings of one of my forebears, Muhammad b. Ibrahim (d.1417). In a poem, he advised his brother, who was a minister, to stay aloof from politics. The 'ulama might choose a ruler but they themselves should only act as moral critics of the holders of power. They should make a living from things such as bookbinding, use the income for their immediate needs, and give the rest to the poor. Honor was derived from asceticism and good deeds, whereas all temporal rulers (*ahl al-mulk*) were to perish.[5] I came to know about the French revolution and its principles which were preceded by Islam in emphasizing freedom, brotherhood, and equality. I also studied Kawakibi's work *Taba'i' al-istibdad* (The nature of oppression). It helped me to understand why the Qur'an has condemned the Pharaohs of mankind whom it describes as tyrannical. They had no regard for consultation (*shurah*) and democracy.[6] I have found that the Qur'an has always emphasized reason (*'aql*), knowledge (*'ilm*), and comprehension (*fiqh*) by which I mean thinking and understanding. If man uses his mind he will be enlightened and no one can stand before the mind with which humans are endowed from birth, and he will understand that human nature is good and that freedom entails responsibility. The human mind can be used for good as well as for evil.[7] I also learnt some love poetry, the Diwan al-Mutanabbi, pre-Islamic poetry, and details about the life of Imam 'Ali.[8] I continued with my studies of the Qur'an and hadith, and I was also trained to give speeches.

When my younger brother Muhammad finished reading the Qur'an at the age of nine, there was a big celebration. The retinue of the governor arrived at the schoolyard which was open to the townspeople. My brothers and I were all dressed in our finest clothes. Our garments were shiny; our headgear was embroidered with gold thread, and our belts were also embroidered. I was carrying one of my father's golden swords. All those who took part in the celebration, the students, the notables, and the soldiers were wearing different kinds of clothes. The celebration began with recitations of the Qur'an, which were followed by speeches from one of the 'ulama and my brothers and myself. I gave a speech in front of my father, the *amir*, the *qudah* and the 'ulama who were with him. I began with a reference to the holy Qur'an (1:12). I talked about the oneness of Allah, saying that there is nothing like him and that the message of the Qur'an was one and unalterable. I said that Muhammad was Allah's final messenger, and that the Qur'an was the proof to mankind that this message was the final one. The Qur'an was an example for all people until the Day of Judgment. All humans were of the same origin, the father was Adam, the mother Eve, and Allah divided them into peoples and tribes so that they could know each other and be diverse. In the eyes of Allah the best of them were those pious ones who endeavored to avoid evil and do good. I recalled the revolutionary message the Qur'an had sent to the world and the omens of the modern reform movement. Islam had generated a civilization which extended from Mecca to the Great Wall of China in the East and to the borders of France and Switzerland in the West.

Ibrahim's speech was followed by recitations (*nashid*) by the students of the school. Thereafter they performed a play which affirmed the primacy of the *ahl al-bayt* over the usurpers of their power. The children re-enacted the pilgrimage to Mecca by the Caliph Hisham 'Abd al-Malik (Umayyad ruler; 724–43) during which he was prevented from touching the pillars of the Ka'ba by other pilgrims.[9] When however Zayn al-'Abidin, the son of Imam al-Husayn, approached the Ka'ba people greeted him and moved aside. When one of the caliph's companions asked "Who is this man who is granted so much respect?," the poet al-Farazdaq who was also present, began to recite verses of poetry in praise of Zayn al-'Abidin. The caliph became angry and ordered the poet imprisoned. According to Ibrahim, the children's goal was to demonstrate how the strong pious souls [among past generations] stood up to authoritarianism, and how their example has influenced the hearts of men.

At the age of fifteen, Ibrahim was to witness the translation of the ideas he had adopted from his teachers into the language of resistance during the uprising against Imam Yahya. Inevitably the young Ibrahim was drawn into the cataclysmic events which divided the elite and two of the leading houses who had worked together to strengthen the Imamate after Ottoman occupation and who were related through affinity.[10] Ibrahim's coming of age coincided with the last contest of this kind in the history of the Imamate. While his father was awaiting execution, Ibrahim sent letters to him comparing Imam Ahmad to Yazid and pleading to fight the oppressor (*al-taghiyah*) for the rest of his life. He invoked the martyrdoms of Imam Husayn and Imam Zayd, saying that once again the principles of law and justice were not honored by the rulers of the time.

During this period Ibrahim's family suffered immense hardship. His mother had to sell milk, traycloth (*quwwarah*) and carryall fabrics (*buqshah*), and to take loans. At the daily women's afternoon gatherings (sing. *tafritah*), the women were made to suffer the consequences of actions taken by their male kin. The women had previously been the focus of attention, but in the aftermath of the uprising they were passed by other women without a word of greeting. Ibrahim's half-brother 'Abdullah fled to India where he died. His older brother 'Abbas was jailed along with other leaders of the movement. He was held in al-Qal'ah, a citadel in the capital San'a where ammunition and grain for the army were stored. When 'Abbas was taken to hospital for treatment and given permission to visit his mother, Ibrahim had to replace him in prison. Ibrahim had been distributing a journal published by the Yemeni opposition in Cardiff (England) called "Peace" (*Al-Salam*), which the Imam attempted to prevent him from doing. At al-Qal'ah he taught the child hostages (sing. *rahinah*) and conveyed to them his views about the regime. In 1954, some shaykhs helped

Ibrahim and his brothers 'Abbas and Muhammad to escape to Aden from where they went to Egypt.

Khadija and Safiyyah

Khadija and Safiyyah grew up during the reign of Imam Ahmad. They were among the first girls who attended school. Khadija's father had opposed Imam Yahya and spent several years abroad; Safiyyah's father and other relatives were loyal civil servants.

Khadija started to veil at the age of nine. This was also the time when she was discouraged from entering her father's *maqyal*. She was told by her mother "you cannot go there, you will be a woman next year." Khadija was never told that she descended from the Prophet.

> I understood that I was a *sharifah* (female descendant of the Prophet) by being in the company of adults. People who came to our house addressed my mother as *sitti* and my father and my older brothers as *sidi*. When you learn that you are a *sharifah*, you understand that you must have good morals. Some of my friends were reminded that they were *banat al-nabi* (female descendants of the Prophet) whenever they misbehaved. My parents only spoke about our family's good name. I was mixing a lot with children who were not *sadah*, but I was not allowed to play in the street. I was told "You are from Bayt F. You are not a girl from the street (*shawari'yyah*)."

When Khadija started to attend the *maktab* (Qur'anic school), her mother and elder sisters helped her with her reading in the evening. Khadija did not worry that she understood little of the meaning of the words she had learnt, for she was at pains to prove herself by finishing to read the Qur'an as quickly as she could. She was looking forward to the *khatm* celebration which had previously been performed for her older sisters, and she was competing with her brother 'Abdullah who was only slightly older than her. For her, acquiring knowledge was as important as it was for him.

> Between nine and thirteen, girls' ability to learn is much greater than boys'. Later society encourages girls to develop *'atifah* (emotion, senti-ment) which is why girls fall behind. I was so eager to finish reading the Qur'an that I skipped some of the lines I was reading when my teacher wasn't looking. I wasn't really interested in grasping the meaning of the verses I read. It took me about two-three years until I finished reading the Qur'an.

The *khatm* ritual took place when Khadija was nine years old. On that day, she went to the *maktab* and returned with all her classmates, about twenty-five girls. She and her teacher went ahead of them; Khadija carried the slate on which a Qur'anic verse was written, decorated with flowers and *shadhab* (rue). Khadija wore a dress of special and shiny material, and she had *naqsh* applied to her hands. (*Naqsh* [drawing, inscription] is an ornamental pattern which is based on gall-ink, *khidhab*. It resembles a tattoo, but it does not last longer than about a fortnight.) Khadija, the other girls and her relatives were given lunch in the lower *diwan*. After lunch, Khadija wore a white dress, make-up, and pearls. This was the only occasion when an unmarried girl was allowed to wear make-up because unmarried girls must not enhance the body's sexual attributes (see vom Bruck 1997a). Khadija enjoyed wearing make-up so much that she wanted to get married only to be allowed to put it on. She was reluctant to remove it and did so only superficially after her mother urged her. Adornment is one of the things women value about marriage; it is appreciated aesthetically beyond what it signifies for sexual relationships between men and women. To Khadija, marriage seemed desirable because it would transform her body into that of a fully adorned grown-up.

Later a *zaffah* (chant) was performed by a woman reciter (*nashshadah*) while everybody moved to the upper *diwan*. Khadija's sisters and paternal cousins were next to her, followed by the others. She was given the best place in the *diwan*, and she was the only one sitting on a chair on a slightly elevated platform. Khadija was to hold a speech; for the last three nights she had been unable to sleep. Lying awake, she had repeated the speech which had been written for her (possibly by her father) again and again. On that special day, she recited these well-rehearsed verses that praised God and spoke of the glory of the Qur'an, and a few verses from the Holy Book. Thereafter the *nashshadah* gave a speech. After the *fatihah* (the opening surah), she praised the girl's success, saying that she was specially gifted, that she came from a good family which had raised her well, and that her mother had taken exceptional care of her. She asked God to preserve her parents. Khadija and the unmarried girls went to another room where they were given raisins, almonds, and chocolate. Later in the evening, Khadija held her speech in front of her father and brothers who gave her a gold ring. She wanted 'Abdullah to acknowledge her success. That night she felt so bold that she went and stole the batteries from his radio in order to annoy him.

The next day a *mawlid* was performed in which only the teacher and married women took part. She remained only for a short period. She was still not entitled to participate in the *tafritah* (pl. *tafarit*, daily women's gatherings) which must not be attended by young unmarried women. Whenever she wanted to enter, she was told "this is not your place."

After the *khatm* ritual, Khadija continued her studies. Those were the late 1950s and one could notice that things were changing. Qur'anic schools were declining.

My *khatm* was less elaborate than that of my sisters. I attended an elementary school where I was instructed by a male teacher. Under his supervision I read the Qur'an again, and I learnt some mathematics and improved my writing skills. Later I went to a school for girls called Bilqis. I was still very thirsty for knowledge. I wanted to study or kill myself. There was hardly anyone who would stop me. My father had already died, and during the 1960s all my brothers were abroad. During this period, I read the Qur'an again. My mother explained the meanings of the verses to me. It was only at the age of sixteen that I felt that I had gained some real understanding.

In 1969 Khadija got married at the age of seventeen. The civil war was over but she did not continue her studies. By that time Safiyyah, whose grandfather was Khadija's cousin, was abroad. Safiyyah began her education with a private teacher (*sayyidatna*) when she was seven years old, a year after she had begun to cover her face when she was about six years old "in order to get used to it."[11] She and other girls would go to the teacher's house every morning. After they had learnt letters and words, they wrote sentences of Qur'anic verses on a slate and then read them aloud. When they made a mistake they would re-read the text from the start. Each day one of the girls said prayers in front of the other girls and was corrected by the teacher. They had to wipe off the words they had written on the slate before they left. The teacher feared that the slate might not be kept clean and the words of the Qur'an might get polluted. After dinner Safiyyah's father explained the meaning of the Qur'anic verses which she had learnt to her, and they discussed them together. He told her that the knowledge of the Qur'an would open her eyes to everything in the world. She did not learn the Qur'an by heart with the exception of two verses, Ya Sin and Tabarak.

Her aunt was not so lucky. When she wanted to learn how to write, her older brother objected by saying "tomorrow she will write letters to men." Nevertheless, her father agreed that she would get some instruction. He asked to see his children's homework every night. Once he came across an especially fine piece of writing. On learning that it was his daughter's, he asked her teacher to stop teaching her. He was worried that she might not find a husband.

When her teacher had fallen ill, Safiyyah was sent to the local *maktab*. There was a boy in her class who looked very grown up and the girls were wondering whether he was still under age. "We said to each other that even while he was wearing a *kufiyyah* (skull cap), he was likely to have an *'imamah* (headgear of adult men) at home." Safiyyah did not study at the *maktab* for long because she could not bear witnessing how the boys were beaten or made to stand in a corner for an hour after misreading a text. She asked her father for permission to leave.[12]

> My father told me that boys were devils (*shayatin*) and that the teacher had few other options than to beat them. However, he agreed to take me out of the *maktab*. I was then educated by one of my uncles. In 1961, I visited my uncle Ahmad who was a diplomat abroad. Under his care, I greatly improved my writing style. I collected all my notebooks so that I could present them to my father on my return to the Yemen. When I learnt that he had been executed I burnt all my papers in the fireplace.

Chapter 5

Performing Kinship

We ought to transform what we read into our very selves, so that when our mind is stirred by what it hears, our life may concur by practicing what has been heard.

Gregory the Great

Dealing with childhood reminiscences, chapter 4 indicated that by focusing on the transmission of knowledge, narrative memory selects from the past what is culturally appropriate. This chapter seeks to illuminate this cultural elaboration by examining more closely the inextricable link between what one knows and what one is and becomes (Bourdieu 1990: 73; Lambek 1993: 6). 'Alid kinship ideology stresses that on its own, generative substance fails to create moral persons. This substance is imbued with moral values and can be acted upon. A person gradually unfolds his or her intrinsic potentials through activities such as the study of the scriptures and speaking with eloquence, as is exemplified by the biographical snippets of Ibrahim and Khadija. Patrilineal descent articulated with knowledge is a marker of status and of boundaries between learned and ignorant *sadah*, who are barely recognized as kin. On the other hand, the learned ones share a common outlook with knowledgeable non-'Alids to whom they render respect. Looking at the teacher–disciple relationship, the chapter asks whether knowledge transactions can relate people to each other in such a way as to produce kinship.

The emphasis that is placed on knowledge, which in conjunction with "blood" serves to produce "proper" persons, encourages a re-examination of the notion of kinship and descent as immutable, which has gone largely unchallenged in Middle Eastern anthropology. These notions are based on the assumption that the ideology of patrilineal descent takes precedence over all other principles of social organization. Following the path-breaking work of Schneider (1964, 1984) and Needham (1971), there has been a shift away from the biogenetic criterion of substance as the defining feature of kinship toward the elaboration of other criteria such as place and nurture or ritual processes.[1] Being a *sayyid* is not simply having been born a *sayyid* but an identity constituted in time.

Ideas about the mutually constitutive role of blood and knowledge foster idealtypical notions of the 'Alid self and are reflected in the Zaydi doctrine of the Imamate. The notion that inherited substance requires animation through learning implies a theory of mind, which is predicated on the assumption that by virtue of sharing the Prophet's substance, the *sadah* are endowed with the potential to better interpret the holy scriptures. This notion is common to Shi'i teachings, yet Twelver Shi'i and Zaydi teachings about the Imams reveal different approaches to the function of inherited substance and knowledge in producing persons capable of ruling. Beyond its function to demonstrate how Shi'i scholars have grappled with the twin notion of descent and knowledge which informs concepts of Shi'i authority, the subject illuminates the significance of state transformation which has occurred in recent times.

Being of the Prophet's *sulb*

The *sadah* explain their descent from the Prophet in terms of sharing his substance (*nahna min sulb al-nabi*).[2] The adjective *sulb* means "hard, firm, solid"; the term also refers to the backbone and its marrow which produces both blood and male and female fertile fluids (*mani*). Both substances are endowed with generative potency. Descent lines are conceived of as vessels of accumulated substance. How do persons come into being? According to the folk model of reproduction which has wide currency in the northern parts of Yemen, men and women are equally implicated in the creation of the embryo.[3] Both possess *sulb*, and *mani* is contained in semen, the ovaries, and breast milk.[4] Because *mani* is transferred through nurturing, people who have been fed by the same woman are prohibited from entering into conjugal relationships. The consumption of a substance that contains *mani* transforms strangers into kin who, like siblings, must not mate.[5]

Yemenis of diverse status affiliation hold the view that although the human embryo is produced by the *mani* of both men and women, either might be "stronger" than the other. If a man's *mani* is dominant, the child will resemble him and have his characteristics. It is likely that these ideas have been adopted from medieval Arab scholars. According to the Syrian Hanbali jurist Ibn Qayyim (d. 1351), "when her [the mother's] semen dominates the man's semen the child will look like her brothers, and when the man's semen dominates her semen the child will look like his brothers." Referring to a hadith, he claimed that when male and female "semen" meets, "and the male semen overpowers the female semen, it will be a male; when the female semen overpowers the male semen, it will be a female" (Musallam 1983: 50–1).

Historically, the notion that both men and women are equally endowed with the capacity to bring forth new generations and are carriers of generative substance had implications for 'Alid claims to a distinct status, for they trace descent from the Prophet through his daughter Fatima. This notion also fosters ideas of appropriate and desirable conjugal relationships, which I discuss in chapter 6. 'Alid concepts of kinship and descent appear akin to those Middle Eastern ethnographies which are predicated on the Fortesian notion that a person's status is acquired at birth and fixed once and for all. Analyzing the nature of descent groups, Fortes (1969: 304) writes that "the living plurality of persons constitutes a single body by reason of being the current representation and continuation of a single founder. The group is one by physical perpetuation and moral identity." Abu-Lughod's (1986) study of the Egyptian Awlad 'Ali Bedouin bears out Fortes's emphasis on the fixity of descent. According to her, genealogy and a specific moral code focusing on honor and modesty are the principles which structure persons' identities as well as their relationships with others. "Blood both links people to the past and binds them in the present. As a link to the past, through genealogy, blood is essential to the definition of cultural identity" (1986: 41).

This approach which defines identity with reference to heredity and the past contrasts with the protean concept of persons according to which they are in flux and potentially mutable (Southall 1986; Fox 1987; Linnekin and Poyer 1990: 6–8; Bloch 1995; Astuti 1995a, b). Arguably the most radical formulation of this anti-essentialist position is Astuti's study of the Vezo. She claims that identity is defined neither with reference to preceding generations nor to successive political events, which might be instrumental in shaping their consciousness as a certain "kind" of people. The Vezo are neither determined by their own nor anybody else's history (1995a: 75). As they move from one historical space into another, they abandon their past, which does not have any impact on the present (1996: 249). "What the person 'is' is made anew and from scratch every day." Identity is inherently fluid to the extent that "anybody can become Vezo" (Astuti 1995a: 79, 15).

The *sadah*'s concept of identity differs from both the approaches referred to earlier. In the first instance, heredity determines whether one is entitled to membership of the House of the Prophet. Yet birth marks only the beginning of the process of becoming a *sayyid*. Through biological kinship, the Prophet's qualities are implanted in the *sadah* like seeds that require cultivation by external factors. Birth transmits potentials but without appropriate sociocultural stimuli, children are not believed to take to preaching and teaching like ducks to water. But even if the inherited potential is not unfolded through learning processes and reinforced through action, it will endure in a subdued, dormant form. However, mismatch between performance and innate disposition cannot last; although the structure of behavioral traits adopted by people over several generations is assimilated into the genetic matrix, the original substance is not erased. Over time, descent lines may be implicated in transformational cycles of genetic reshuffling. This means that uneducated *sadah* are not "proper" *sadah*, but their descendants can reactivate their potentials by becoming knowledgeable once again. The "good" qualities of generations which preceded the slackers are redeemable.

Moral attributes are activated in social relationships rather than automatically passed on through biological reproduction. In the process of learning the scriptures, the body uncovers inborn potentials that gradually give a person definition as a *sayyid*. This twofold, intertwined bio-cultural activity is different from Bourdieu's (1977) analysis of the body's "reading" of its immediate environment—for example, the house—which serves as a mnemonic. Bourdieu's theory of bodily self-historicization, which is a development of Henry Bergson's (1970[1911]: 197, 299) earlier work, centers on those embodied rituals of everyday life. Bergson argued that the body acts as a repository of its history. Bourdieu (1990: 73) insists that "the body . . . does not represent what it performs, it does not memorize the past, it *enacts* the past, bringing it back to life." His notion of the body as the repository of incorporated history (Thompson 1991: 13) is commensurate with 'Alid kinship ideology.[6] However, in the case of the *sadah*, embodied history crucially includes inherited substance which must be animated through morally informed praxis, thus animating the process of becoming 'Alid.[7] As Bloch (1996: 216, 229) has pointed out, for the *sadah* the significance of "mnemonic objects" like the Qur'an and the corpus of ancestral verdicts as markers of the past can be understood only if due consideration is given to the processes which create persons.

In accordance with the theory of procreation to which the *sadah* subscribe, there is a natural affinity between a certain kin group, the Prophet's descendants, and a certain type of knowledge (*'ilm*) which centers on religious truth. *'Ilm* is the knowledge on which Muslim scholarship is based (see Sabra 1971). Eickelman (1978: 489) aptly defines *'ilm* as "the totality of knowledge and technique necessary in principle for a Muslim to lead the fullest possible religious life." Zaydi 'ulama conceive of *'ilm* as internally ranked in accordance with its divine or mundane status, including such disciplines as Arabic grammar, principles of jurisprudence, and astronomy. Only divinely inspired *'ilm* is characterized as *hurmah* (inviolable, sacrosanet). During Imam Yahya's reign, it was referred to as *al-'ilm al-sharif* (the honorable science) (Taminian 1999: 204). Some scholars considered poetry as *'ilm*; others did not. Nowadays the older generation distinguishes between different disciplines by speaking of philosophy as *'ilm al-falsafah* or literature as *'ilm al-adab*. *'Ilm* which is objectified and debatable, notably that which is referred to as *al-hikmah* (wisdom), for instance history and algebra, is distinguished from both the divinely inspired *'ilm al-kalam* (a discipline which focuses on the notion of the unity of God)[8] and "practical" knowledge such as that held by builders. In describing the processes that lead to becoming a Master Builder of minarets in old San'a, Marchand (2001: 238) argues that an apprentice achieves competence through acquiring an intuitive and non-objectified knowledge that is neither propositional nor amenable to scrutiny. Within the traditional value scheme of knowledge categories, this type of knowledge is not classified as *'ilm*. Builders do not possess *'ilm* because they do not have expertise in engineering (*handasah*). In accordance with local concepts, I define *'ilm* in terms of a myriad of ethical principles and knowledge categories.

Generally, the study of the various Islamic disciplines is considered a means of establishing a connection with the teaching of the Prophet (Brown 1976: 76). In accordance with their self-images, for the *sadah* this kind of study has an added dimension because religious knowledge was first transmitted by their putative eponym. Their duty to become the divine paragon to which they are predisposed by descent, and to uphold the message conveyed to mankind by the Prophet, is thus greater than that of people who merely belong to the community of believers. The emphasis they place on learning derives from the Shi'i view that 'Ali was the most eligible successor to the Prophet because he was his closest blood relative and because he was taught by the Prophet who grew up in his household. Learning is conceived as a *sine qua non* for the creation of moral persons. *'Ilm*, then, is a constituting element in 'Alid kinship: it provides nurture to the growing body that is to become a *sayyid*.

Thus, the *sadah*'s view of identity combines elements of the analytical approaches to kinship and identity which I referred to earlier. Heredity determines identity, but one ultimately becomes a *sayyid* through performance. Indeed both factors are mutually constitutive. A local proverb which says that "a *sayyid*'s injuries are sown up by a *faqih*" corroborates the notion of *'ilm* as a naturalized substance that harmonizes with the *sayyid*'s body texture.[9] The proverb implies that if a *sayyid* suffers from misfortune or has acted with impropriety, his condition can be alleviated by a *faqih*. The *faqih* is a metaphor for knowledge. The proverb suggests that *'ilm* is a kind of natural substance which runs through the body like a tree of life—a substance which must be incorporated so as to create a complete person.

Plato's Legacy and Shi'i Doctrine

The notion that the Prophet's progeny are best suited to interpret God's message by virtue of their inherent capacities is reminiscent of the Platonic idea that persons possess inborn potential to develop particular types of knowledge. For Plato, seeking knowledge and learning is merely remembering. The essence of learning, which nonetheless requires diligent effort, is recovering and eliciting the knowledge which is already in the mind (Platon 1957: 21–2; Coleman 1992: 5–6, 9). Twelver-Shi'i teachings about the Imam share Plato's idea that what the body "knows" derives from its inherited knowledge. In broad terms, the Twelver-Shi'is support a strongly Platonic version, the early Imamis and the Zaydis a weak one. Early sources of the Imami school which predate the disappearance of the twelfth Imam in 874 (the pre-ghaybah period) define the excellence of the Imam in terms of the knowledge he possesses.[10] The Imam is portrayed as the *'alim* (learned) *par excellence*, and the terms Imam and *'alim* are used interchangeably. Excellence is mainly defined in terms of the possession of knowledge. The Imam is more knowledgeable than the prophets, with the exception of Muhammad. The knowledge of the Imam, which is inaccessible to other humans and superior to that of all others, is the main constituent of the Imami theory of the Imamate.

According to Hisham b. al-Hakam (d. 795), a disciple of Imam Ja'far al-Sadiq, the Imam's knowledge is based on transmission from the Prophet and passed on to other Imams via 'Ali. Hisham stresses that the Imams have perfect knowledge of the Qur'an, but rejects the idea that the knowledge of its interpretation may be acquired without learning (*'ilm bidun ta'lim*). For him infallibility (*'ismah*), which is attributed to the Imams, is the divinely instilled ability to acquire and preserve unaltered the knowledge transmitted from the Prophet. In his eyes, the Imam is no more than a perfect transmitter of revealed knowledge (Daou 1996: 37, 56; personal communication).[11]

Hisham's position was different from that held by his adversaries whose theories about the knowledge of the Imam gained widespread acceptance two centuries after the ghaybah (the occultation of the twelfth Imam). There were diverse views about the kinds of knowledge possessed by the Imams. According to some sources, they possess exceptional knowledge of the law. Others speak of their access to divine mysteries, of their knowledge of future events—such as the time and circumstances of a man's death, and of men's innermost thoughts; their command of all languages of humans, animals, and plants; and their having been "spoken to" by the angels, that is, knowledge was revealed to them through the medium of angels (Daftary 1990: 561; Daou 1996: 15; Madelung 1996: 422). Despite these attributions, in post-ghaybah Shi'ism the Imams' role as teachers of law and other ethical principles was not played down. However, the crucial point about these concepts of the Imams is that learning represents recollection of the knowledge that is already inside them; it does not increase it.[12] The twelve Imams were born with perfect bodies which required hardly any refinement such that in the process of learning, their bodies need only "remember" what is in store.

The Zaydis take a somewhat different point of view. Early Zaydi sources stress that the 'Alids are endowed by God with a special capacity to acquire knowledge (Madelung 1965: 48). If properly activated, this potential aids them in apprehending divine knowledge. Because unlike the twelve Imams the Zaydi Imams (those who came after 'Ali's sons) are not reckoned to be infallible, their bodies become gradually moralized through disciplined learning until they are fit for their difficult office. Their innate potential is as important as their understanding of the revealed law gained by being taught. As noted in chapter 1, the classical Zaydi doctrine of the Imamate stresses that descent from 'Ali and Fatima is a necessary prerequisite for legitimate leadership. However, it states equally clearly that this condition is insufficient. Candidacy for the highest office is reserved for a *learned* descendant of the Prophet. According to al-Qasim b. Ibrahim, the intellectual godfather of the Zaydi school, the candidate must be the most erudite, abstemious, and pious among the *ahl al-bayt* (Madelung 1965: 142–3).

Zaydi doctrine requires the holders of power to be of superior moral character which is why they must be *mujtahids* and display piety. At the core of Zaydi theory is a concern for the ethical foundations of power. Ideally, knowledge of the scriptures serves to moralize and tame power, but the doctrine recognizes that it can be put to evil ends which is why it stipulates that it is the duty of the believers to rise against corrupt rulers. In attempting to wed power to morality the early Zaydi thinkers, notably of the Hadawi school, required the ruler to be of 'Alid descent and endowed with the highest qualification for leading the community of believers. Zaydi-Hadawi doctrine assumes that in conjunction with a cultural disposition to learn, a specific inborn disposition best guarantees the exercise of moral judgment. The Imams are afforded moral agency by reason of being the inheritors of the Prophet's biomoral substance and the recipients of the knowledge transmitted by him. Some Imams regarded the divine message as their patrimony. According to Imam al-Mu'ayyad Muhammad (r. 1620–44),

> We are heirs to the Book and the Wisdom, and We honor the birth of His Prophet Muhammad, the Seal of the Prophets, and therefore We take his place in his Community, in the same way as Abraham . . . and may He [God] make our love for Muhammad—God bless him and his family and grant them salvation—a wage in the worlds and for them good tidings in Paradise . . . And the People of the House of Muhammad . . . will continue to preserve for us his religion and his legacy, stand up for the right of his call to all mankind, in order to bring the proof for him; they will draw the sword of Holy War in the face of those who are opposed to his command; among his servants they will give good advice (quoted in Haykel 1997: 32).

The idea that the most eligible from among the *sadah* is best disposed to interpret the holy scriptures was explained to me by a *sayyid* with reference to the Yemeni saying that "the owner of the house knows best what's in there" (*sahib al-bayt adra bi-alladhi fihi*).

> It makes a difference whether I quote someone to whom I am not tied by blood or whether I quote my father, and my father quotes his father and so on. I know better than others what has been said in my own house, and I shall not tell a lie about the words of my ancestors.

Heredity is also believed to dispose the *sadah* to have a greater capacity to acquire wisdom (*hikmah*). As it was put to me by a friend:

> God gives *hikmah* to anyone he likes. It means that somebody is divinely inspired. People possess *hikmah* only as a potential. A person must work for it to develop. It reveals itself in good intentions, good works, and humility. In the political sphere, it means acting with moderation. These traits are more readily shown in the *sadah* than in others.

The notion of *'ilm* was inextricably connected with the religious character of the Zaydi state and central to its political culture. It led to the aestheticization of politics. According to a saying, "ink is the perfume of the 'ulama." Enshrined in this culturally "valued cognitive style" (Eickelman 1985: 5) acquired by the scholars was a logic which contrasted different social categories in accordance with the knowledge they held. The teaching of religion was perceived as a way of civilizing the tribespeople who, once saved from their ignorance, would be loyal subjects and become integrated into the Zaydi state.[13] As a result, the subjects would respect both the message of the Holy Book and those who presented themselves as the natural repositories of that knowledge. As it was phrased by a contemporary Zaydi scholar, "the Zaydiyyah itself was a school" (*Al-zaydiyyah nafsha kanat madrasah*).

In view of the doctrinal emphasis placed on both descent and learning, it is not surprising that those Yemeni scholars who were convinced that heredity is a person's destiny did not accept that education per se could produce moral persons. Among them were *qudah*, many of whom resented being ineligible for the leadership by reason of lacking the appropriate descent credentials. Thus al-Shawkani (1979: 128–30) argued that some people who became educated did not use their knowledge to improve themselves morally. How, he asked, can one distinguish good students from bad students? The good ones are of pure descent, and they aspire to learn about their religion and to belong to a respectable social stratum. By contrast, those of lowly birth who pursue manual work are self-interested and not concerned about their reputation. If they become learned (*talib 'ilm*), they will not invest in their moral education and try to emulate their teachers (*ahl al-'ilm*). They will have regard only for their own kind, and will "end where they started." Hence non-'Alid scholars, who like the *sadah* claimed to be of pure descent, embraced certain ideas about the connection between descent and knowledge, even if they failed to espouse Zaydi-Hadawi doctrine. Paradoxically, al-Shawkani endorsed heredity as a criterion for access to knowledge but disagreed that the supreme ruler must necessarily be of 'Alid descent and a *mujtahid*. Indeed, he claimed that obedience must be shown even to a ruler who is a slave because of the Traditions which state "Obey and be obedient even if he is an Ethiopian slave whose head is like a raisin" (Haykel 1997: 99).

The emphasis Zaydi-Hadawi doctrine places on the descent affiliation of the holder of supreme authority indicates that it is based on the implicit assumption that people are ranked according to the potential inherent in their bodily constitutions. It puts forward a theory of the mind–body which centers on wholeness, excellence, perfection, and refinement. Enshrined in the Zaydi-Hadawi theory of the Imamate is the image of the quintessential *sayyid* who is devout, erudite, generous, modest, courageous, patient, fearless, and has an intact body. Even the loss of a finger or the inability to smell call into question a man's eligibility to rule (Sharh al-azhar Vol. 4: 520; Haykel 1997: 4). Zaydi doctrine which is predicated on hierarchicized descent also constructs a gender paradigm which privileges men over women. Irrespective of their knowledge and piety, women, like non-'Alids, cannot compete for the supreme leadership.

Imams were declared "complete" by their peers. For example, the fourteenth century Imam al-Nasir was tested by 1300 'ulama who verified his qualifications and found him "complete" (Haykel 1997: 20–2). On the other hand, in the twelfth century a descendant of Imam al-Hadi claimed a "limited Imamate" because he could memorize only a third of the Qur'an. He did, however, gain the support of the 'ulama for his candidacy (Madelung 1965: 210). This pretender endorsed the doctrinal stipulations regarding the leadership, and acknowledged his lack of competence vis-à-vis his more erudite peers.

Moralizing the Body

Thus far I have examined *'ilm* as a constituting element in 'Alid kinship and as a means for the validation of claims to rule in Zaydi-Hadawi doctrine. There is a series of acts oriented toward acquiring knowledge of the ancestors, thus articulating kinship with them. Knowledge as performance can be conceived as an act of remembering or, put differently, remembering occurs performatively "in and by and through the body" (Casey 1987: 147). Many of these acts possess a performative quality, "so that 'having' knowledge is very closely linked to 'performing' it." Knowledge is something "that comes to be embodied, an intrinsic part of the self . . . [it] has an indexical, personal function; what one knows is not fully distinguishable from what one does or who one is" (Lambek 1993: 6). The biographical sketches of Ibrahim and Khadija drawn in chapter 4 convey the notion of *'ilm* as a source of nurture—a person without religious knowledge is a person without morality, without humanity. The process of inscribing and encoding memory in somatic and somatized forms (Strathern and Lambek 1998: 13) begins immediately after birth and marks the beginning of the long process of engaging with the collective memory—the Prophet's revelation. The *adhan* (call to prayer) is pronounced into the right ear of the infant, and the *iqamah* (a short *adhan* that initiates prayer) into the left ear. The Prophet Muhammad is said to have done so after his daughter Fatima had given birth to her sons al-Hasan and al-Husayn. The infant is made to "memorize" the name of God even before it is capable of (conscious) remembering and understanding.

In her account of the *khatm* ritual (chapter 4), Khadija conceded that she studied the Qur'an without understanding the meaning of the words. In Falih, I also observed this phenomenon among small children who played near the mosque where their elder siblings were taught. The boy of the family I stayed with returned home reciting fragments of Qur'anic verses he could memorize without any tonal control or understanding of their meanings. His parents tried to teach his infant sister the name of God ("*quli Allah*") before they taught her kin reference terms.

Several practices serve to locate the child in the moment of genealogical time which she or he is to inhabit. The child's name is recorded on the genealogical chart and invocations (*du'a'*) are written down next to his or her name on the first blank page of the Qur'an. Men enter the dates of birth and the personal names of their children on this page, a precious personal belonging which is hardly ever taken outside the house.[14] The invocations focus on the child's moral conduct. God is invoked to "give boys the honor of *'ilm*" and to guide the girls to become virtuous women (*al-qanitat*).

Mawlids which commemorate the Prophet's birthday provide further occasions for "performing" knowledge. They are held mainly by women as part of the post partum celebrations following the birth of a child, a wedding, or someone's recovery from a serious illness. Even very young children are brought along. A female reciter (*nashshadah*) narrates the events in the Prophet's life while incense is burnt. She talks about the *ahl al-kisa'*, those who figure in the "Mantle hadith," about Muhammad's life as an orphan who, when he had become a man, fought for the right and refused to surrender. She narrates how he encouraged people to fight for their own beliefs, an activity through which they would gain merit (*ajr*). These rituals, whose prime goal is remembering, may be perceived as "landmarks in genealogical knowledge, and at the same time they constitute genealogy *in action*" (Bahloul 1996: 115). However, the *sadah* insisted that the rituals had no special meaning for them and were not performed more often by them.[15] Once however a young girl of an 'Alid family burst into tears at a *mawlid* whilst listening to the narratives about Muhammad's hardship—though the extent to which she identified with him is difficult to ascertain.

Prior to 1962 boys were expected to learn "patrifiliative name chains" (H.Geertz 1979: 343) by rote. Some of the girls also acquired this knowledge. Children started to memorize about four to five names at an early age and continued to learn more until they were able to recite them without mistake. Nowadays few San'ani children are encouraged to do so. However, a family which had recently migrated from Shaharah to San'a took their children's memorization of ancestral names seriously. Whenever I came to the house, their five-year-old boy was asked to demonstrate the progress he had made. (I felt uncomfortable whenever his memory failed him and he disappointed his father; I suspected that he resented my visits.)

Learning one's genealogy is a way of pronouncing the bond between the living and the dead. Speaking the names of the dead is actualizing their presence in the lives of their descendants. The names memorized as a child become mnemonic devices for genealogically saturated histories that have become one's own. Once I asked a man from Bayt al-Wazir about one of his ancestors, a well-known *mujtahid* called Muhammad b. Sarim al-Din. Without consulting the family annals, he told me that he was his ancestor in the fourteenth generation, that he had lived in the sixteenth century, and that he was killed when his wife was pregnant with his daughter. The father of the boy who had moved from Shaharah to San'a a few years ago, a real estate agent of Bayt al-Mutawakkil, explained his location on the genealogical map to me.

> All Bayt al-Mutawakkil trace descent to al-Qasim b. Muhammad. They are about 30,000. You will find Bayt al-Mutawakkil in Jiblah, Ibb, Ta'izz, Shaharah, Anis, and Bayt al-Abyadh. The patronymic (*laqab*) al-Mutawakkil derives from al-Mutawakkil 'ala-'llah Isma'il who united north and south Yemen. His rule extended up to the borders of Oman. He ruled until 1087H [A.D. 1676]. He was the son of Husayn b. al-Qasim whose mother was from Bayt Sharaf al-Din of Kawkaban. Those Mutawakkil living in Shaharah, Ta'izz, and Ibb are descendants of Isma'il; al-Mutawakkil Muhsin (San'a, al-Sudah) are related to Bayt Hamid al-Din. We belong to that branch. Bayt al-Mutawakkil of Shaharah and of San'a have little in common, they only meet in the sixth generation. Bayt al-Mutawakkil San'a think that we are conservative.

The acquisition of this genealogical knowledge—which is also historical knowledge—is seen as a substantial part of the process of becoming 'Alid. However, during early childhood the major focus is the text which was revealed to the Prophet and which is "copied, taught, memorized, enunciated, listened to, [and] performed as prayer . . . Each of these acts . . . is mediated by the identical text" (Lambek 1995: 265). The performative qualities of knowledge practices analyzed by Lambek and also Carruthers (1990) can be brought into relationship with Bourdieu's notion of the body's enactment of the past. Yemenis share with ancient Greek and medieval Christian thinkers an awe for trained memory. As Carruthers (1990: 10) explains, "a work is not truly read until one has made it part of one-self—that process constitutes a necessary stage of its 'textualization.' " In Islamic learning the aim of "mnemonic domination" or "possession" (Eickelman 1978: 489; 1985: 57) is the conquest of the self by the sacred text. This molding process links 'Alid children with their ancestors: the somatization of knowledge (Bourdieu 1977) generates a textualization of the body such that the incorporated scriptures meet innate potentials, thus providing the largest possible scope for moral agency.

In Yemen, a well-trained memory was (and is) considered to be morally virtuous in itself. Beyond its moral force, speaking the words of God and the ancestors from memory generated an authority which marked the Imamate's political culture. Men who wanted to take up positions as judges, scholars, and governors in the Imamic government had to study and memorize the Qur'an (in parts) and other standard works. Grammar books had to be memorized so that these men could use the language of the Qur'an correctly. This disposition toward rote memorization has been aptly described by Carruthers (1990: 13) in her study of the ethos of memory among medieval Christian writers. "The choice to train one's memory or not . . . was not a choice dictated by convenience: it was a matter of ethics . . . Memoria [trained memory] refers not to how something is communicated, but to what happens once one has received it, to the interactive process of familiarizing—or textualizing—which occurs between oneself and others' words in memory."[16] Carruthers (1990: 13) stresses memory as an ethical concept. However, when the memorization of key texts was part of the job description of Yemeni senior scholars and administrators, memory training had practical advantages. It provided access to prestigious positions for those who by reason of Zaydi stipulations and specific socio-moral codes were obliged to stay aloof from mercantile activities (see chapter 8). A well-trained memory was (and is) important for issuing fatwas, and it would lower a scholar's reputation were he to consult books before pronouncing his judgment.[17] This is illustrated by the biography of a *mujtahid*, Sayyid 'Abd al-Qadir b. 'Abdullah 'Abd al-Qadir (figures 5.1 and 5.2), who died in 2004.[18] The data record the stages of his academic career and his engagement in politics, and offer clues as to how knowledge was produced and reproduced.

> I was born in 1908 in al-Rawdah [the garden suburb of San'a]. I was brought up in close contact with my father. I studied and completed the Qur'an with *al-Ustadh* Ahmad 'Abdullah al-Kharush. I learnt by

Figure 5.1 Sayyid 'Abd al-Qadir b. 'Abdullah b. 'Abd al-Qadir (right) with his father, 1940s

Figure 5.2 Sayyid 'Abd al-Qadir (center) with his youngest daughter (right) and other children and grandchildren and a retainer, 1986

heart how to recite the Qur'an properly (*al-'ilm al-tajwid*). My father sent me to study at a scholar's house where I was taught together with my paternal cousins and some other children of well-known people and relatives. I learnt how to write, mathematics, and learnt the *Matn al-azhar* (corpus of Zaydi law) by heart. I also studied *talkhis* (summary), legal principles (*fara'id*), a work on grammar (*al-Mulhah*), and summary (*khulasah*), but I did not complete the legal opus by al-Husayn b.al-Qasim (d.1640) known as *al-Ghayah*.

I was still under age when I finished my study of grammar (*'ilm al-nahw*) with *al-Ustadh* Ahmad b. 'Ali al-Anisi. [At that time] I began to join *al-'Allamah* al-Husayn b. Muhammad Abu Talib at the Fulayhi mosque. I studied grammar and learnt most of *al-Ghayah* by heart. With *al-Sayyid al-'allamah* Muhammad b. Isma'il al-Kibsi I studied *fara'id*. Then I started to study the branches of Islamic law (*al-furu'*) with *al-Sayyid al-'allamah* Ahmad b. 'Ali al-Kuhlani in al-Wushali mosque during the period between the sunset prayer (*maghrib*) and the last evening prayer ("*'asha'*").

By that time I still had not reached puberty. When I completed studying the *Matn al-azhar* at the age of about thirteen, I asked my father to honor my teacher *al-Faqih* 'Abdullah b. Husayn al-'Abadi and to celebrate the occasion. My father referred the request to *al-Qadi al-'allamah* 'Abd al-Karim b. Ahmad Mutahhar while they were reading the Qur'an in the Qubbat Talhah mosque during the month of Ramadhan. The *qadi* wrote a poem and gave it to my father . . .

After learning *talkhis*, *al-fara'id*, a work on grammar (Ibn Hajib), *mulhah*, *khulasah* and the other disciplines, Muhammad b. 'Abd al-Rahman Kawkaban, a man of letters and poet (*al-Sayyid al-adib wa sha'ir*) praised me in his poetry. He said that I was blessed by God because I was a brilliant student. Addressing me as "you, noble son from 'Ali's lineage" (*ya ibn al-karim min sulalat haydar*), he said that I had achieved a level of knowledge which could "neither be matched by the one who begins nor the one who finishes his studies." I learnt the summaries (*mutun*, pl. of *matn*) by heart.

I completed my studies with *al-'Allamah al-jamali* 'Ali b. Muhammad b. Isma'il Al-Fidah. I specialized in *nahw*, *fiqh*, hadith and a branch of rhetoric; kinds of sentences and their uses (*al-'ilm al-ma'ani*). I always joined my teacher at the Qubbat al-Mahdi 'Abbas and al-Taqwah mosques. *Al-Sayyid al-muhaddith* ("the one knowledgeable in hadith") *al-'allamah al-'abqari* ("the genius") al-*shaykh al-qurra'* (master of Qur'an recitation) 'Ali Ahmad al-Sudumi instructed me in the compendium to the *Matn al-azhar*, the *Sharh al-azhar*, for a short period during the "days of the grapes" [the grape harvest] in al-Rawdah. Then I studied *furu'*, *'ilm al-ma'ani*, *usul al-fiqh* and *nahw* under *al-Sayyid al-'allamah* Ahmad b. 'Ali al-Kuhlani in the Great Mosque and al-Wushali mosque.

Then I entered the *Madrasah al-'ilmiyyah*. I was among the first students who enrolled in it. I studied under various *mashaykh* (pl. of *shaykh al-Islam*, a scholarly rank), among them my father *al-'Allamah* 'Abdullah b. 'Ali 'Abd al-Qadir who taught me in hadith and the method of identifying the correct or incorrect hadith (*'ilm al-hadith*). *Al-Sayyid al-'allamah* Ahmad b. 'Ali al-Kuhlani instructed me in *usul al-fiqh* and the basic principles of religion (*usul al-din*), and exegesis (*tafsir*), and *furu'*. I studied grammar under *al-Sayyid al-'allamah* 'Abd al-Khaliq b. Husayn al-Amir, and *nahw* and hadith under *al-Sayyid al-'allamah al-zahid* ("the ascetic") Ahmad b. 'Ali al-Kibsi. I also accompanied him to the Sayyad mosque where I listened to his lectures. I studied *tafsir* and *furu'* with *al-Sayyid al-'allamah al-muhaqqiq* Zayd b. 'Ali al-Daylami, *furu'* and *sarf* (a branch of grammar) with *al-Qadi al-'allamah shaykh al-Islam* 'Ali b. 'Ali al-Yamani, *tafsir* with *al-Qadi al-'allamah* 'Abd al-Karim b. Ahmad Mutahhar, hadith with *al-Ustadh al-khatib* ("the speaker") *al-'allamah* Muhammad b. Husayn Dalal, *usul al-fiqh* and *sarf* under *al-Qadi al-'allamah al-muhaqqiq* ("the inquirer") 'Abdullah b. Muhammad al-Sirhi, and *usul al-fiqh* and hadith under *al-Qadi al-'allamah* Lutf b. Muhammad al-Zubayri. In his house and the Taqwah mosque I learnt the most famous work of hadith (*Ummahat*).

I received a general certificate (*ijazah 'ammah*) from him and my father. I asked *al-Qadi al-'allamah al-hujjah* (a religious title) Husayn b. 'Ali al-'Amri for an *ijazah*, and I received others from *al-Muhaddith al-madinah al-munawwarah* (someone who narrates the Prophetic traditions in Madina) *al-Shaykh al-'alim* al-Salah b. al-Fadl al-Tunisi and from *al-Shaykh al-'alim* Mahmud b. 'Ali Shuwir al-Masri during my first pilgrimage in 1354 (1935 A.D.). I was also given one by *al-Ustadh al-'allamah* 'Abd al-Wasi' b. Yahya al-Wasi'i.

After I had graduated from the *Madrasah al-'ilmiyyah* I became a teacher there. I taught all classes and I had many students. I also taught at the Talhah mosque from *fajr* (dawn prayer) until sunrise, and then at the *Madrasah al-'ilmiyyah* until noon. I continued teaching at home until the late afternoon prayer (*'asr*). Afterwards I went to listen to lectures myself until the evening prayer. My time was busy with obedience to God's orders (*ta'ah*).

One day I received a letter from the crown prince (*wali al-ahd*) Ahmad b. Yahya Hamid al-Din who asked me to become his secretary. I worked with him for a few days, and resigned when the prince moved from San'a to Ta'izz. A year later Imam Yahya assigned me the post of *'amil* (governor) in the sub-province of Zabid. I declined the offer but later I accepted to become the head of the ministry of endowments (*awqaf*) for Mecca and Madinah in Yemen (*nadarat awqaf al-haramayn*). I fulfilled my duty with honesty and interest. It was an easy job which did not obstruct my studies. I continued this work until the 1962 revolution.

In 1941 I was assigned the post of chairman of the *Majlis al-niyabi* [a legal council which answers questions to the public on behalf of the Imam] by Imam Yahya. I had this job until the Constitutional movement (*al-harakat al-dusturiyyah*) [of 1948] began and the Imam was assassinated. I was a member of the new cabinet which was appointed by the leaders of the uprising. I became Minister of Endowments. I was imprisoned for two and a half months by Imam Ahmad. After my release from prison I became a member of the Court of Appeal (*Majlis al-isti'naf al-shari'ah*) which was under the presidency of *al-Sayyid al-'allamah* Muhammad b. Husayn al-Wada'i. I was his assistant and acted as his deputy in his absence. At the same time I attended the *maqam* (royal court, also referred to as *al-maqam al-sharif*) where I assisted the Imam's deputy. Later I was appointed Deputy Minister of Education. Many times I was asked to supervise the work at the Ministry of Education and the Ministry of Work. In all my jobs I was well-known for my good behavior, sincerity and truthfulness.

In 1962, I was in prison again for one and a half months until I was appointed deputy minister at the Ministry of Interior. Later I worked as a judge (*al-hakim al-awwal*) in San'a, and then I joined the Court of Appeal (*Mahkamah al-isti'naf*) in 1972. Then I became Minister of Justice. I also was a member of the first National Council (*Majlis al-watani*) whose chairman I became. I became the President of the High Court in 1972. I was asked to draw up documents for the Parliament (*Majlis al-sha'ab*). At present I am involved in the codification of the *shari'ah* (*taqnin al-shari'ah*).[19]

Equivalence and Difference

The majority of the *sadah* feel obliged to acquire the revealed knowledge and have had at least some basic training. However, *sadah* of learned houses who are troubled by the existence of ignorant ones refuse to recognize them as equals. To recapitulate, according to their self-stereotypes, shared substance with the Prophet is the defining characteristic of the *sadah*'s identity. Those who fail to activate their intrinsic potential are not the same kind of person. This calls into question the equivalence of all *sadah*. In specifying the conditions for the supreme leadership, Zaydi doctrine acknowledges diversity among Muhammad's progeny: they comprise devout and impious, learned and ignorant, generous and mean, with complete or incomplete bodies.

As noted earlier, inherited potentials which remain undeveloped are never entirely lost. The Prophet's substance is thought to be variously present in the bodies of his offspring. The status of houses which have abandoned learning for centuries is somewhat ambiguous. As pointed out by Lambek (1997a: 132), "knowledge provides a main vehicle for both social action and social differentiation." In medieval times, Hijrat Waqash had subscribed to a religious movement which was disapproved of by the ruling Imam and was subsequently demolished (chapter 3). The majority of the members fled to other parts of the country where they continued their tradition of learning. Those staying behind no longer invested in *'ilm* and lived as farmers. Referred to as *qaba'il* of 'Alid origin, *fallahun* (farmers), or simply as *qaba'il* by their educated urban peers, these black sheep were denied a space in genealogical time. Distinguishing themselves from the others, the educated ones described themselves as *al-tabaqat al-raqiyah* (elevated stratum) of the Prophet's House. Once there were no more 'ulama in Hijrat Waqash, the names and activities of their descendants who changed their occupations were not worthy of being recorded and memorized for posterity. "Why would I want to know about them since I was interested in the *buyut al-'ilm*?" I was asked by their erudite kin. This was an instance of "forgetting" among learned *sadah* who excluded their uneducated kin from their genealogical charts. While their social biographies were eventually erased from memory, those of mothers and sisters, women 'ulama, and of women marrying into that branch who were daughters of scholars were recorded. In one case, the date of a man's marriage to the daughter of an Imam was recorded, but the year of his death was not. He was described to me as "not so important"—that is, he was a scholar of minor reputation. The notion of a dormant genetic potential only became fully clear when I sought to visit those who had been willfully forgotten. As to how I would find them—since I was not sure whether they had maintained their original patronymics—it was suggested to me (somewhat humorsome) that if I were to read a hadith in front of all the villagers, those of 'Alid descent would understand it better than anyone else.

The descendants of men who were not scholars tend to gloss over their occupations by saying that they were "ordinary people" (*nas 'adiyyun*). Family annalists of past centuries expressed in no uncertain terms their disappointment about those who abandoned knowledge. Their writings suggest that the history of the scholarly houses manifests itself in the kinship of the bearers of knowledge. The silence about the descendants of scholars who were not learned reveals the link between male personhood and knowledge. Indeed, among the *buyut al-'ilm* knowledge is fundamental to the realization of the masculine self. Consider the history of Bayt al-Wazir (chapter 3 and appendices II and III). The names of the descendants of Yahya b. Uthman b. 'Ali (Uthman lived 1642–1717) and of Yahya b. 'Abdullah b. Zayd are not listed because they were not 'ulama. Notes on the dates of birth of the male members of houses are an indicator of their social worth because they are only provided for men who made a reputation for themselves as scholars. For example, we know when al-Hadi IV and his father Salah b. al-Hadi III were born, where they studied and so forth, but we learn near to nothing about Muhsin b. al-Hadi IV, Salah's grandson.

The annalist only mentions that he had two sons, Ahmad and Muhammad, but omits his date of birth and data about his studies and character traits. Indeed we are not told how he made his living. One contemporary scholar commented on those *buyut al-'ilm* who neither studied and taught nor fought unjust Imams, "if they were not 'ulama, they were dead, even if they were rich."

Talking about 'Abdullah b. Hasan al-Wazir and his children al-Baqir and Yahya, one of the annalists of Bayt al-Wazir who lived after the seventeenth century bemoans the fact that this scholar's sons had failed to seek knowledge.

Al-Baqir had four sons, 'Ali, Yahya, Muhammad and Ibrahim who lost their wealth and did not continue their education. They had a lot of land which could have brought them prosperity. But they lost the nobility that inhered in their origin and wealth (*sharaf al-asl wa-'l-mal*). They found themselves in bad condition. May God help us (*Allah al-musta'an*).[20] (Using this formula the writer places blame on them and indicates that he has no more to say about them.)[21]

These men who had disregarded the moral imperative of their noble station were stripped of their honor, and the names of their descendants were not recorded in the annals. The subtext reads like a censure for their reluctance to study which threatens the integrity of the 'Alids as a collectivity: there is fear of being leveled and of losing the fundamental moral vision which differentiates them from others.

Another chronicler of Bayt al-Wazir refers to the son of Qasim b. Yahya b. Mansur who was called Ahmad (d. 1386). Ahmad had two sons, al-Mahdi and 'Ali. All that is said about them is that some of Qasim's descendants were not interested in *'ilm* even though their grandfather was a well-known scholar.[22] Again, family annalists did not write about these men's descendants, and would only do so if they dedicated themselves to knowledge once again.

Sometimes a failure to study caused a rift between close relatives. 'Ali al-Farisi, a cousin of Ahmad al-Farisi (chapter 3), was well educated and a respectable civil servant. His brother 'Abbas was a belt and cap maker. Ahmad al-Farisi's son could not answer my question as to what had become of Yusuf, 'Abbas's only son. "No contact" was his short reply.

Sayyid 'Ali lived up in San'a and rarely went down to his brother's house except on special occasions. They hardly saw each other. (One of 'Ali's friends)

Neither the ignorant nor those who are quarrelsome or lack propriety are fully recognized as kin by other members of their house.

In one case, a woman asked her husband to divorce her because he did not allow her to pursue her education. (Prior to the marriage he had agreed that she could attend university.) He demanded that she compensate him for all the expenses he had made on her behalf. These included the wedding expenditure and gifts he had given her during their marriage. His kin were outraged, arguing that he was not one of them (*mush min sulbna*).

One man, a grandson of Imam Yahya, argued that he would break off relations with a kinsperson or any *sayyid* involved in reprehensible entertainment. "If a *sayyid* opens a cabaret, he is no longer a *sayyid*." He explained that if he did so, others were better than him. "One's actions must be consistent with one's social location (*sharaf al-nasab yatazawwij sharaf al-'amal*)."[23] A *sayyid* who is ignorant, negligent, and of weak character is referred to as *sayyid haqq al-jum'ah* ("Friday *sayyid*"; that is, born on a religious holiday rather than of 'Alid parents). Even a *sayyid*'s misconduct is explained with reference to Muslim history, using the Prophet's uncle Abu Lahab as an example. He was opposed to Muhammad and was singled out for divine curse in the Qur'an (Madelung 1997: 16). No claims are made that the Prophet's progeny is inherently good.

The fact that all *sadah* have a common status by virtue of their descent but some might fail to realize their innate potential renders problematic the notion of the equivalence of all. In a different cultural context, namely the Indian caste system, Parry (1989: 494, 501–2) has dealt with an apparent paradox of this kind which might serve to illuminate the Yemeni case. The cultural logic according to which all members of the same caste are commensurate, conflicts with the notion of the mutability of their inherited substance through different kinds of transactions. In his seminal article about the potential mutability of bodily substance, Marriot (1976) argues that the notion of persons as indivisible, bounded units is alien to South Asian thought. Through manifold transactions ranging from natural reproduction to food exchanges, "dividual persons" might absorb inappropriate substances which then transform their bodies. Parry (1989: 513–14) concludes that this contradiction between the notion of the immutable substance inherent to members of specific castes and that of the potentially alterable body has a moral rationale. Improper transactions threaten both the integrity of the individual's body and the collective body of the caste. Therefore, as Marriot (1976: 111) points out, persons preserve their own integrity and that of their caste "by admitting into themselves only what is homogeneous and compatible." According to Parry (1989: 494), in the final analysis the cultural representation of the transformational potential of substances serves to maintain and reinforce the static ideology of caste, justifying rigid boundaries between castes. In the Yemeni context, it might be argued that the notion of the potential mutability of biomoral substance constitutes a motivating force in the continued quest for knowledge. The *sadah* will endure as a unique social category only as long as there is a steady, harmonious interaction between bodily substance and performance in accordance with the moral imperatives defined by their ancestors. The work of memory undertaken by family annalists and genealogists is concerned with preserving what is worth being remembered. Writing only about those who were learned, they set the terms of inclusion and exclusion. Rather than reconstructing the past, these works define for future generations what they ought to become: pious seekers of knowledge.

Teacher and Disciple

If *sadah* who are ignorant or behave improperly are denied full recognition as kin by their peers, how do they conceive of relationships with non-*sadah* to whom they are tied through learning? As Sayyid 'Abd al-Qadir's biography made clear, learning occurs in close relationships with kin and teachers. Relationships with teachers of any extraction are morally charged and cast in the language of kinship.[24] Systematic teaching is conceived as an act of nurturing which creates permanent and irreversible links between teacher and student. But do teachers become kin in the process of sustained nurture?

Learned *sadah* acknowledge the contribution *qadi* scholars made to their moral growth, and they are therefore treated like older kin. Sayyid 'Abd al-Qadir places as much stress on the subjects he studied as on the identity of his teachers. Pedigrees based on the transmission of *'ilm* co-exist with those based on blood descent. In the case of Sayyid 'Abd al-Qadir, the line established through teaching is partly identical with his patrilineage because he was taught by his father and other relatives.[25]

The bonds between a teacher and his disciple are considered to be permanent, and the teacher shares some of the father's authority.[26] The disciple is expected to have the same loyalty and respect for his teacher as he does for his father. The student is to offer any kind of service to his teacher, and he is expected to pray for him, asking God to reward the teacher's efforts, to forgive his sins, and to preserve his health. A student's obligation and attachment to his teacher is expressed in proverbs such as *Man 'allamuni (qarani* in Yemeni dialect) *harfan, kuntu lahum 'abdan* (who taught me a word, enslaved me). Teachers are addressed by the honorific *sidi/sidna* (my lord) which is used for close agnates, and in the past students kissed their knee. The teachers closest to the student are referred to as *'ammi* (paternal uncle). In turn, the teacher calls his student *walad* (boy, son) irrespective of his age and social standing. Once the student has become a scholar himself, his teacher uses forms of address such as *al-walad al-'allamah* (the learned son), a combination of kinship terminology and scholarly title. Teachers are also asked by their students to accompany them when they are presenting their marriage proposals to a girl's kin, and marriages between students and their teacher's daughters are not uncommon. In one case, the obligation toward the teacher had grown out of their joint experience of imprisonment after the 1948 uprising, and was particularly strongly felt. According to one of the students, "we have never argued with him because he cleaned our cells and cooked our meals so that we could study."

The answer to the question whether kinship relations are brought into being through sharing knowledge is not straightforward and depends on the parameters which are made salient in the analysis of these relations. Writers like Weismantel (1995) and Carsten (1997) seek to overcome distinctions between biological and metaphorical or symbolic "relatedness" by stressing the significance of performative criteria in the production of kinship. Within the context of the teacher–disciple relationship in Yemen, category distinctions based on descent are played down and blurred, but ultimately not obliterated. In other words, difference is not eliminated to the extent that boundaries become entirely permeable. Non-'Alid teachers may be addressed and treated like kin, but they do not enjoy inheritance rights.[27] The *sadah* are adamant that non-'Alids cannot become *aqarib* (kin) and be incorporated into the House of the Prophet. Unlike non-'Alids, uneducated *sadah* share the Prophet's substance, if only in its dormant state, and are granted their share of family waqf if they are entitled to it. The creation of proper 'Alid persons centers on consubstantiality founded on different components, notably blood and knowledge, but neither can succeed on its own. Nor is one substantial element (for example, knowledge) convertible to another (blood).

The inheritance factor may also be worth examining in the context of Islamic family law which does not rigidly define kinship in genealogical terms. Other kinds of relations which create kinship, for example, those based on shared breast milk, are also recognized. In her work on India, Lambert (2000) is at pains to identify nurture as a "form of relatedness" equivalent to other substantive relations based on shared blood. She argues that flows of substance and sustenance which have a prominent place in the anthropology of India have failed to influence understandings of Indian kinship (2000: 83). As mentioned before, it is because this link is recognized in Islamic law that persons who have shared bodily substance with one mother are prohibited from mating with each other. And yet milk ties are considered to be different from blood ties. Milk siblings have no right to inherit from each other's parents nor are they included into the sibling's descent category.

Furthermore, Islamic law leaves scope for people to express their attachments to others with whom they may not be consanguineally related by allowing them to set aside a third of their patrimony to "nurture" them. Different forms of sharing which create enduring bonds—for example, through milk, knowledge, respect, and affection—are encompassed within the paradigm of genealogical kinship, but they are of a different order. Carsten (2000) proposes to replace the ambiguous notion of kinship with "relatedness" to include genealogical and performative elements in the creation of enduring relationships. However, by arguing that the use of this term broadens the scope for cross-cultural comparison, she nevertheless acknowledges that it "begs the question of whether like is being compared with like" (2000: 33). This and the next chapter show that because substances or components may or may not be subsumable to each other, and shifts in emphasis are historically contingent, the task of comparing relationships involving these components may indeed be as problematic as ever.

* * *

This chapter discussed the kind of knowledge which in the past centuries has buttressed political authority and to which is attributed the "capacity to *do* things" (Weiss 1997: 165). This knowledge, which informs idealtypical notions of the 'Alid self, is more potent when wedded to descent. In 'Alid descent ideology, learning is performing kinship. Ways of moral knowing and reasoning are embodied in performative acts which are inseparable from procreative ones. Learning involves a process of gaining knowledge of the ancestral scriptures and of unearthing potentials for comprehending them. The body itself, then, becomes a kind of memory, not only a metaphor for memory. In the case examined here, memory is not just produced in bodily practices as Bourdieu (1990: 73) has argued. The capacity for what the body enacts is innate, so that what the body learns is not merely an imprint on a blank sheet. The theory of the mind which derives from these notions is predicated on mental templates which aid the understanding of the holy scriptures.

This theory, which centers on predisposition, illuminates Shi'i concepts of authority. Despite the heterodox views held about the sources of knowledge of the Imams among Shi'i scholars of different periods and schools, the very emphasis on blood descent from the Prophet, which is a marked feature of Imamic authority, demonstrates that this authority is not based only on the "gift of knowledge" received from eminent forebears. Regarding the attribution of a special capacity to expound divine knowledge to certain persons, I have argued that Zaydi interpretations constitute a form of weak Platonism. Taking the preceding analysis a step further, it would appear that the emphasis on the cultivation of a disposition to generate virtuous conduct has much in common with the Aristotelian concept of phronesis or reasoned moral practice. For Aristotle, morality is striving for human good, and the exercise of judgment in personal practice constitutes virtue. Ethical knowledge is a form of *connaissance* rather than *savoir*, it is integral to who one is as a person (Lambek 2000: 314, 316). Practice within the framework of the *taqlid* is such ethically informed practice which does not discourage personal reasoning. In Zaydi theory the *mujtahid* has the greatest capacity to interpret the scriptures, thereby applying reason with the goal of establishing and maintaining a moral social order.

Descent and knowledge were also explored as a means of social differentiation. The various relationships dealt with in this chapter may be analyzed from a continuum based on the combination and separation of key substances. Ties of blood may prove to be weak and those that bind may be based on the transmission of knowledge as in the teacher–disciple relationship. Regarding the diversity within the House of the Prophet, it would be too simple to infer that 'Alid concepts of descent are based on the assumption that difference is rooted in bodies. Descent in terms of substance transmission is not divorced from ethical principles on the basis of which similarity is founded. This means that within limits, generative substance is mutable and thus has an historical dimension. Biology is not one-dimensional, nor strictly ahistorical. Among the members of the Prophet's House there are people who are equivalent to each other and those who perceive each other as different kinds of kin. Through learning, the ignorant can revitalize bonds of kinship and reenter the annals. Family annals reveal the extent to which both men and women derive their status from knowledge. References to women who are either 'ulama or daughters of scholars indicate a submerged matrifocal element in the overall patrilineal scheme. Both men and women can enhance each other's status as is illustrated by the case of a man whose wedding day to an Imam's daughter rather than the year of his death was recorded.

This chapter stressed that actualizing kinship with the ancestors through learning serves to produce continuity over time. For the *sadah* one aspect of maintaining themselves as a unique category has been the practice of endogamy, a subject which will be taken up in the chapter 6. The body is conceived as the locus of inherited substance and of moral relations which are best created and reproduced through learning and marriage within the Prophet's House.

Part III

Self-Fashioning in the Idiom of Tradition

Chapter 6

The Politics of Motherhood

In the seventeenth century, the Zaydi jurist Salih b. Mahdi al-Maqbali expressed his indignation at a verdict of the ruling Imam al-Mutawakkil Isma'il b. al-Qasim which prohibited the *fatimiyyat* (sing. *fatimiyyah*, 'Alid women) from marrying *'arab* (sing. *'arabi*; non-'Alids) (al-Maqbali 1981: 440–1).[1] The Imam argued that a marriage of a *fatimiyyah* and an *'arabi* violated the sanctity of the *ahl al-bayt* and should not be allowed. Al-Maqbali protested that those who desired to become close to the Prophet by marrying his female offspring were prevented from doing so. Although these women were fortunate to be desired as wives, most became old without getting married and some might misbehave. Where can we find *fatimiyyun* ('Alid men) to marry those women he asked?

In the twentieth century, the subject still aroused passions among *sadah* and non-*sadah* alike. Marriages between 'Alid women and non-'Alid men were no longer illegal, but nonetheless few were contracted. In the Imamate, the close link between political authority and heredity had repercussions for conjugal relationships. 'Alid descent was the essential requirement for both the supreme leadership and for unions with women belonging to the Prophet's House.

The idiom of descent was used in a politics organized around status and gender differentiation. The patrilineal principle at the core of Zaydi doctrine excluded non-'Alids from the leadership, and it privileged 'Alid men over women. Whereas men were able to arrange hypergamous marriages, women were obliged to marry men of equal descent so that their offspring could adopt their status. The discourse about preferential or enforced endogamy of 'Alid women was predicated on the purity of the *ahl al-bayt*. Endogamy, which was at certain times enshrined in law, defined a "true" 'Alid woman in accepted cultural terms. She and her agnates thus demonstrated that they put the interests of her children and of the Prophet's House first by guarding their 'Alid status and sensibilities. Cross-status marriages inevitably threatened to destabilize both the images of self and the categories of ruler and ruled.

Beginning with the debates among Zaydi 'ulama over what constituted the equivalence of spouses (*kafa'ah*), this chapter examines the articulation of the descent principle in the marriage system. Marriage is understood as part of a wider system of political relations which are the outcome of historically specific processes. The chapter explains the rationale behind the preoccupation with motherhood by the Imamate's leading authorities and its *asl*-bearing subjects. Because a woman transmits her *sulb* to her progeny, at both the top and the bottom of the social hierarchy women were prevented from "marrying out."

However, descent (*nasab*) was not the only criterion which determined matrimonial relationships among the Prophet's descendants. As indicated in chapter 5, low ranking *sadah*, who had left the path of *'ilm*, were neither treated as equals by their more privileged peers nor reckoned to be suitable spouses for their daughters and sisters. Descent was the main status attribute that structured conjugal relations between 'Alid women and non-'Alids, but unions among members of the Prophet's House were contingent upon nonhereditary attributes. Once the descent principle became disjointed from state authority, marriage regulations were relaxed. Chapter 7 will deal with the transition from Imamic to republican rule, showing that the emphasis on heredity has gradually shifted toward nonhereditary factors in the definition of conjugal eligibility. The chapter also considers how women as wives and daughters have accommodated to the realities brought about by the revolution. Chapter 11 deals with a case of a marriage contracted by *sadah* of different rank.

What Constitutes the *kafa'ah?* The Debate Among Zaydi 'ulama

There have been major differences of opinion concerning the *kafa'ah* among Zaydi 'ulama.[2] Early Imams of the Zaydi School considered either descent (*nasab*) or piety (*din*) as the sole conditions of the *kafa'ah*. Imam Zayd b. 'Ali and Nasir al-Utrush, the Imam of Tabaristan, held the view that piety and manners (*ahlaq*) were the main criteria for the equivalence of spouses. According to Mutahhar (1985: 161), Imam al-Hadi Yahya applied the same principles, the proof of which was his approval of his daughters' marriage to the Tabariyyin who were Persians. According to the "followers of the doctrine" (*ahl al-madhhab*), both piety and descent are conditions for the *kafa'ah*. The Quraysh (the Prophet's tribe) cannot enter matrimony with the

Banu Hashim, nor can the latter marry the *'arab*. A group of *'ulama* gave all women except the *fatimiyyat* the right to disregard rank disparity between spouses based on wealth (*mal*), occupation (*sina 'ah*), and free descent (*huriyyah*, here referring to the status of a non-slave). In the eyes of those *'ulama*, nothing could make up for lack of *'Alid* descent for the *fatimiyyah* even if she and her guardian agreed to the marriage.[3] Imams like Mutahhar b. Yahya (d. 1297) and his son Muhammad (d. 1327) contested this view arguing that the Prophet had married his daughter Umm Kulthum to the caliph *'Uthman*. Imam Sharaf al-Din (d. 1558) held the opinion that there were no legal grounds for a prohibition of marriage between the *fatimiyyat* and non-*'Alid* men, but such unions would lower the women's social worth and be an insult to them (Sharh al-azhar Vol. 2: 301–4). In the fifteenth century, a woman *mujtahid*, Safiyyah bt. al-Murtada al-Wazir, wrote a strong-worded treatise against those *'ulama* who sanctioned marriage of a *fatimiyyah* and a non-*'Alid* (Z. al-Wazir n.d.).

In the controversies over the *kafa'ah*, occupation was not of paramount concern. However, the Faqih Muhammad b. Sulayman b. Nasir b. Sa'id b. Abu Rijal ruled that a *sayyid* who had a low occupation was not eligible to marry a *fatimiyyah* (Sharh al-azhar Vol. 2: 302).[4] Like his brother Imam al-Mutawakkil Isma'il, Imam al-Mu'ayyad bil-lah Muhammad b. al-Qasim (r. 1620–44) decided that women—with the exception of the *fatimiyyat*—were allowed to ignore status discrepancy in marriage. However, he saw the injunction of marriage of the *fatimiyyat* and the *'arab* as a divine instruction. At first, the Imam had not emphasized the rule prohibiting this kind of union, but later he charged all those who rejected the rule with unbelief (*kufr*). In his view, the prohibition was a divine decree that could not be negotiated (*al-haqq li-'llah laysa li-ahad an yusqitahu*). As already mentioned, al-Maqbali questioned this verdict, arguing that many *fatimiyyat* suffered from this injustice. In some areas of the Yemeni mountains *fatimiyyat* had married low status *'arab* because they were poor, and no one had made a fuss. Yet the *'ulama* had no choice but to follow the principles established by the Imam. Referring to the Prophet as saying that "every *nasab* and *sabab* (reason, here: mission) comes to an end except his *nasab* and his *sabab*," al-Maqbali makes no secret of his anti-Hadawi stance (al-Maqbali 1981: 440–1).

His writings reveal that the endogamous marriage rule which was imposed on *'Alid* women was already opposed by Yemeni writers long before the issue became highly politicized during the 1962 revolution (see chapter 7). His work is also of interest because it shows that by no means all *'Alid* women approved of this rule. Al-Maqbali introduces the reader to a woman who tried to bypass the law. The woman, a wealthy lady, was eager to marry but unable to find a spouse among the *sadah*. When she met a pilgrim who was passing through Luhayyah, a western town, she asked him whether he was a *sharif* because no one knew him. He did not reply, but when she repeated her question the answer was "no." She suggested to him that if he were not, he should pretend to be so, but he refused to do so. Expressing her disapproval of the Imam by saying "let God do with you [what he will] (*fa'al allahu bika ya Mu'ayyad wa fa'al*)" (al-Maqbali 1981: 439–41), she also indicated that injustice had been done to her.

The uncompromising attitude some Imams held toward the *kafa'ah* should be seen in the context of the political economy of their time. I am unable to establish whether the legally prescriptive discourse which required 'Alid women to marry men of their own kind had much effect before the seventeenth century, but there is no doubt that overt control was exercised over their sexuality during a period of national reconsolidation, expansion, and economic prosperity. Under the rule of Imam al-Mutawakkil Isma'il (1644–76), for the first time in its history the Zaydi Imamate expanded southwards to Aden and the Hadramawt, and even aspired to intervene in the affairs of Mecca and the Hijaz (Blukacz 1993: 47–8).[5] His influence was such that his name was even mentioned in the *khutbah* (the allocution at Friday prayers) in the Red Sea port of Yanbu' in the northern Hijaz (al-Shahari 2001). The Imamate's missionary zeal combined with an ambition to expand its trade activities: the coffee and horse trades were at their height. After the Ottomans were ousted from Yemen, the Imam also introduced the ritual of Ghadir Khumm which centers on the righteousness of the Prophet's progeny ('A. al-Wazir 1985: 314). During this period, doctrinal distinctions gained more social and political significance (Würth 2000: 26). In Sunni towns such as Ibb where the Ottomans had had a strong presence, appointments to judgeships were contingent on the candidates' knowledge of Zaydi law and the verdicts of Imams had to be applied in the local courts (Messick 1993: 41). In some parts of the north, shari'ah law was enforced and efforts made to collect the zakah (Blukacz 1993: 43). Thus, the assertion and reassertion of Zaydi supremacy after foreign occupation manifested itself in an exclusive marriage rule and regional religious, economical, and political expansionism.

In the twentieth century, the Imams did not pose legal obstacles to marriages between the *shara'if* and non-'Alids. Local officials enquired at the court of San'a, or of the Imams, about the legality of proposed unions. On one occasion, a local judge (*hakim*) asked Imam Yahya whether the marriage of an *'arabi* to an orphan 'Alid girl was legal, informing him of the possible objection of the girl's brother, and the need to have a judge to act as her guardian. The Imam agreed on the following conditions: (1) that the girl should have reached maturity; (2) that she was unable to obtain a *sayyid* in marriage; (3) that she was in need of provision and of being prevented from having to leave her house; (4) that the suitor was pious, prosperous, honorable, recognized her honor, and was able to prevent her from leading a reprehensible life. The Imam ordered a judge to ask the girl's brother to contract a marriage on her behalf with either a *sayyid* or an *'arabi*. The Imam sought to secure the girl's moral integrity through marriage, a matter to which he gave priority over her marrying a *sayyid*.[6] Like Imam Zayd b. 'Ali, Imam Yahya ruled that piety constituted the *kafa'ah*. According to Qadi 'Abdullah al-Shamahi, Imam Yahya's verdict was inspired by the Prophet's saying that "a *sharif* is the one who is *sharif* (noble) in his deeds." He argued that "a man is the son of today rather than yesterday, defined by his work rather than by his patrilineage." God had sent the Prophet to unify the *ummah* and to inform its members that they were all equal, and there was no privileging of black over white or *'arabi* over *'ajami* (non-Arabic speakers, mainly Persians) (al-Shamahi 1937: 45–7).[7] However, the Imam's judgment in the case of the orphan girl reveals that his view toward the question of matrimony was nonetheless ambivalent. He consented to the girl's marrying an *'arabi* should she be unable to find a *sayyid*.

In another case, a non-'Alid asked a *sayyid* for his daughter's hand. Before he had received an answer, a *sayyid* proposed to the girl. Her father gave preference to him, but the first suitor went to court claiming that the other man had no right to propose to the girl before a decision had been taken. The judge, who was employed at the royal court (*hakim al-maqam*), argued that the timing of the second suitor's proposal was inappropriate. On being asked who she wanted to marry, the girl favored the litigant. The judge then enquired about whom she would have chosen had both men proposed to her at the same time. She again confirmed her choice and the first suitor was given the right to marry her. The case was transferred to Imam Yahya who agreed with the judgment.

The Transmission of *sharaf* or How to Marry Safely

Irrespective of specific 'Alid concerns in Yemen, elsewhere on the Peninsula ruling houses tended to engage in non-reciprocal marriage relations, too. Al-Rasheed (1991: 198) writes of the Rashidi dynasty of Hail (1836–1921) that "although the amirs [the Rashidi leadership]. . . took brides from outside their *beit* [*bayt*], lineage and even tribe, they did not reciprocate with their own daughters and sisters." The interests of these houses notwithstanding, hypogamy is usually considered to be incompatible with a woman's dignity. For a Yemeni 'Alid woman, any union with a non-'Alid was by definition a hypogamous marriage. In marriage, it is said, a woman "becomes" her husband's *sharaf* (honor). An 'Alid woman who "marries out" cannot transfer the *sharaf* which is intrinsic to her *nasab* to her progeny. During the (twentieth century) Imamate, even poor families were disinclined to give their daughters to more prosperous non-'Alid men. During Imam Yahya's lifetime, the majority of *sadah* living in the northern part of the country ignored his verdict concerning the *kafa'ah*. With few exceptions, women entered into endogamous marriages. As al-Maqbali's writings indicate, this was common practice at least during the last two centuries of Imamate rule. Mufti Ahmad Zabarah (personal communication) claimed that prior to Imam Yahya's time, 'Alid women could not marry non-'Alids for a thousand years. This assumption is confirmed by the histories of 'Alid houses which I was able to consult, but it is of course possible that marriages with non-'Alids were not recorded.

Yemenis of diverse social location also argued that hypogamous relations were likely to upset domestic harmony; a woman's superior social status would conflict with her inferior position in marriage. "If there is a problem between the couple," I heard it said of an 'Alid woman, "she will remind him who he is and ask him how he dares to tell her to listen to him." Although there is much more tolerance of hypergamy, many people consider it to be degrading for women. Indeed, marriage does not neutralize status divisions. The offspring of hypergamous marriages adopt their father's status and regard their mothers as different from themselves.[8] One 'Alid woman did not categorize her mother, a tribal woman, as a member of the family (*usrah*) ("*ummi qabiliyyah*"). Once I asked a shaykh's daughter, who was married to one of the Imam's deputies, what life was like for her after she had moved into his household. Was it different from what she was used to? This she denied, but her daughter exclaimed "How come? Among the *sadah* men and women do not mix; the *qaba'il* do not pray, and they are not educated." "Of course the *qaba'il* pray," her mother protested.

Marriage involves material and moral exchanges, part of which are transmissions of biomoral substance between men and women. Because Yemeni women, like men, are conceived as carrying and transmitting the *sulb* of those men and women from whom they descended (see chapter 5), those who consider themselves to be of pure descent are concerned about their *nasab*. This is why even today marriages with those formerly labeled *qalil asl*, irrespective of their achievements, are rarely contracted (see chapter 7). A woman of a "good" family is not only pure but also deemed to inculcate moral dispositions and, moreover, the values associated with her specific social location, in her children. A mother's successful instruction of her daughter is acknowledged by her son-in-law who offers her a small gift (usually money) after the wedding night which is called *haqq al-salam* ("that which belongs to peace"). The gift communicates the groom's satisfaction. It is a symbolic recognition that conflict between the affines could be avoided because the girl has been raised well.[9]

The link between lowly birth and moral deficiency has implications too for ideas about a person's ability to exercise authority. According to one *sayyid*,

> 'Adnan is superior to Qahtan because he is a descendant of [the prophet] Isma'il.[10] Isma'il is superior to Abraham [another prophet]. Isma'il had leadership qualities. The reason why nobody wants to marry the *mazayinah* is because they have no potential for leadership, they are lacking roots (*asl*). The Prophet married Safiyyah of the defeated Jewish tribe because she was the offspring of Moses and hence had leadership qualities. It was the sons 'Ali had from Fatima who became Imams rather than his other sons who were born of ordinary women. [The Caliph] 'Umar heard the Prophet say about 'Ali "this man's blood will not vanish, others' will," whereupon 'Umar married 'Ali's daughter.

In the Imamate the notion that *sulb* is transmitted by both men and women provided one rationale for preferential 'Alid endogamy which I define broadly as marriage within the House of the Prophet. Even though Zaydi doctrine does not stipulate that the mothers of Imams must be of 'Alawi-Fatimi stock—quite a few of them were slaves—concerns about motherhood figured in scholarly debates and were fundamental to Imamic state building. Many of these debates centered on who provided eligible partners for 'Alid women—in other words, whose patriline were they permitted to reproduce? A high number of marriages with men of Qahtani descent would have undermined the logic of descent-based political control that privileged the 'Alids.

Before I explore the implications of endogamous marriages, let me say a few words about how people were married. Like women, men were (and are) urged to marry so that their sexuality be properly channeled. Both conjugality and parenthood are attributes of adult status, and contracting a marriage is also a fulfillment of one's religious duties for which one is credited in this and the other world. This is why the onset of conjugal relations is the most highly ritualized event in a person's lifetime. Marriage was arranged by the couple's parents, but older siblings and close kin also played a role. Women acted as marriage brokers and looked for suitable spouses for their sons and brothers, but men usually had the final word. Even after contracting the marriage, the couple did not meet before the wedding night.[11] Girls married between the age of eleven and fifteen; men were a few years older. They were told that they would learn to love in the course of their marriage.[12]

> I was married when I was thirteen. I only agreed because I wanted to be close to my sister who was married to my husband's brother. I was unhappy. I went back to my mother every day and cried. She told me that after a while love would come, and my father praised my husband's good qualities. (A scholar's daughter in her fifties)

The endogamous unions of 'Alid women served to maintain the prestige and independence of distinguished houses. Marriages, which aimed at reproducing "the social relationships which made them possible" (Bourdieu 1977: 53), were quite often contracted among people of the same house.[13] As noted by al-Rasheed (1991: 195), these marriages blurred divisions between maternal and paternal kin. Unions between houses of similar rank were just as common; these houses often stood in fixed and symmetrical affinal relationships to each other over several generations. Some of them shared both a secondary ancestor (such as Imam al-Qasim) and political interests, and they were landed.[14] Affinal ties between members of the same house also strengthened relations between the hijrah and the towns. Both men and women tend to consider marriage within the same house to be desirable because "one knows them well," but no special preference is given to unions with the paternal parallel cousin.[15] The question of motherhood also concerns the relationship between brothers and sisters. A brother has a particular stake in his sister's marriage because relations between the maternal uncle and his sister's children are expected to be close, and he might wish them to be of equal status. In a sense all marriages within the Prophet's House, particularly those within and between well established houses, were (and often still are) seen as "close" and "safe," not least with regard to women's interests. For a girl to marry "safely" meant that a man would respect her as he would his sister and provide for her; his social position would not be lower than that of her family. According to a high ranking *sayyid*,

> As you know, women are always looking for safety. For a girl to marry a *sayyid* means that she feels safe in every respect. A woman knows she

will be treated better by her husband if she has a powerful father and brother [assuming that she comes from an influential family], and she wants her daughters to have brothers so that they will be safe, too. You marry a *sayyid* to be sure that you will be treated well, it's like marrying from the family.[16]

Transacted Men?

Women's "safe" marriages within the Prophet's House fostered 'Alid claims to political pre-eminence, but women were not "a sort of symbolic money allowing prestigious alliances to be set up with other groups" (Bourdieu 1977: 66). Bourdieu's assumption that a woman derives her social worth from the affinal links she engenders for her group is commonplace (for example, Vieille 1978; Tapper 1991). The data on the Yemeni old elite suggest that sons were more important "political instruments" (Bourdieu 1977: 54) than daughters for strengthening ties between (sometimes culturally and geographically remote) 'Alid houses and for creating alliances with non-'Alids. Using the example of women at the highest echelon of the twentieth-century Imamate, I suggest that the marriages contracted by the women were of less strategic value for their senior male kin than those arranged by men. Even when these women did not enter conjugal relations with close kin, their marriages usually reinforced pre-existing ties between prominent landed houses whose members held political offices. Moreover, the notion at the core of alliance theory that "women are for men to dispose of, they are in no position to give themselves away" (G. Rubin 1975: 175; compare Yanagisako and Collier 1996: 235–6) must be qualified. Although a son's refusal to marry a girl his parents had chosen for him was more likely to be accepted, he was cautious not to alienate them.[17] Men were just a few years older than their brides and hardly in a better position than them to assess the proposals made by their mothers and sisters.

The affinal relations men established with women of the royal House had little material advantage for them. The marriages enhanced their prestige, but the men were well placed to assume leading positions even without these affinal ties. These marriages had political value in so far as they cemented relationships between ruling 'Alid houses. A further point to be made is that at the top of the social hierarchy, women's status was defined less through marriage than through kinship with the ruler. Usually the category "woman" is identified with wives and mothers whose status is considerably higher than that of unmarried daughters and sisters (vom Bruck 1997a). The latter hardly ever gain as much control over the daily affairs of the household as do women who have produced children. However, some of the Imam's kin acquired authority and influence even while they were young and unmarried. Most women of the royal dynasty married, but conjugality and motherhood were not necessarily the prime source of their self-worth and status. One of Imam Yahya's daughters who never married did not enjoy less prestige than those who did. Women of the royal dynasty and eminent houses could afford to turn down proposals and to pursue their own interests. Some refused to return to Yemen from exile after the civil war had ended in 1970 even though they were aware that their husbands were going to take other wives after repatriation. Royal women's status was defined primarily by their kinship with the ruler rather than through marriage. Their brothers invested their trust in them, and some women brought their husbands into the royal household in spite of the general preference for virilocal residence. To a man, a daughter's or sister's (or other female relative's) loyalty was of greater importance than the marriage she contracted. Brothers entrusted their sisters rather than their brothers or affines with the supervision of property while being abroad.[18]

As far as women of distinguished 'Alid houses were concerned, the "genealogical legitimacy" (Bourdieu 1977: 35) of their marriage had to be complemented by the suitor's education, wealth, or political office. Bourdieu (1977: 37) assumes that the actors themselves will always pose "the problem of marriage in strictly genealogical terms," thus denying its orientation toward the pursuit of material and symbolic interests. However, the genealogical records and annals of 'Alid houses reflect rather than conceal the rationale that underlies affinal relations within the Prophet's House. The very exclusion of those *sadah* referred to as "*qaba'il*" or "*fallahun*" from the documents produced by their higher ranking kin defined them as ineligible partners of their daughters. "Official" and "practical kinship" merged and were not opposed to each other as suggested by Bourdieu. The records mirror the concerns of "practical kin who make marriages" (1977: 34). Just as the verdicts of Zaydi-Hadawi 'ulama on the principles of rule disfavored the uneducated scions of the Prophet's House, exclusionary marriage practices *within* it were legalized by scholars like Muhammad b. Sulayman who forbid unions between the *fatimiyyat* and 'Alid men exercising lowly professions. Those who did not represent masculine strength and honor (*sharaf*) in terms of erudition and other rank-related practices were feminized and hence ineligible partners for women of more distinguished houses. Thus, while in most cases *nasab*-based distinctions served as impediments to marriages of *shara'if* and non-'Alid men, poor or badly educated

'Alids were also limited in their choice of wife. Because of these restrictions, it was not uncommon for high ranking 'Alid women to remain unmarried. The intrinsic link between marriage and rank within the Prophet's House contributed to the estrangement of its less privileged members. The *sadah* who participated in the revolution did not have access to the daughters of those collaterals who exercised political authority, some of whom carried the same patronym. A few years after the revolution, the officer who explained his motives for engaging in anti-regime activities (chapter 2) asked for the hand of the daughter of a leading scholar, who did not refuse.

Unlike women, men had the option of taking wives of lower status without endangering the purity and prestige of the House. Some high-ranking *sadah* preferred a high ranking non-'Alid woman to a low ranking *sharifah* because of the latter's lack of sophistication. These men realized two ideals—that of endogamy and asymmetrical alliances—by contracting multiple simultaneous marriages.[19] They married the daughters of important shaykhs, *qudah*, and *sadah* (some of them rivals) from different parts of the country. The Hamid al-Din Imams contracted marriages with both the 'Alid and non-'Alid elite and commoners and slaves.[20] By marrying the foster-sister of the Hashid Shaykh Nasir al-Ahmar who was a *sharifah*, Imam Yahya at once reinforced relations with a closely related house and one of his main supporters.[21] Alliances with the shaykhs were especially important because they commanded large numbers of armed men. Some of the governors and judges who were assigned posts in different areas married up to thirty times during their professional careers.

The *sadah* were able to simultaneously establish links with the tribes and to marry *shara'if* even without contracting multiple unions. They did so by arranging marriages with 'Alid houses in the eastern part of the country who are referred to as *ashraf*. According to San'ani *sadah*, the *ashraf* became "tribalized" after abandoning their tradition of learning two or three centuries ago. They began to engage in warfare like their neighboring tribes. Several *ashraf* are rich landowning houses. They were appointed army commanders under the Imams who did not need to fear them as rivals because they were not learned (for example, Bayt al-Dumayn). Urban *sadah* refused—and still do—to give them their daughters because *ashraf* women work in the fields and their lives might be affected by violence caused by feuds. Urban *sadah* also point out that marriages among the *ashraf* do not last, a notion that is held by the *ashraf* themselves. An *ashraf* woman who had moved to San'a after getting married to one of Imam Ahmad's commanders turned down her brother's son's proposal to her daughter because she feared the marriage would be terminated sooner or later. She explained that her father had married about ninety times, and she wanted to spare her daughter the experience of serial marriage.

Boundary Transgression and Its Sanctioning

Just as definitions of endogamy within the Prophet's House have to be qualified by reason of considerations of rank disparity, hypogamy was not a straightforward matter either. In either case, there was ambiguity over what constituted "marriage between equals." On the one hand, those who were of equal descent but uneducated and poor were often ineligible partners. On the other hand, well situated non-'Alids who married 'Alid women who were no longer virgins were not inevitably inferior to them. The marriages which were contracted between high-ranking *shara'if* and non-'Alid men during the period of Imam Yahya and Ahmad were not hypogamous unions because most of the women were widows or divorcees. The husbands' lack of a genealogy as eminent as theirs was balanced by the women's lower status as non-virgins. As pointed out by Bourdieu (1977: 68; emphasis his) because of the "price" put on virginity, for a woman the "*depreciation* entailed by previous marriages is infinitely greater" than for a man. Contracting these marriages did not undermine the *sadah*'s position of supremacy because "giving out" women in marriage did not arise from the necessity of creating alliances. Most importantly, these unions demonstrate that the categories which derived from the hierarchies of rule were negotiated relations and disputed classifications.

Since the number of women's exogamous marriages has increased steadily after 1962, one particular case is often cited by *sadah* to show that 'Alid women had always married out, and that this was not just an effect of the revolution. In the 1950s, Qadi Ahmad al-Sayaghi, Imam Ahmad's governor in Ibb, married a widow from the famous Sharaf al-Din family. Several members of her house had reservations about the marriage but did not try to prevent it. According to the woman's kin, the Imam was angered and reproached them for their consent. Other *sadah* suspected political motives for the Imam's displeasure, arguing that he had shared his father's opinion on the *kafa'ah*. From a legal point of view, he had had no reason to object. He had never declared the marriage to be unlawful, and if indeed this had been his opinion, the governor would have had to divorce his wife. They suggested that he had been concerned about the alliance between an important *qadi* house and the Sharaf al-Din who were closely related to the al-Wazir who had played a role in his father's assassination.

Moreover, the Imam had not been on good terms with his governor and tried to create problems for him. The writer Muhammad b. 'Ali al-Mahaqiri said kind words about the couple.[22]

> The pure, erudite, pious *Sharifah* Taqiyyah bt. (daughter of) Muhammad b. Ahmad b. 'Abdullah b. Yusuf b. Muhammad b. 'Abd al-Rabb b. Muhammad b. Husayn b. 'Abd al-Qadir b. al-Nasir b. 'Abd al-Rabb b. 'Ali b. Shams al-Din b. al-Imam Yahya Sharaf al-Din, was born about 1923 in Makhadir [between Ibb and Yarim] where her father worked. She was brought up under his care and studied the Qur'an, arithmetic and literature, and learnt to write. Before she had reached puberty, she was married to Ibrahim b. Husayn b. 'Ali 'Abd al-Qadir who died very soon after.[23] Then she married the great scholar 'Ali b. Hamud Sharaf al-Din[24] who loved her and looked after her because she was beautiful, intelligent, and had good manners. When he died, she returned to Makhadir. I got to know this when I was staying in Makhadir. She was a woman of splendid intelligence, and she had a fine handwriting and a good style. She used to ask me for a book to study and would return it very fast and would then ask for it again. I married my daughter to her father through her mediation. She looked after the house and managed all his affairs. She offered alms and comfort to the needy with all her abilities until the governor of Ibb, the Qadi Ahmad al-Sayaghi, married her. He was a very good, able, and generous man who looked after her and valued her. He took her with him to Aden when he escaped from Imam Ahmad. When he died in the Jawf where he was hit by an Egyptian bullet, she moved to Jiddah. She, her son Khalid al-Sayaghi and her three daughters were supported by King Faysal. She suffered from diabetes and was treated in London. She died in Jiddah and was buried in Madina in 1979.

The marriage of the divorced daughter of a judge to a Shafi'i student was interpreted along similar lines. The judge had participated in the Constitutional Movement and was not looked at favorably by the Imam. The judge himself had hesitated for some time, before he gave his consent after receiving the approval of a respected relative, then living in Cairo, to whom he had sent a long letter which is still in his possession. His daughter had also written to this man explaining that she concurred with the proposal. In her view, the man was a believer (*mu'amin*) like themselves. Her relative, who had been asked for help by the suitor as well, convinced her father to agree, arguing that some pious Imams had also given their daughters out in marriage. The woman's father asked her for a written statement of approval so that she could not complain later. When I met him in his seventies, he said that he had agreed because it was his daughter's decision and because he did not believe "in these things," that is, restrictive marriage regulations. He had received letters of protest from members of his family and other 'Alid houses. In later years, when the woman's husband became an ambassador, her status as a wife was surely preferable to that of a divorcee in her father's house.

In another case a minister in Imam Ahmad's cabinet was willing to agree to his daughter's marriage to a non-'Alid. The girl was also in favor, but the decision was opposed by their relatives who claimed that "he who agrees [to the marriage] is lowering us." Several friends and 'ulama also protested, and the Head of the Court of Appeal (*ra'is al-isti'naf*), who disapproved, interfered in the affair. The woman's father contemplated complaining to the Imam or going to court to secure permission to marry his daughter to the man of her choice, but in the end he gave up. A friend of mine who had known him well commented that "he was a true Zaydi (*Zaydi ḥaqiqi*) because Zayd b. 'Ali himself did not believe in endogamy."

The politics of motherhood discussed in this chapter expose the intricate pattern of (self-)identification and difference which is one of the book's central themes. The debates Zaydi scholars have engaged in over the centuries leave no doubt about the profound political dimension of the question of whose mother women were to become. Women have always been both subjects and objects of the politics of motherhood which have shaped their experience as wives, mothers, and sisters. Chapter 7 examines how in the post-revolutionary epoch marriage reflects as well as creates new political realities. Marriage is analyzed as a means of conquest and reconciliation, and as a relationship not just between husbands and wives but between sisters and brothers as well.

Chapter 7

Marriage in the Age of Revolution

As chapter 6 indicated, some of the scholars who opposed the rule prohibiting marriages between 'Alid women and non-'Alid men in previous centuries tended to cultivate anti-Hadawi sentiments. Since the 1962 revolution the discourse of those who had reluctantly accommodated to this rule focuses on their dual exclusion from the supreme leadership and 'Alid women. Indeed, the passion the subject arouses leaves one with the impression that in important respects the revolution is conceived in terms of its challenge to elite marriage practices. Whilst many non-'Alids—above all the *qaba'il*—were prone to accept the rule of a just Imam (*imam 'adl*), they were less willing to acquiesce in these practices. In the debates about the moral foundations of the society, which was to be built on the ashes of the ancien régime, motherhood has remained a contentious issue.

Opposition to the *sadah*'s marriage practices is voiced most emphatically by those who themselves had aspirations to the leadership, the *qudah* and the shaykhs. In the Imamate marriage was an idiom through which status distinctions were constituted in gender terms. Hypergamous unions are predicated on the inferiority of the wife-givers and modeled on the relation between the groom and the bride. On marriage a man assumes authority over his wife who is expected to be subservient to him. By virtue of being the principal wife-givers, non-'Alids symbolically became the "brides" of the *sadah*. This caused resentment particularly among shaykhs who provided protection for the *sadah* who lived in the hijrahs. In accordance with their hijrah-status the *sadah* were considered worthy of protection like women, yet they asserted their masculinity as rulers and by refusing their women to their protectors.

Men of lower status, who never had ambitions to marry 'Alid women, are less likely to challenge the *sadah* on these grounds. The upwardly mobile among them desire unions with higher ranking non-'Alids. The number of 'Alid women who have "married out" has risen continuously in the past four decades. These marriages are of great symbolic significance because they are interpreted as a renunciation of 'Alid claims to ascendancy and exclusive status. As noted in chapter 2, the strength of the pre-revolutionary elite was demonstrated in part by their non-reciprocal marriages even under Turkish occupation. Therefore, Amat al-Karim al-Farisi's marriage to a descendant of a former colonial official in the early 1980s was a story about shifting relations of power (see chapter 3).

The Morning After

Men who took up prominent posts in the revolutionary government were the first who asked 'Alid women in marriage. Shaykh Amin had joined the revolution because his close relatives had been executed by Imam Ahmad after being accused of rebellion.[1] He asked for the hand of the granddaughter of one of the Imam's officials who lost his life during the revolution. Like his predecessors, the shaykh granted hijrah-status to the official who was at once his protégé and his superior in the political hierarchy. Many of the woman's kin objected to the shaykh's proposal, but those who argued that a rejection would rule out a reconciliation with the new holders of power got the upper hand. The woman, Amat al-Malik, became the shaykh's second wife. Affinal relations had previously been established between the two houses. However, in the early 1960s the shaykh's family became wife-takers for the first time. The children of the shaykh's first marriage with his paternal cousin were called after pre-Islamic tribal ancestors, but the first son born to the *sharifah* was called after one of the Prophet's kin. The shaykh explained that the *qaba'il* expected baraka from a marriage to a *sharifah* and that it was an honor to marry one even if she came from an impoverished family.[2]

The empowerment of the new elite manifested itself in marriage proposals to 'Alid women as well as in a refusal to create affinal ties with the *sadah*. Ahmad al-Husaynat, a young *sayyid*, asked for the hand of a girl of a *qadi* family which had served in the Imam's government and has continued to prosper. In the early 1960s, Ahmad had difficulty in reconciling the pride he took in his descent with the abusive language directed at the *sadah* in the streets. He was mystified by the notion of being a *sayyid*. He decided against marrying an 'Alid girl in order to come to know more about those who had been described to him as different from himself. The girl's kin turned down his proposal, informing him that they "had finished with the *sadah* once and for all," and that they would "rather burn the girl than marry her to a *sayyid*." After his rejection, Ahmad married a *sharifah*.

To this *qadi* family, his proposal seemed reminiscent of former asymmetrical marriages between the *sadah* and non-'Alid women. Other *qudah*, among them a wealthy landed family, which holds prominent positions in the government, welcomed a *sayyid*'s proposal to their daughter in spite of their strong resentment toward the *sadah*. They had named one of their sons Yazid after the Umayyad ruler who was Imam al-Husayn's enemy, and often talked about the *sadah* in derisive terms. The girl had met her husband-to-be, a student of law, at the university. She had always worn a light-colored overcoat and headscarf; after the wedding she began wearing the *sharshaf* (pl. *sharashif*), a black garment that covers the whole body, even though her husband did not require her to do so. Although her family never conceded the *sadah* any special status by virtue of their pedigree, they now spoke of them as adhering to religious principles more strictly than others. According to the girl's brother, "the *sadah* like that sort of thing. They like their women to be covered."

In the mid-1990s, one of the young men of the family proposed marriage to a girl of a reputable 'Alid house. Both families have held influential positions in the Imamic as well as the republican government. The girl's family was concerned about the family's anti-'Alid attitude rather than their descent. The girl's sister, a confident professional, demanded that all the elders of renown belonging to the suitor's family, including those in exile, give their approval before they agreed. In another case, the son of a famous judge of *qadi* extraction married the daughter of a high ranking *sayyid* who had fled the country in 1962. He talked about his proposal to the girl as if it was an act of charity, thus affirming claims to his own superiority. He explained that the girl's father had returned from exile in 1973, finding himself discredited and jobless. "He was then an ordinary man, he was nothing."

Men like this young *qadi* discussed the *sadah*'s marriage practices in terms of their claims to social precedence (*al-tabaqah al-'uliyah*) rather than with reference to religious sources. Few pointed out that there was no Qur'anic injunction against marriage between the Prophet's female descendants and men like themselves. Even elderly *qudah* who were well versed in Zaydi jurisprudence refrained from invoking past rulings of Zaydi 'ulama on this issue. According to an old judge, in the Imamate the social distance between the *sadah* and others discouraged the latter from contemplating marriage to 'Alid women. "In the past some men thought they did not have the right to marry a *sharifah* because the *sadah* were ruling. Now, if a *sayyid* refuses, he will be told that he is still attached to the old regime."[3]

In the Imamate, *qaba'il* who moved into the professional class of the *qudah* no longer gave their women in marriage to *qaba'il*. The son of one of Imam Yahya's secretaries of *qadi* descent argued that "a *qadi* married the daughter of a shaykh, but he could not give him his daughter because she could not have lived in the countryside." His grandson, who trained as a doctor, agreed with him.

> A Qahtani did not ask a *sharifah* in marriage because he was afraid of being rejected. No one ever thought that marrying a *sharifah* was a violation of the principle of *kafa'ah*. The *sadah* thought they were better than us and hence could not give us their daughters in marriage, but they took ours. I have no desire to marry a *sharifah*. She may be imperious (*mutakabbir*) and tell me that she comes from a better family than mine. The *shara'if* are not as conservative as our girls. It's much better to marry someone of the same rank (*tabaqah*).

Women who described themselves as *qabiliyyat* (of tribal descent) generally approved of marriages with the *sadah* but, depending on these men's personal circumstances, felt uncertain about their future. One elderly woman told me proudly that her husband was a Qurayshi from Mecca (*zawji qurayshi min makka*). Another woman, a shaykh's daughter, told me that she was gratified having been proposed to by a man who was closely related to the Imam. Her pride notwithstanding, she had reacted with apprehension because he already had two wives, and she did not know what her life was going to be like. She was still very young and had burst into tears when her father informed her about the man's proposal.[4]

In the rural north attitudes toward marriages of this kind differed widely. In Sa'dah province, a Zaydi stronghold where the *sadah* were prominently represented as government officials and scholars, there is a network of closely interlinked hijrahs. Some *qaba'il* who were farmers living in the vicinity of Sa'dah town argued that they did not wish to marry a *sharifah* because of the respect they had for the Prophet's descendants. These men, who in the 1980s still referred to 'Alid women as *shara'if*, were also concerned that they might be looked down upon by their affines. One argued that since a man could not help swearing at his wife from time to time, it was better not to marry a *sharifah* because it was improper to insult the Prophet's offspring.

The *sadah* in the area hardly ever referred to religious sources when defending women's endogamy. Some, who might have been unfamiliar with these sources, conceived of the rule as *muharram 'urfi* (habitually unlawful). Often it was explained with reference to the practices of the *qaba'il*. As a *qabili*'s wife, they said, a *sharifah* would have to work outdoors and would not be asked to veil.[5] The *qaba'il* rejected this suggestion, saying that the wives of the *qudah* and shaykhs also lived secluded lives. The sons of shaykhs with urban aspirations seek to marry *shara'if*, considering their fine manners and education. They hold images of these women as delicate and beautiful because they do little outdoor work.[6] A young *sayyid* living in Hijrat Falih where there was strong opposition to women's out-marriage was aware of the moral dilemma. "The *qaba'il* who come to ask for our girls will be told that they are not believers (*ma-fish din 'andakum*). The *qaba'il* will then say we are racist, reactionary, and arrogant (*mutakabbirun*)."

This was confirmed by a young *qabili* called Muhammad (see chapter 10) who works in San'a as a clerk.

> I myself do not recognize the *sadah*'s superiority by virtue of their descent from Muhammad, neither do I endorse their reluctance to give me one of their daughters. They say their daughters' children will not be *sadah* if they were married to us, but so what? This is just another assertion of the superiority they claim for themselves. Endogamy is tragic for the *sadah*. It means that their women have few choices to find a spouse, and it prevents them from integrating into the society. Once a *sayyid* told me that some *qaba'il* might not want to marry a *sharifah* because she will not work in the fields. Yes, perhaps. But that is not a sufficient reason to deny us their women. The *shara'if* don't do any hard work, they always stay at home and cover their faces. Even the poor ones never fetch water from the well or go and fetch wood. The *shara'if* think that they must never be seen. They have very fair skin.[7]

When in the early 1990s the first 'Alid woman of Razih in northwest Yemen entered into matrimony with a wealthy non-'Alid merchant, the crowds of men who enthusiastically celebrated the event sang a song:

> Oh sayyids, you tricked us
> With your turbans, remedies and charms
> Whenever we proposed marriage, you said
> "With a *sharifah*, a sayyid's daughter? It's not allowed."
> God only knows whose book you studied!

(Weir 1997: 26)

The derisive tone of the song demonstrates lack of respect for the bride-givers. Yemeni wedding songs usually flatter and praise the new couple and their families. By exclaiming "God only knows whose book you studied!" the men failed to argue from within the *taqlid* and instead denied the existence of that "book" which renders such unions unlawful. The rhetorical question of "whose book" rather than "which book" insinuates the fabrication of a religious text and even heresy. The knowledge derived from *that* book has lost the status of authoritative memory. This case is paradigmatic of situations where literacy becomes more widespread and power changes hands. It is not clear whether lay persons are able to read the religious sources as well as the clergy, as Fischer (1990: 140) has argued for Iran, but there is certainly a notable denial of the latter's specialized judicial authority, even if this denial is only politically motivated.

In the north-eastern part of the country, the *qaba'il* had less desire to enter matrimony with 'Alid women. There 'Alid control was weaker than in the north-western provinces, and there were few learned *sadah*. One of the important shaykhs, who was married to his paternal cousin, claimed that he would only have married a close relative. Talking about the endogamous marriages practiced by the *sadah*, Shaykh Hamid said "I also have my principles (*'andi 'asabiyyah aydan*), I would not marry a *sharifah* or anyone who does not belong to my house." He dismissed the *sadah*'s arguments in favor of their women's endogamy as *kalam fadi* (empty words, nonsense), and denied that people could approach the Prophet through marrying his offspring.

> For a *sharifah*, marrying a *qabili* was not like marrying a *qarawi* (person of lowly birth). The *sadah* created an atmosphere such that it was *haram* (religiously unlawful) for a *qabili* to marry a *sharifah*. Those who were ignorant believed that it was *haram*. Marriages of a *qabili* and a *sharifah* do not express disrespect for the Prophet. (Shaykh Hamid)

The shaykh's vision of the world was one in which the *sadah* and the *qaba'il* were equals whereas farmers, craftsmen, and traders were inferior by virtue of their heredity and occupations.[8] "If we were all princes, who would take the donkey to the drinking trough?" Until recently, the shaykh had had the right to prevent a *qarawi* from marrying a *qabili* woman. To him, marriage with a *qarawi* was abhorrent, but restrictions on marriage between the *qaba'il* and the *sadah* seemed unreasonable to him because both are of "pure" descent (*asli*). When the nephew of one of the lesser shaykhs of his tribe married the granddaughter of a barber, Shaykh Hamid declined to attend the wedding. Soon after the wedding rumors spread that the groom's family was from the Hadramawt and not "proper *qaba'il*."

The discrimination against the *mazayinah* decreases slowly. Mahmud, a *qabili* from the Jawf who had settled in San'a in the late 1970s and worked for a development agency, held the view that "oneself and one's children would pay the price if one would marry a *muzayyinah* because one would become isolated." He quoted a saying according to which "a child comes after his maternal uncle or his first ancestor" in order to support his claim. He was adamant in his opposition to matrimony with the *mazayinah*, but he had little sympathy for *sadah* who refused their daughters to the *qaba'il*. "The *qaba'il* think of themselves as gods. They could not accept being treated like that."

Men like Mahmud have redefined hypergamy in their own terms. They do not acquiesce to the *sadah*'s hesitation to assent to women's out-marriage but reckon those they stigmatize as *qalil asl* to be inappropriate spouses for their daughters and themselves. The following example illustrates again the concern of those claiming pure ancestry about their child's mother's identity.

> My brother 'Abd al-Rahman was engaged to a girl whose family we did not know well. The engagement party took place in her house. We had all gone there and were chewing qat with her relatives. One of the guests who entered was Ahmad al-Hallaq [a *muzayyin* who has achieved high political status]. 'Abd al-Rahman asked one of his fiancée's relatives what he was doing there. On learning that the guest was the girl's uncle (*khal*), he threw his qat on the floor and left. He was angered that he had not been told and was determined to break the engagement. My family was too embarrassed to inform the girl's family of his decision. Finally my sister picked up the phone and told them that 'Abd al-Rahman was not yet ready to get married. (A shaykh's son)

About the time the shaykh's son broke his engagement, there was talk of a politician of *muzayyin* descent who at the house of Shaykh Amin had expressed his grievance about people's refusal to marry their daughters to his sons. He had argued, it was said, that since the *sadah* were marrying Christians (*nasara*), there was no good reason for the *qudah* and *qaba'il* to reject his sons' proposals to their daughters.

Whither "Our" Tradition?

Among the *sadah* the question of women's out-marriage was debated with reference to religious orthodoxy, but political concerns were also expressed. In doing so, they were not exposed to accusations of wrongdoing as long as others were prepared to pursue a dialogue in such terms. There does not seem to be a correlation between opinions on this subject and occupation or age. Gender, however, does shape their views because women are less familiar with religious texts. Senior kin tend to be remarkably tolerant toward the opinions of the young.

The *sadah* were acutely aware of the political dimension of their marriage practices and were deeply divided over the issue. Those who opposed out-marriage referred to Imams such as al-Mu'ayyad billah Muhammad b. al-Qasim or to hadith according to which one should "choose carefully whom you marry because the roots of your spouse will slip in"; "take into consideration alliance and descent because some of the roots (of your spouse) will thrust themselves forward (*tawakhkhu al-hasab wa-'l-nasab fa-inna ba'ad al-'irq dassas*)." One woman told me an anecdote relating to the early Shi'is.

> Imam al-Husayn's daughter Sukaynah, a poet, turned down a proposal from a man who did not belong to the *ahl al-bayt*. She was reprimanded by her friends who told her that she needed some excitement in her life. She replied "if you want excitement, you will get it." A letter which informed the suitor that she agreed to marry him was sent to him. When he arrived with his entourage, they were resisted by the Banu Hashim, and a hundred and twenty men were killed. Sukaynah then told her friends "you wanted excitement, there you have it."

The *sadah* who give preference to women's endogamous marriages nonetheless conceded that it was unreasonable to refuse all marriage proposals until a *sayyid* of equal rank was found. "Since Allah implanted desire in both men and women," they reasoned, "it would be unjustified to prevent anybody from marrying." Some men admitted that they had agreed to their daughters' marriage to men who were not *sadah* because no *sayyid* had asked for them, and they might have become too old. Even today there are *shara'if* in their thirties who have not married because their families refused to accept a non-'Alid spouse. According to Waterbury (1970: 103), in the Moroccan town of Fez there were until recently "sorts of boarding houses . . . where old-maid daughters of the shurafa were allowed to bide their time, often futilely, until one of their own came along to seek their hand."

The growing number of *sadah* who approve of women's out-marriage can refer to as many religious sources as their adversaries to authenticate their decision. Referring to women's endogamous marriages, they hold that "our God did not order us to do so," and "a man is his work (*al-rajul 'amaluh*)." Thus, a suitor should be judged according to his morals. Imam Zayd b. 'Ali, who did not forbid exogamy, and the marriages of the early Imams were also referred to. Imam al-Hadi was said to have married his daughters to his Tabaristani supporters who emigrated from Iran to Yemen. (However, the question of whether or not they were *sadah* was not raised.) As on other occasions, historical memory did not fail the actors. An anecdote referring to the time of the caliphs was quoted to me to illustrate Imam 'Ali's judgment on the issue of marriage.

'Umar, the second caliph, repeatedly asked 'Ali for the hand of his daughter Umm Kulthum. Some of 'Umar's followers told him there were other beautiful women of good families he could espouse, and he should not wait for Umm Kulthum. In reply, 'Umar said that the Prophet had declared 'Ali's offspring to be the best and purest, and he considered it worthwhile being patient. Later he took 'Ali's daughter as a wife.

The poet who told me the story about 'Umar's marriage to 'Ali's daughter vehemently opposed women's endogamy as violating Islamic principles. He insisted that piety (din) was the sole criterion of the $kafa'ah$, judging all other opinions as $taghut$ (idolatry).[9] This man's uncompromising view was uncommon. There is usually tacit agreement among the $sadah$ that all verdicts of the Zaydi $mujtahids$ have their own truth value and that their validity must not be denied. His attitude is likely to have been influenced by the trauma of the revolution during which he lost relatives and friends, and the resentment the $sadah's$ enemies expressed toward them thereafter. He encouraged his young relatives and students to marry foreign women in order to continue the "tradition of the ahl $al-bayt$." "If someone like al-Maswari [man of $muzayyin$ descent who acted as governor of San'a until April 2001] came to ask for my daughter, I would give her to him as long as he was educated and would pray and fast. There is no hierarchical order ($tabaqat$) in Islam" he would say. In spite of his convictions, he got upset when acquiring 'Alid brides seemed to non-'Alids like a triumph over the defeated rulers. A man of $qadi$ background, a government employee whose father had held a prominent post in Imam Yahya's government, visited an old friend who had maintained contact with the House of the Imam. The man had asked him "do you think they [the Hamid al-Din] have some girls for my sons?" When the poet heard about this, he remained silent for several minutes; the sarcastic undertone of the question had not escaped him.

Several $sadah$ still felt bitter about marriages such as that of the Shaykh Amin and his 'Alid wife which had been contracted soon after the revolution. That union healed some old wounds but also opened new ones. The relatives of women who entered marriages with influential non-'Alids soon after the revolution felt that they were intended to humiliate them. During the 1960s, the $sadah$ had often been forced to give their consent to their daughters' marriage with the new rulers, hoping to effect their relatives' release from prison and to prevent their executions. Indeed, these unions could hardly be conceived as "alliances" like those previously formed between the *sadah* and non-'Alid women. The women's kin had little left but the memory of their past glory. The official propaganda labeled them "oppressors," their property was confiscated, and they were insulted in the streets. However, the defamatory rhetoric might have camouflaged the esteem in which the non-'Alid wife-takers still held the Prophet's descendants. The marriages which were arranged with those disgraced families were seen as perfectly respectable by the wife-takers, thus revealing a deeply felt ambivalence toward the *sadah*.

The *sadah* are cognizant of the social damage that might result from a *sayyid*'s refusal of a suitor's proposal who in the past may have been told that he lacked the appropriate descent credentials; the latter may claim that it was motivated by this consideration. Ahmad al-Husaynat, a professional in his forties, was born into a prominent family of 'ulama and Imams (see earlier and also chapter 11). His father was a senior government employee. He had vital memories of the insults of other children he had suffered in the early revolutionary years. One of the guiding principles of his adult life has been the avoidance of these experiences. He believes that the *sadah* should neither stress their distinct status nor have their own organizations, either political or humanitarian, for this would merely nourish the prejudice of those who accuse them of aspiring to restore Imamate rule. Ahmad agreed to the marriage proposal of a non-'Alid friend to one of his sisters even though he was neither wealthy nor from a well-known family (*usrah ma'rufah*). He was convinced that the man was good-natured and would treat his sister well.

Ahmad's sister had no objections to the marriage. Her brother had to reassure the suitor that he had actually agreed; a few days before the wedding, members of his own house and his future in-laws repeatedly asked him whether he had the approval of his father who was living abroad. His father had declared his son to be his deputy, and did not interfere with his decisions. His maternal kin had come from their village to protest but were conciliated by the fact that the suitor's grandfather had been a shaykh. Ahmad wished to prove that the *sadah* no longer had pretensions to superiority based on birthright. He held the opinion that guilt was the driving force behind the *sadah*'s decision to accept proposals like this, and conceded that the *sadah* could not afford to alienate those who had defeated them. A few years after the wedding he told me that his sister dominated her husband. "He never tells her to do this or that because she is a *sharifah*, and he does not take any decision without consulting her. I am worried about this; she might destroy his masculinity."

Hanna al-Farisi married a government employee who holds a Ph.D. in engineering from Peking University. His sister is a friend of Hanna whom he met when he came to collect his sister from her house. Hanna's family knew little about the man's family but accepted her wish to marry him. According to Hanna's paternal uncle, the family's main concern was the suitor's education. "We did not even enquire whether he was a *sayyid*. Nobody was interested in that. In the past this was important because the girl's family was concerned about her future. A *sayyid* had better chances to get a job in the government."

Engagements between a *sharifah* and a non-'Alid are hardly ever welcomed unanimously by the girls' relatives. Usually her kin receive letters accusing them of violating "our tradition," of "stepping out of Islam," and of affronting all *sadah*. These letters are usually ignored and ill-feeling persists only temporarily. The liberty with which the *sadah* often take decisions in this matter is quite different from the picture drawn in the literature on kinship in Middle Eastern cultures which stresses the constraints placed on men to comply with the wishes of their kin. Men follow their convictions even when such behavior threatens their rapport with their brothers.

Ignoring his brother's concerns, a man sided with his niece, a girl called Amat al-Latif. The girl had fallen in love with a neighbor of an unassuming family who had just begun his career in the army. When her father rejected his proposal, she took refuge with her uncle who informed his brother that she was staying in his house. (The girl's father was aware of the reason for her running away.) She remained there for two months, and her father made no effort to take her back. Meanwhile her uncle acted as the girl's guardian (*wali*) and married her to the soldier. In the following years there was no contact between Amat al-Latif and her father until people in the neighborhood declared some of his land their own. Help could not be expected from the government. Amat al-Latif's father was advised by his friends to turn to his son-in-law who by then had gained influence in the army. There followed a reunion between father and daughter. Her husband quickly solved the problem by threatening his father-in-law's neighbors that force would be used against them. After the incident, Amat al-Latif's father was teased by his relatives. The wife of the girl's uncle, who had arranged her marriage, and whose daughter had also married a non-'Alid, told her brother-in-law that he readily sacrificed his principles for material advantage. Amat al-Latif's husband is now his favorite son-in-law.

A substantial number of marriages contracted by the *sadah* after the revolution have involved Sunnis of Ta'izz and the Hujariyyah some of whom, until the mid-1980s, constituted the commercial elite. Of the four out-marriages of women of one of the great 'Alid houses, all spouses were Sunnis. Giving women in marriage to Sunnis rather than their Zaydi rivals might have been reckoned to be less degrading because in the south anti-'Alid sentiments have been far less virulent than in the north. Since unification, marriages between north Yemeni and Hadrami 'Alid houses have also occurred, as well as with foreigners. The *sadah* were among the few who had already married foreign women, of predominantly Turkish and Egyptian origin, before the revolution. Since then they have married German, Hungarian, Yugoslav, Russian, and American women. For the first time, names such as Thomas have appeared on the genealogical charts (see Bayt Zabarah in appendix III.1). In conjunction with a European education, such unions allow men to represent themselves as cosmopolitan and enlightened. Some of these foreign brides built their reputation as good wives by spending their afternoons educating their children rather than chewing qat.[10] In the early 1970s, one of the sons of Imam Ahmad's deputy in San'a returned with his German wife from Italy where he had studied medicine. His father did not object to the marriage because the Prophet had married a Christian (see chapter 8). One man, a writer, explained that marrying a foreigner was no problem because the mother of Imam Zayd b. 'Ali had come from Sind, currently a province of south-east Pakistan.

> I married a foreigner to avoid conflicts with my relatives over whom to marry, about how I should conduct myself as a *sayyid*, and how I and my wife should live together. (A doctor married to a Danish woman)

In the aftermath of the revolution, for these young men to marry women who have no place in the Yemeni status hierarchy might have provided a sense of aloofness from the conflicting demands of family and wider society. Deracination, of course, also provides a force on its own. Today some of the descendants of Imam Yahya, who thought that Christians might have a corrupting influence on his subjects, are brought up by French, Spanish, Mexican, Japanese, and English women.

Women's Voices

Unlike men, very few women of the old elite have married foreigners. Regarding marriages with non-'Alids, however, women are as divided as men. Women who oppose out-marriage are more likely to invoke religious precepts than those who approve. They explained their reluctance to marry a non-'Alid in terms of the pride they took in their ancestry and their desire not to deprive their children of membership of the Prophet's House. As proof of the inappropriateness of marriage to non-'Alids they referred to invocations such as "Oh God, bless (*salli 'ala*) Muhammad and his Family." These words are spoken in prayers and whenever someone wishes to invoke the Prophet's name, for example, during a moment of silence at a gathering. According to the women, those who ask God to bless the Prophet and his progeny recognize the *sadah*'s distinct status. Therefore, it was improper for anyone who did not belong to the Prophet's House to marry one of his "daughters." Furthermore, since in the Qur'an (33:32) the Prophet's wives were reproachfully reminded of their special status ("You are not like any of the [other] women, provided that you remain conscious of God"), this would certainly apply to his female offspring (*dhurriyyah*). A young woman from the *ashraf* of al-Jawf, who had married a San'ani, quoted a hadith saying that "a *sharifah*'s sweat cannot go to a shaykh's children." "How can she labor for his children? It used to be *'ayb* for a girl not to marry from the same house. She was supposed to strengthen the house from inside. Nowadays girls become educated and get new ideas."

Several women who had entered endogamous marriages simply commented upon them by saying "*lazim*" or "*wajib*," implying duty-bound behavior. Most women are keen to avoid alienating their families when conflicts over the choice of spouse arise. Very few instigate law suits if their families oppose the partner of their choice. Once cases are brought to court, a woman is usually given the right to marry the man she wants.[11] The elite tends to be more tolerant toward women's rejection of proposals or their decisions to remain unmarried. Most are able to provide for them and wish to be seen as adherents of religious precepts. Men who force their daughters or sisters to marry men against their will risk tarnishing their reputation as pious Muslims.

I came to know a man who was the only breadwinner in his large household which included his wife, his own unmarried children (seven) and those of his deceased brother (two), his mother, his mother's co-wife, his unmarried aunt, and his divorced sister. It was evident that he would have been delighted had his aunt and sister accepted one of their many suitors, but he never tried to pressure them. He insisted that Islam required women to give their consent and therefore accepted their decision. Among those with adequate means this attitude was by no means unusual.

It is not uncommon for women whose families refuse to accept the suitor of their choice to seek the support of 'ulama. Men who have been rejected by an 'Alid family because they are poor or not *sadah* also ask learned men to intervene. If a reputable scholar sanctions a proposed marriage, the women's parents may be persuaded to accept it. His support reduces their vulnerability to ill fame and is likely to prevent their relatives from accusing them of endorsing a misalliance. If they were to make such an accusation, it is they who lose countenance.

> Women approach me for help because their families disapprove of their suitors who are Qahtaniyyun or Shafi'is. It is difficult for them to stand against their parents on their own. I usually ask my wife or sisters to find out from the women whether they wish to marry those men, and then have a word with their families. These people do not put forward an argument about what is *haram*. They simply feel superior to others. (A scholar in his sixties)

On one occasion, he acted as a woman's guardian and drew up the marriage contract on her behalf. Suha's family had spent a few years in exile in Lebanon. She wanted to marry a Christian Arab who converted to Islam on her account. In her father's view, he was "not a *sayyid*, not a Yemeni, and not a Muslim." He conceived of the suitor's conversion as opportunism rather than conviction. After Suha had approached the scholar, her father agreed reluctantly to the marriage. He accepted the scholar's argument that it did not violate Islamic principles and appointed him as his deputy. The wedding was held at the scholar's house but the bride's father did not attend.

Women who approve of unions with non-'Alid men evaluate types of marriage in terms of the personal qualities of suitors. Amat al-Latif was an elderly *sharifah* who had been brought up to believe that the *sadah* ought to marry their own kind. All the women of her house had respected this ethical principle. Later in life she had given up her reservations toward women's out-marriage. "Nowadays even *sadah* are found drinking and neglecting their prayers, so why should we prefer them to others?" she would ask. With the exception of her youngest daughter Bushrah, all her sons and daughters were married to *sadah*. She judged the suitability of her daughter's spouse in terms of his allegiance to his religion: he should pray and fast, and read the Qur'an in his free time—then he would be worthy of her. Bushrah, a girl of eighteen who was expected to get married in the next few years, held the view that "marrying out" was *'ayb*. Whilst listening to our conversation, she asked: "All my sisters got married to *sadah* and you want me to marry a Qahtani?" For a moment, this idea seemed unacceptable to her. Then she said, "if I were to marry a Qahtani, he would have to look up to me."[12] A little later, she jokingly rebuked her small niece who was playing with a knife: "If you don't put the knife away, we'll marry you to a Qahtani."

Unlike Bushrah, her mother, a very pious woman, held the opinion that a man asking for her daughter should be judged by his propriety. She knew that her husband and sons would be reluctant to give their consent to Bushrah's marriage to someone of non-'Alid descent. (In the 1990s, she got married to a relative.) Amat al-Latif was not educated well enough to justify her arguments by reference to religious texts, but her view was congruent with that of men who claim that piety constitutes the *kafa'ah*. Based on her comparison between the wives of her nephews of whom one had married from outside the Prophet's House, Amat al-Latif was even more convinced that there were no good reasons for either men or women to give preference to endogamy. She described the 'Alid wives as haughty and argumentative, and the other as a reliable and caring wife (*maqtubah*).[13] This one, she said, adhered best to her domestic duties, served her husband well, and was modest and kind.

Farida al-Nasir, a university graduate in her late twenties, expressed her disapproval of status barriers between *sadah* and non-*sadah* in a provocative fashion. She was critical of the Imamate because its leaders had not even provided sufficient health care in order to prevent her father's premature death. Farida had been brought up by her paternal uncle Yahya who, although his brother was executed in the revolution, was opposed to Imamate rule. A relative of Amat al-Malik who had married Shaykh Amin in the early days of the revolution, he had pleaded with his close kin to accept the shaykh's proposal. Farida seemed somewhat tired of the conflicts over the issue of marriage which she had witnessed in many 'Alid houses. Of her future spouse she said "may he, with the help of God, be a butcher" (*Insha 'llah sayakun jazzar*). Her paternal uncle showed no reaction when she expressed such views. Her maternal uncles, conversely, were opposed to her taking up a profession, and disapproved of women's out-marriage.[14] Finally Farida got married to a colleague from the Hujariyyah who was neither a *sayyid* nor a Zaydi. Her marriage to a man she had met at work, who came from an area of Yemen which had been under the rule of her ancestors who were Imams, was conceived as an indiscretion by those who conceived of *nasab* as the chief principle guaranteeing equivalence in marriage.[15]

For Farida's paternal uncle, Amat al-Malik's marriage to the shaykh provided the opportunity to portray the *sadah* as opponents of the old hierarchical order. He argued that marriages such as these imposed constraints on the *sadah* because they had to take into consideration their affines' concerns whenever they contracted a marriage. Since Amat al-Malik had married the shaykh, none of her kin were able to marry a person of disreputable descent. If this should happen, the shaykh might return his wife to her family for fear of losing the respect of his tribe (compare Meissner 1987: 183). Thus the *sharifah*'s kin were obliged to observe social rules they might otherwise be tempted to ignore.

In San'a, the first marriage of a *sayyid* and a girl of *muzayyin* descent who now works for the government occurred in the early 1990s.[16] Depravities had cost the man's family much of its former prestige. His bride had received a good education and was employed at a government ministry, but her name did not betray her extraction. A *qabili* from the Jawf who had migrated to San'a commented on her family's confident and assertive demeanor: "Sometimes I ask myself whether their name is not spelled differently." Her background was not discussed disapprovingly by the groom's kin. One of his close relatives, Iman, came to like her and spoke about her favorably, though I did ask myself what she really felt about the marriage. "She really is a nice girl, she always asks about me and invites me to their house." Her sisters had already contracted marriages which were resented by several *sadah* at the time. I was working in Falih when word spread among the women that Ahmad al-Farisi's daughter Amat al-Karim had married "a *qabili*"—a term they used to refer to anyone who is not a *sayyid* (see chapter 3). They clearly thought of the woman as unfortunate and were astonished that her male kin had given their consent. In their eyes, women's marriage within the Prophet's House is a rule sanctioned by Islam, and transgressions are said to be *haram* (unlawful). When Iman learnt about their disapproving attitude, she said "we want to finish with these things."

* * *

Chapters 6 and 7 were concerned with the ongoing controversy over eligibility in marriage among both religious scholars and men and women who "make marriages." There are two conclusions I wish to draw. First, I argued that during the past three centuries, at least, sexual sanctions and prohibitions related to rank and gender were central to Imamate politics. The controlled flow of substances in sanctioned sexual unions was not just an expression of these politics but served to define and produce boundaries within the ruling elite (that is, *sadah* and *qudah*) and between rulers and ruled. In other words, positions of power were demarcated as much by Zaydi-Hadawi doctrine as by gender-specific sexual control. The argument about descent as a principal criterion of the *kafa'ah* was less about the survival of the *sadah* as a social category than about their political viability and cultural reproduction.

Second, I have used the example of marriage to show that memory continuously negotiates between religious sources and present-day social and political concerns. Elements of the *taqlid ahl al-bayt* are imported into disparate social realities and reinterpreted in novel contexts in ways that are unpredictable. This process of transformation is perhaps best understood as modulation for it is rooted in continuities. Rationalizing their activities with reference to the verdicts permits the actors to argue in an idiom of static time whilst commenting upon changing events; it is a way of *not* talking about new relations of authority and domination.

However, as this chapter has demonstrated, specific historical circumstances are also referred to in order to account for changing marriage practices. Pragmatic reasoning is often entwined with the citation of religious verdicts, thus emphasizing their relevance to present-day realities. The salience of specific traditions or elements of these traditions is contingent on certain historical contexts and on the strategic interests of actors and groups who make use of them. But it is equally important to bear in mind that it cannot be explained exclusively in these terms. For example, among *sadah* who are not prepared to marry their daughters out, concerns of purity and dignity override considerations of favorable marriages to the new elite. Nor does the prospect of social damage caused by the refusal of a proposal by a non-'Alid provide sufficient motivation to alter cherished positions. Like many of their Hadrami peers (Freitag 1997: 15), these *sadah* are happy to say farewell to gestures of deference such as the kissing of their knee or hand (*taqbil*) and being addressed by the title *sidi*, but they cannot accept that their grandchildren should not be members of the Prophet's House.

A further point to be made is that because status is always elusive, the principles that constitute equivalence in marriage are not as clearly defined as they might appear. As I have argued, rigid restrictions on 'Alid women's reproductive choices served to articulate the politics of Imamate rule, but these were lifted in exceptional cases. Prior to the revolution, the marriage of a non-'Alid to a virgin *sharifah* of a prominent family could be judged as hypogamous, but once her value in the marriage market was diminished, criteria other than descent were given greater consideration.

Nowadays those *sadah* who downplay the significance of descent in marriage can recognize those who used to fall into the category of inferiors as equivalent to them. Yet many *sadah* and non-*sadah*, however much they may disapprove of the derivation of privilege from the facts of birth, are prepared to accept that within a religious scheme of values, the *sadah* rank highest (see chapter 10). Thus, even when descent is made light of, the conflict between the woman's superior religious status and the couple's equal social status may not be fully resolved. If, as Mundy (1995: 125) suggests, marriage belongs to a history "structured by understandings about property," then this history must include understandings about people's intrinsic properties. It is only then that we can apprehend why even today some 'Alid women are forced to grow old without partners, and why well-to-do men of *muzayyin* background face great difficulty in marrying those who think of themselves as *asli* (of pure descent). Writing about upward mobility in Hureidah, Bujra (1971: 112) says that newly acquired status becomes legitimized by marriages with other groups of similar social location. However, he acknowledges that "those . . . who have improved their economic status have not been able to translate this into a higher descent status" (1971: 114).[17]

However much people are concerned to depreciate the descent criterion in marriage, the ongoing desire of men of the new elite to take 'Alid brides tells another story. Like in the past, marriage creates and reflects new realities. Some four decades after the revolution, the pattern of pre-revolutionary hypergamous marriages appears to have been reversed. These marriages ensured the loyalty and benefit of the bride-givers, and indicated the power afforded by rulers and officials. Nowadays, some 'Alid women, whose social standing has declined after the revolution, are proposed to by non-'Alid men of highest status. It is these brides who now secure prestigious jobs for their relatives in the administration. In the post September 11 world, where tensions between the ruling elite and elements of the Islah party are rising, marriages with the women of the deposed holders of power highlight a new dimension of the relationship between them and the new ones. These marriages could give further impetus to the *sadah*'s sense of national belonging, but the increase in anti-'Alid rhetoric in the aftermath of the Husayn al-Huthy crisis may put it in jeopardy.

Chapter 8

" 'Ulama of a Different Kind"

Focusing on marriage, chapter 7 showed that explanations of practices in terms of religious verdicts are often intertwined with references to present-day realities. This pattern also emerges from the analysis of the *sadah*'s professional engagements over time. Beginning with an analysis of the professional ethos among the scholarly elite in the first part of the twentieth century, this chapter concentrates on changes in the attitudes toward knowledge. Since the mid-twentieth century, the *buyut al-'ilm* have adopted the "instrumental" sciences (Taminian 1999: 204) which in the traditional value scheme of *'ilm* categories constitute a kind of "subjugated knowledge" (Foucault 1994: 203) which is inferior to divine knowledge. I look at the conflicts over this kind of knowledge within the elite and their reasons for adopting it in the past decades. Dealing with the Iraqi *sadah*, Batatu (1978: 210) says that he could not say how they fared economically after the 1958 revolution. It is due partly to their investment in the "new *'ilm*" that their Yemeni counterparts have been able to maintain and even to improve their standard of living.

The Ethos of Teaching and Trade

In contrast to the *sadah* of South Yemen, among those who were subjects of the Imams, emigration was almost unheard of before the civil war. The reason might be sought in their greater involvement with the state and lack of proclivity for economic activity. Prior to the revolution, a moral distinction was drawn between the Zaydi establishment who taught, judged, and ruled, and those involved in trade and commerce. As Weber observed, the most privileged strata often "consider almost any kind of overt participation in economic acquisition as absolutely stigmatizing" (Gerth and Wright Mills 1977: 193). Trade was considered to be vulgar by the learned nobility. Scholarly activities were incompatible with the self-centered and profit-oriented spirit which governed the transactions of material goods. Trade was thought to inevitably involve some kind of illicit practice such as bribery. Unlike *'ilm*, trade could not establish a person's links to the "hereafter," and it was believed to inhibit the acquisition of *'ilm*. Self-respecting scholars were expected to carry out their divinely imposed duty of transmitting knowledge without compensation in this world, for they were to be rewarded in paradise. Learned men who did not accept payment for their services ranked higher than others. Although the pursuit of material gain was not condemned as inherently immoral, trade was seen as violating the spirit of the gift that knowledge ideally constitutes. For a Muslim scholar, teaching was and is a moral obligation; failure to transmit knowledge is *haram* (unlawful). Ideally those of elevated birth would commit themselves to the transmission of religious knowledge (*'ilm*), the pursuit of which was perceived as the most honorable activity (*ashraf 'amal*).

Indeed, until recently the roles of the 'ulama, many of whom were *sadah*, and of the merchants have been conceived as mutually exclusive. In Muslim cultures a multiplicity of idealtypical constructs of the scholar exist, ranging from the "trader-scholar" to the more retracted who is entirely devoted to the pursuit of knowledge. Yemeni 'ulama fell into the latter category but, significantly, erudition entitled and obliged them to exercise political authority. Elsewhere in the Middle East, the 'ulama have engaged in economic ventures, and they and the merchants tend to share cultural and political orientations.[1] Geertz (1968: 68) notes that "mosque and market have been a natural pair over much of the Islamic world." The close ties between Shi'i religious and commercial classes as well as their own trade activities in Iran have often been noted.[2] Regarding other societies with Shi'i majorities, notably Iraq, local views differ from the findings of historians. According to Nakash (1994: 233–4), the 'ulama and the Shi'i merchants have neither held shared values and interests, nor have the latter constituted the financial backbone of the religious establishment. Litvak (2000: 54), examining the financial sources of the 'ulama in the Iraqi Shi'i shrine cities in the nineteenth century, notes that only a few were engaged in trade. However, the religious circles of Najaf do not unanimously subscribe to this view. They argue that it is by no means morally inappropriate for the 'ulama—with the exception of the grand *mujtahids* (*maraji'-i taqlid*, sources of emulation)—to engage in commercial enterprises. In fact, as some 'ulama assert, these activities verify their morally sound attitudes for they demonstrate that they refrain from spending the *khums* on themselves. As one Najafi scholar commented, "the Prophet was a merchant, 'Ali was a merchant, and the Imams owned land and were involved in buying and selling. A substantial amount of the *khums* which was received by Grand Ayatullah Abu'l Qasim al-Khu'i (d.1992) came from Iraqi merchants."[3]

In the late Yemeni Imamate the relationship between the 'ulama and the merchants was rather like that described by Nakash for Iraq. Unlike the Moroccan *shurafa'*, who became wealthy through marriage with merchant families or economic enterprises (Waterbury 1970: 97), we can assume with some certainty that at least in the twentieth century, few men who belonged to the religious establishment either through descent or profession took up a trade.[4] For example, M. Zabarah (1979: 447) refers to a relative, Jamal al-Din b. 'Ali b. 'Ali Zabarah (b. 1888/89), who was engaged in trade and later worked as a treasurer in al-Sudah.[5] Cases like these were probably exceptional.

Yemen presents a peculiar case because the negative attitude the Zaydi elite held toward trade was above all buttressed by the Zaydi injunction according to which rulers and governors must not engage in economic transactions in the areas under their control.[6] They are to be prevented from using political institutions for economic advantage and from becoming corrupt. In the eyes of other nobles who did not hold senior posts in the government, the rule which required the governing elite to abstain from the market place was interpreted as noblesse oblige which even they could not afford to disregard. Hence one of the characteristics of distinction and nobility was abstention from trade and commerce.[7] There were very few big merchants in San'a, among them al-Hajj Lutf Aslan, 'Abdullah al-Sunaydar, and Isma'il Ghamdan. The latter, who was a *sayyid*, sought to demonstrate his concern for knowledge by establishing a formidable library and by cultivating friendships with the 'ulama. Some merchants offered support to the 'ulama for which they expected *ajr* (remuneration; divine merit) in return.[8]

Trade and commerce were held in low esteem by both the religious elite and the majority of the *qaba'il*. It would appear that Zaydi doctrine and other moral codes upheld by the *qaba'il* reinforced each other. Thus, the conspicuous distance the *buyut al-'ilm* maintained toward the market place was not merely an elitist preoccupation. Among the *qaba'il* of the central highlands and the eastern planes, traders were looked at as inferiors because they had to rely on their protection (Dresch 1989: 120; compare Gerholm 1977: 79, 132, 141; Adra 1982: 64). However, according to Weir (personal communication), in Razih in north-west Yemen trade did not carry a stigma among either elite *sadah* or the *qaba'il*. High-ranking *sadah* acted as merchants and worked in the local suq. However, as elsewhere, the 'ulama did not trade. Messick (1978: 275) writes that before 1962, the southern

Sunni town of Ibb had commercial ties with its surrounding hinterland and beyond, but none with the next major town to the north, Yarim. "It is said in Ibb that Zaidis of the northern tribal areas regarded merchants as low status (*naqis*); no trader could pass the night in a Zaidi *qabili* village." In the Sunni areas south of San'a trade was not as stigmatized as among the Zaydis. During the Imamate, the Sunnis were most active in trade but were politically underrepresented. Their religious affiliation disqualified them from high office, and mercantile activity served as an effective mechanism of exclusion from political power.

Biographical works on the leading Yemeni personalities of the nineteenth century reflect the hierarchical valuation of occupations among the elite. Merchant families are never spoken of in these books. Among the families better known for their economic rather than their scholarly activities only those who were devoted to knowledge are referred to. For example, among Bayt al-Sunaydar, a well-known (non-*sayyid*) merchant house, M. Zabarah (1979: 584) mentions only the Faqih Muhammad b. Muhammad. To some extent, then, social differentiation was based on a distinction between social status and economic power deriving from trade. Big merchants, among them Sunnis and Jews, were excluded from political power; *sadah* and *qudah* who held political offices were not always wealthy but some were rich landholders. By abstaining from trade and commerce, the majority of the *qaba'il* could identify with the Zaydi elite and distinguish themselves from the low status tribal and non-tribal traders and the Sunnis.

Those who represented religious learning through either descent or profession—including women—were supposed to avoid even the physical space of the market place. This picture contrasts with Gilsenan's (1982: 174, 177) description of the market as "public space *par excellence* . . . [which] might be seen in its turn as an extension of life in the adjacent mosque." The market place was considered an indecent place where people would shout, negotiate prices, and be prone to spontaneous pinching and grabbing. I came to appreciate this ideology when a man of an eminent family told me about his first visit to the suq of San'a in the 1950s. He had spent most of his youth in the more liberal Sunni town of Ta'izz, but entering the suq still caused him tremendous discomfort. Anxious not to be seen, he kept looking around in case there was anybody there who knew him. "I was so apprehensive that I was sweating as if I had had (illicit) sex." Likewise, in the highlands it was inappropriate for the tribal elite to spend time in the suq.

> When I accompanied my grandfather to the weekly market in Kuhlan, we would just quickly buy a few things and leave immediately. (A shaykh's son)

Because the Imamate was rationalized in terms of the Zaydi doctrine which insists on the irreproachable character of the holders of power, violations of its central precepts were taken seriously. For example, in 1946 the 'ulama demanded that Imam Yahya put an end to the economic transactions of some of his officials and the princes (Douglas 1987: 118). According to Mufti Ahmad Zabarah (personal communication), the princes' involvement in trade tarnished the reputation of the royal House and contributed to their downfall in 1962.[9]

Imam Ahmad attempted to dissuade young men of distinguished houses from pursuing careers as merchants. He took a personal interest in their aspirations, judging them according to their abilities and family histories. In the 1950s the eldest son of Ahmad al-Kuhlani, one of Imam Ahmad's governors and a renown scholar, decided to establish a business. Expressing his displeasure, the Imam named some of his forebears ("you are the son of . . .") who had been 'ulama, suggesting that his chosen profession was unsuitable for him. The young man's motives were purely pragmatic. He explained to the Imam that his father only ate 'asid (a dish made of wheat or sorghum flour and water which can be enriched with beef stock, onions, and clarified butter), and that he himself desired a better life.[10]

Even a man who was well off had no more than two to three zinnin (sing. zannah, a full or calf length garment). People borrowed things from each other when they had a wedding. Chicken was eaten only by sick people and women who had given birth. A chicken cost 1/2 RF (riyal fransi), a sheep 1 RF.[11] My father had chicken at lunchtime but he was the only one in the family. There were few wealthy Hashimite houses. Some of the poor were helped by my father. (A sayyid from a landowning family)

Some sadah were driven by necessity to work in the suq. In some cases, market trading provided their primary source of income (compare Weir 1986: 227). In San'a, these sadah were frowned upon unless they were bookbinders and thus dealing with religious texts. In spite of the differential degrees of honor inherent in the various types of labor ranging from engagement in 'ilm to blacksmithing and streetsweeping, the reputation of men of learning who were involved in craft activities to make a living was not tarnished. Some of those who neither owned land nor had access to funds from endowments (waqfs)—which many rejected because they thought it ought to be used for charitable purposes—were involved in craft production. The production of caps or dagger belts (which were often embroidered with Qur'anic verses) by those who had no other sources of income enabled them to live up to the ideal of unremunerated teaching.

Even men of the highest rank were involved in such activities at some stage of their careers.[12] Within the field of craft production, book binding carried more prestige than, for example, carpentry. The shops of the bookbinders, some of whom were well educated and used to teach a few hours a day, were located near the Great Mosque. This location and the nature and purpose of their craft served to neutralize the polluting effects of labor in the suq. They were symbolically set apart from the suq while physically interacting within it. Thus, paradoxically, some noblemen produced goods in a place that was looked down upon by the majority of their peers whilst maintaining their good reputation.

The present day *sadah* who live in the countryside do not consider agricultural labor beneath their dignity, but working as tenant farmers in the service of a *qabili* is less agreeable. When during the early period of my fieldwork I asked Sayyid Isma'il al-Mukhtafi, a scholar residing near the town of Sa'dah, whether it was *'ayb* for a *sayyid* to be a farmer, he replied: "What a ridiculous question! 'Ali b. Abi Talib dug twenty-four wells during his lifetime!"

Embracing the "new *'ilm*"

In the early twentieth century, young generation urban *buyut al-'ilm* were influenced by developments in countries such as Egypt and Lebanon where educational reforms had been already implemented. North Yemen was free from colonial rule but nonetheless influenced by the regional and transregional processes involved in changing the meanings and institutional arrangements associated with learning and knowledge in all spheres of life. As pointed out in chapter 2, the educated elite was not as insular and inward looking as is often made out. Some never even visited places a 100 miles away from home, yet others completed their education at religious colleges in Cairo and Mecca (Snouck Hurgronje 1931: 192). During the reign of Imam Yahya and Ahmad, demands were made for the introduction of a parliament, a constitution, freedom of speech, and access to knowledge which was available in some other Arab and European countries.[13]

From the 1930s onwards there was thirst for knowledge beyond the classical education in Yemen, and a realization by some of the *buyut al-'ilm* that their future would not lie in religious education. A passage from a book written by al-Shami (1965: 135–6) five years after the 1948 uprising testifies to the discomfort felt about the status quo.

The [conservatives] see life awakening while it is still sleeping, and shining while it is in darkness. They have build a wall of superstitions, fanaticism, and ignorance between themselves and progress. [In their view], anything which is not mentioned in their books is unbelief (*kufr*), misguidance, and disobedience. Everything people enjoy from the fruits of civilization is forbidden (*haram*) because it has been created by the infidels (*kuffar*), and to seek wisdom from books other than those of the forefathers is misguided and sinful. And good deeds, reform, and following Islam serve only to freeze the status quo; [they do not want] to think of the future; they follow the path of our forefathers blindly . . . The [reformers] do not attach significance to the new things just for their sake; their yardstick is benefit and usefulness. Whatever is useful we like, whether it is old or new.

By sending students abroad, the Imams reluctantly recognized what they refused to admit: they were nostalgic about a classical Imamate whilst discerning that the colonial encounter had irrevocably changed the Middle East, and that a modern state apparatus required staff who were trained abroad. Imam Yahya's educational policies aimed at controlling knowledge; surely Imam Ahmad was aware that were he to allow greater numbers of students to study abroad, he would lose that control. He desired a traditional education for his wider family and resented their wishes to study in other Arab capitals, but was lenient toward their defiance. In 1953, one of the sons of his sister Amat al-Rahman escaped to Cairo via Jiddah while he was on pilgrimage. At the request of the Imam, he was returned to Yemen by the Egyptian government. Later on the Imam allowed him to continue his studies abroad, and he and nine princes joined the Yemeni students already in Egyptian schools. Revolutionary Egypt made a tremendous impression on them and some, among them the late Yahya b. al-Husayn, demonstrated against their uncle Imam Ahmad in front of the Yemeni embassy.

'Abbas b. 'Ali Zabarah, the son of Imam Ahmad's deputy in San'a, received an education in the Islamic sciences. In the 1950s, a sensitive period during which the Imam's rule was contested, he aspired to advance his education abroad.

Although I had the opportunity to be taught by the best teachers, I wanted to expand my horizon and continue my studies abroad. My father disagreed. When he visited Cairo, he was taken to a nightclub, and he did not want me to be exposed to this kind of thing. He told me that outside Yemen, people were *kuffar* (unbelievers) and that I would be corrupted. However, I was determined to go. I went to Ta'izz to see the Imam. There were other youth who had come to see him. He encouraged us to sit on his right side. On seeing Muhammad 'Abdullah al-Wazir [the son of the interregnum Imam who ruled in 1948] entering, he uttered "*La rahim Allah abuk*" (May God have no mercy on your father). On noticing the son of the governor of Hajjah, Muhammad b. ʿAbd al-Malik al-Mutawakkil, he exclaimed "*Ah, al-mutaharrir!*" (Advocate of liberty).[14] Turning to me, he asked "*Wa anta malak?*" (and what's the matter with you?) I told him that I was ill; my head had been aching severely for some time, and I wanted to seek treatment abroad. When Muhammad b. ʿAbd al-Rahman al-Shami [a scholar and poet] entered the room, the Imam asked him whether he could confirm that I was ill. He was slightly embarrassed but said yes. The Imam was pleased to hear that I wanted to go to America or Italy rather than Cairo where many of his opponents lived. In order to leave the country, I needed a written permission of the Imam. I noted on a piece of paper that I was ill and had to go abroad. When I asked him to sign it, he advised me that "the pen is already closed." I then offered him my own pen, and he signed. Muhammad al-Shami urged me to leave the country as soon as I could before a telegram of my father would reach the Imam, asking him to refuse me permission to leave. I first went to Lebanon, but the ensuing war forced me to leave and to go to Damascus instead. There my Arabic was praised. The teachers said that it was better than that of the other students, but my general knowledge was not sufficient to enter the university. I first had to finish high school. Because of my fine Arabic I was often asked to give official speeches. Meanwhile my father had written to the Imam bemoaning the loss of his son. He asked him to help him to bring me back to Yemen. Later he accepted my decision to go abroad and was pleased to hear that I had received a scholarship to Italy where I studied medicine. He did not insist that his sons would become ʿulama, but he wanted them to be intellectuals—for instance poets or historians. It was only after an increasing number of Yemenis pursued their education abroad that he agreed that they would study medicine or attend the military college.

By the time ʿAbbas returned to Yemen with his German wife in 1972, he joined the small group of professionals who had been educated abroad (figures 8.1, 8.2, 8.3, 8.4, and 8.5). His brother had been trained as a computer specialist in the United States. Mosque teaching had declined, the *Madrasah al-ʿilmiyyah* was closed down, and religious learning was marginalized. By reason of their families' long-standing dedication to education, the young *buyut al-ʿilm* adjusted easily to the new system and received scholarships from foreign governments to study abroad. It was not before the 1980s that economic activities gained respectability among the *buyut al-ʿilm* and the *qabaʾil*. In the first decade after the reconciliation between royalists and republicans, it was the Shafiʿi merchants who profited from the peace agreement, trade opportunities, and the growing demand for a wide range of consumer goods. Many who had left their villages in the Hujariyyah at the turn of the century seeking fortunes abroad and in Aden had supported the opposition against the Imams from

Figure 8.1 Sayyid 'Ali b. 'Ali Zabarah, Imam Ahmad's deputy in San'a, Cairo 1952

the 1930s onwards.[15] Once Imam al-Badr was ousted, the merchants hoped that in a newly created community of consumers the old social divisions and the ideology of otherworldly piety would be obsolete. They were the major driving force of consumer capitalism in the new republic. They developed the private sector and until the early 1980s, they had a quasi-monopoly on trade in the Yemen Arab Republic. Their experience, international trade networks, and lack of prejudice against trade and commerce put them at an advantage vis-à-vis the Zaydis (vom Bruck 1998a: 279–82).

Until the 1980s Zaydi merchants were less successful than the Sunnis and somewhat similar to the Javanese businessmen Geertz described in

Figure 8.2 'Abbas b. 'Ali Zabarah (left, behind his brother Mutahhar), his maternal uncle 'Ali al-Kibsi and his brother 'Abd al-Malik

Figure 8.3 'Abbas (left) with his friends Ahmad b. 'Abdullah al-Zubayri and Yahya b. 'Ali Zabarah

Figure 8.4 'Abbas (right) with Muhammad b. 'Abd al-Malik al-Mutawakkil in Beirut, 1959

Figure 8.5 'Abbas with his family, late 1980s

Peddlers and princes. What Javanese businessmen lack, Geertz (1963: 28) says, "is the power to mobilize their capital and channel their drive in such a way as to exploit the existing market possibilities. They lack the capacity to form efficient economic institutions; they are entrepreneurs without enterprises." During the first two decades following the revolution, newcomers to the world of commerce from well-known houses, among them tribal leaders, did not conduct their enterprises under their own names.

> When a man from our tribe [the Abu Ras of eastern Yemen] opened a shop, it was seen as an insult to the whole tribe. After the revolution many shaykhs involved in land speculation. The government tried to make business attractive for them so that they would not trouble it. (A member of the Abu Ras who settled in San'a in the 1970s)

Toward the end of the twentieth century, few Zaydis considered high status and political office to be incompatible with economic activities. Nowadays shaykhs and other men of tribal extraction represent a great variety of companies dealing in commodities such as gas, oil, and tourism (see figure 8.6).[16] Like many other well-known Zaydi

Figure 8.6 'Abid b. Husayn al-Surabi, a shaykh's son and businessman, San'a 2005

families, the *sadah* are no longer aloof from commercial activities. By virtue of their traditional dedication to knowledge, they have been able to adapt to the recent economic innovations with considerable ease, diverting this commitment into an excellence in business enterprises. Few pursue careers based on religious learning and the number of *mujtahids* is dwindling. The increasing emphasis placed on economic achievement as a determinant of status and the developing commercial sector provide incentives toward economic engagement. Once the border to South Yemen opened and unity between the two Yemens became a realistic prospect, 'Alid businessmen, alongside Sunni merchants and others, took their first opportunity to explore the 'Adeni market. The professional choices of young generation *sadah* are radically distinct from those of their forebears. One man, the son of a religious scholar whose great-grandfather was an Imam, studied aeronautical engineering. His father commented on his son's field of study saying that "Imam 'Ali said 'Instruct your children for they have been created for a generation other than your own.' " The most successful merchant from among the Zaydis, who like many of his Sunni counterparts is engaged in large-scale entrepreneurial activities which involve taking risks, is the son of an Imam.

In spite of the commitment to practical knowledge among the majority of the urban *buyut al-'ilm*, there is still a certain unease about abandoning careers in the religious sciences. Consider the following joke:

> On the Day of Judgment people who had been friends during their existence on earth were separated from one another. Some were sent to heaven and others were sent to hell. Bemoaning their separation, they got together and decided that each group would build part of a bridge which linked heaven and hell so that they could meet in the future. The inhabitants of hell were very diligent, and they finished building their part of the bridge in a short period of time. Conversely, those on the other side made very slow progress. This was because there were no engineers among them.

In the joke engineers are placed into the category of merchants who are said to greedily suck people's money. I have heard the joke being told in the presence of men who have taken up a business or engineering and are sensitive to charges of social drift. One young *sayyid* who had obtained a degree in French literature commented disapprovingly upon the fact that many of the great houses had gone into business. "It's not good for their identity and the *sadah* collectively. All they are concerned about is money."

Dispute about professional inclinations hardly ever develops into serious rupture between the young and the old not least because the latter have understood that the traditional education cannot guarantee a successful future. As Waterbury (1970: 87) observed, among the "first westernized generation" of Moroccans, few "have renounced their families, and many have become the willing or reluctant accomplices to family interests despite a superficial scorn for filial piety." Unlike the *buyut al-ʿilm*, who stress the rift in their professional histories, a young solicitor (*muhami*) and relative of Ismaʿil b. ʿAli Ghamdan, one of the few great ʿAlid merchants who lived during the time of Imam Yahya, pointed out that members of his house had always pursued diverse occupations.

> During the reign of Imam Yahya, most were teaching and working in the domain of jurisprudence (*al-qada'*). One was employed by the Ministry of Endowments (*awqaf*). Another, Muhammad b. Ismaʿil, was *amir al-hajj*. He led groups of Yemeni pilgrims to Mecca and also had a few commercial interests. His sons helped him in his job. Ismaʿil became a merchant like his brother ʿAli. In 1948 he gave his allegiance (*bayʿah*) to ʿAbdullah al-Wazir hoping that he would develop the economic sector. His three sons ʿAbdullah [died 2000], ʿAbd al-Rahman, and Muhammad were all merchants. Later in life, ʿAbd al-Rahman gave up trade in order to study. ʿAli's son Ahmad was also a merchant; his grandson ʿAli works with Muhammad b. Ismaʿil. [Muhammad was one

of the highly successful Zaydi merchants after 1962; he died in the early 1990s.] Ahmad's sons are also merchants, and one works for the Yemen Airlines. Since the early 1970s, many of our family have pursued their studies abroad. At that time, President al-Hamdi [1974–77] came to power. He was very good. He made a serious effort to eliminate the old social distinctions, and to give everybody equal access to scholarships and jobs. Before he came to power, the *sadah* were facing many problems. Things are looking good now. Some of our family are in the armed forces, the police, the Foreign Office, and the Press Office; others work for oil companies and at the courts. Several have their own business. My brother is an officer. It is not good if all work in the same profession. Our family was well off in the past, but we are doing even better now. In the days of Isma'il six families shared a house; now everyone has his own.

Some *sadah* have done less well. For example, one of the grandsons of the former Supervisor of Endowments (*nazir al-awqaf*) has become a money changer. Some of the sons of previously influential men pursue jobs as minor employees or do not work at all. As one of my friends explained,

> After 1962 they felt they were no longer important. That ruined them. Me and my brothers only survived and did well because my father was very strict about our education. When we wanted to say something to each other during our studies, we would whisper so that he would not hear us. Consider Muhammad 'Abd al-Khaliq who held a high position in Sa'dah. After the revolution he came to San'a and built a house there. He lives off a small pension. He has ten children who are all looking for jobs. One of his sons studied in Czechoslovakia, but when he returned in 1995 he could not find a job. Another is the driver of a high ranking officer (*dabit kabir*), and one works for the army as a lieutenant and clerk.

Economic assets certainly contributed to sustaining social positions in the years after the revolution, but university education was required in order to compete with the new elite some of whom are well educated and have acquired remarkable wealth. Looked at from another vantage, some *buyut al-'ilm* whose property remained confiscated for several decades, but who had already obtained university degrees during the 1960s and 1970s, nonetheless prospered.

Since the 1970s there has been a steady rise in women's education and employment. Before the revolution, women of the *buyut al-'ilm* taught girls at their homes or—sometimes clandestinely—worked as nurses. The majority of professional women are government employees. Many work as teachers in private or government-run schools. Several attended university during years of exile in the 1960s and 1970s and were highly successful on their return to Yemen. The granddaughter of the Imam's governor in Hajjah became the first woman engineer in the Yemen Arab Republic. Women of the House of the Imam have obtained university degrees in physics and English literature; one obtained training in French cuisine while her brother was studying in Paris. Rather than expressing their reservations, her uncles joked about the amount of weight they would put on after her return. Another woman who studied in Beirut and Cairo during the family's exile opened a private school which is attended by the new elite.

> My father supported my studies, but he wanted me to work in an intellectual field (*markaz 'ilm*) rather than a bank where I might have a bad experience [in a male-dominated surrounding]. That's how I became a schoolteacher.[17]

Old generation *sadah* who are critical of their children's professional choices argue that they depreciate their descent (*aslhum*). One of these men, who had little praise for his nephew's recently established advertisement agency, commented upon it by relating his uncle's fate. "We shall see what will happen to him [his nephew]. The last one in our family who became a trader died young." The man's uncle, who had been a bad student, had moved far away to an area in south-eastern Yemen where he traded in horses. He died in his twenties but the cause of his death remained mysterious. The trader's nephew indicated that a *sayyid* who chose to engage in trade rather than to serve others by transmitting knowledge might attract misfortune and even find a premature death. This man regretted that none of the young scions of his house which had produced a great number of excellent scholars had devoted himself or herself to religious studies. "We now have 'ulama of a different kind" he would say. Nonetheless, he was proud of his daughter who had just finished a degree in Political Science. "She will be like Safiyyah bt. 'Abdullah," he commented, referring to one of her ancestresses in the fifteenth century who was a scholar. He was satisfied that his sons and daughters who work as engineers and computer specialists have a sound basic religious education, do not flout religious rules, and are successful in their careers. Indeed, he has come to appreciate his children as channels to a world from which language barriers separate him. They translate English and French newspaper articles, old European academic and travel literature on his behalf which he considers to be authentic descriptions of his country.

This man's ambivalence was shared by a poet in his seventies. The way he glossed over the link between traditional elite status and particular types of occupation revealed his disappointment about the aspirations of young generation *sadah* and the decline of *'ilm*. He argued that all Muslims were equal, and no one should despise any type of labor which is compatible with Islamic values. Conscious of the fact that the *sadah* had been accused of maintaining rigid hierarchical divisions, he suggested that even the work of a *muzayyin* was not beneath a *sayyid*'s dignity. He said of his son who had taken up hotel management that he was to become a *muqahwi* (traditional innkeeper) who had no knowledge of Yemeni history.[18] When in the mid-1990s a young man of the famous Mutawakkil family opened a Lebanese restaurant in San'a, the poet exclaimed: "Splendid! A *muqahwi!*" (*'Azim! Muqahwi!*). His niece, a woman in her early forties, commented "it's good that the *sadah* involve in all professions. It's wrong that some people think that the son of so-and-so (*ibn fulan*) should not work in certain domains." When the poet learnt that the daughter of a former minister in Imam al-Badr's cabinet had became a landscape architect, he said "This is the new *'ilm*. She is a *qashshamah* (vegetable grower). There are no more Imams in Yemen, only *qashshamiyyun* (pl. of *qashsham*) and *muhandissun* (engineers)." He compared engineers to the *qasahat*, "those who built Imam Yahya's palaces—*rijal 'ummiyyun* (illiterate men)." His speech had a bitter sense of irony, yet he wanted to demonstrate that the *sadah* were committed to egalitarian values and that even the pursuit of formerly despised occupations carried no stigma. His commitment to these values notwithstanding, terms like *muqahwi* do not belong to the vocabulary of the business world of Yemeni cities. Once tourism will prosper again, the poet's son will have good prospects in the hotel industry. Young men like him concede that a religious education no longer provides either prestige or a good income. Some argue that one needs money to be respected and to pay the high dowries required for one's sons' marriages. Access to trade and commerce is also valued as a liberation from socio-moral codes which used to oblige the descendants of Imams and 'ulama to pursue careers in the field of religion.

> Under the Hamid al-Din, we could neither take up a trade nor the jobs of the *mazayinah*. Poor *sadah* were forced to make a living through Qur'an recitation which was more respectable than trading. In the countryside many waited for the farmers' gifts at harvest time because the zakah was forbidden to them. Many *sadah* supported the revolution because it gave them a chance to work. Many of us, who now work as doctors and engineers, might not have had any job in the past. We have more independence. Those who leave university no longer have to hope that their father, older brother or uncle will employ them as assistants. (A businessman in his sixties, 1986)

The *sadah* also wish to avoid nurturing the prejudice republican hardliners hold against those whose commitment to religious learning is interpreted as their continuing adherence to the values of the ancien régime. They often refer to the discrimination they have suffered since the revolution and their fear of being accused of aspiring to restore the old order.

> Many *sadah* work at the Foreign Office. They think that the foreign governments they deal with might protect them were they to be persecuted again. For the same reason, some entered the Ba'th party. They left it when Saddam lost influence after the Gulf crisis. (A diplomat)

Several *sadah* decided to engage in commerce because of their de facto exclusion from political posts which involve decision-making. Others spoke of the social benefits of people's dedication to new professions. The Mufti approved of his grandson's choice to become an engineer by saying that society needed people like him. As mentioned, these utilitarian considerations are juxtaposed to doctrinal ones and are produced as if either were intended to add weight to the other. 'Alid businessmen refer to religious authorities such as the Prophet and Abu Hanifa, the teacher of Imam Zayd b. 'Ali, both of whom were merchants. A judge in his fifties, whose documented family history does not record a single merchant, reflected on the economic activities of his kin by saying "they only follow the tradition of the *ahl al-bayt*." Thus, involvement in trade and commerce is represented as an expression of loyalty to the antecedents. By simultaneously committing themselves to Muslim orthodoxy and occupying innovative professional niches, the *sadah* at once maintain and recreate their identities.

Cases: Bayt Zabarah and Bayt al-Wazir

Centering on the twentieth century, this section deals with the professional histories of two houses spanning over a range of about three generations. (They represent only one section of these houses.) During the Imamate, the overwhelming majority of adult men (fourteen) were employed by the government. Some had studied in the famous teaching mosques. In the government sector, they worked as treasurers, customs officers, in the Ministries of Finance and Economics, and in the diplomatic service. 'Ali b. 'Ali, the Imam's deputy in San'a, was the most prominent member of the house who acted as the "head of the family" (ra'is al-'a'ilah). Two of his sons worked for him as his deputy and assistant. Another member of the house was a craftsman who produced dagger belts (sing. hizam). During the reign of the Hamid al-Din, two men were among the students who were sent abroad for further studies. Ahmad b. 'Ali attended a course in diplomacy in Egypt, and Amin, one of the sons of the historian Muhammad b. Muhammad, was trained at the Military College in Cairo and became the first parachutist in Yemen. By the mid-1980s and 1990s, the majority had continued to be employed by the government (twenty-eight of forty-six). Those who are self-employed work as computer specialists, engineers, and doctors; one is a shopkeeper in the suq. Some work in both the government and private sector. At least six hold doctoral degrees from foreign universities. Several women work in the educational and health sector.

The second example relates to two branches of Bayt al-Wazir (appendix II.2). During the Hamid al-Din dynasty, many held posts as governors, judges, administrators, and as employees in various ministries. Since 1962, men and women have worked for the government in various capacities (twenty-seven of seventy-two, many of whom live abroad). Many of those who were jurists before 1962 continued to work in this field after the revolution. Isma'il b. Ahmad al-Wazir, who served as Minister of Justice (1993–2001), received modern legal training.

*   *   *

The examples of Bayt Zabarah and Bayt al-Wazir, which are fairly representative of the prosperous San'ani buyut al-'ilm, demonstrate several points made in this chapter. First, before the background of the social and political transformations which occurred elsewhere during the 1950s, those who were eager to pursue mercantile activities or to seek advanced education abroad were aware that the established distinction between the scholar cum politician and the merchant would soon be obsolete. Their forefathers who had pursued their studies abroad attended Islamic institutions such as al-Azhar. Instead, "the university," a generalized institution of knowledge was now featuring in the vocabulary of the young, and soon became linked to the growing cultural dislocations emerging in Yemeni society. The enlightened young generation (and astute observers among the old) of the Zaydi establishment understood that maintaining authority required a considerable readiness to accept a re-definition of the political bases of power. Their turning away from the old teaching institutions and their growing criticism of autocratic rule which they shared with many others, anticipated the decline of the Imamate as a culturally viable political system of governance. Their new orientation was part of broader shifts in the bases of power which were now occurring. Those who provided a critique of the Imamate's fundamental institutions inadvertently undermined the few positions of privilege they had. However, in regard of the acquisition of knowledge required for new professional careers in the republic, they have by no means been disadvantaged and able to compete with those who were educated in Aden and abroad.

These developments have been accompanied by a much greater emphasis on secular knowledge. The biographical data provided by the two houses examined here suggest that in the nineteenth and early twentieth century, a specific kind of knowledge, *'ilm*, was linked to distinct types of scholars within a distinct type of political order. Since *'ilm* no longer provides the rationale for the state, it has been assigned different meanings and has become recontextualized. *'Ilm* has never been rigidly defined and so is not opposed to the "new knowledge" which has gained unprecedented authority in the past decades. The *buyut al-'ilm* continue to regard a basic religious education as an essential component of self-constitution, and religious notions are invoked in their everyday conduct. Some of the memory work which has been performed by the *sadah* over the centuries may be pursued in different idioms of knowledge. Because secular knowledge can be represented as a kind of *'ilm* and religious knowledge is still being acquired, current pursuits of knowledge do not conflict with notions of *'ilm* as an analogue of biogenetic substance which is instrumental in generating inborn potentials (see chapter 5).

The professional histories presented earlier demonstrate that the *sadah* had contact with the outside world before the revolution as students and in the diplomatic service; that the majority no longer conceive of trade and commerce as incompatible with their status honor; that women are increasingly involved in higher education and earning a living; that the majority are still engaged in public service. There is a noticeable trend toward engineering, medicine, business administration, accountancy, and computer science. Many houses are engaged in both government service and business. Since the 1960s they have continued to be well represented in the domains of law and education. In the last decade the trend toward nonintellectual professions has increased because academics had to augment their salaries.[19] In spite of the high percentage of *sadah* who are employed by the government and have joined the ruling Congress Party, they do not wield real power. Since the revolution they have held ministerial portfolios but high positions are not necessarily an indication of political power. Like other contemporary Middle Eastern elites who hold political positions (Zartman 1980: 3), they are primarily bureaucrats and technocrats rather than politicians. Some *sadah* who have abandoned government service in favor of private business, argue that economic independence guarantees surviving the vagaries of the political system (see Waterbury 1970: 97). The professional histories also reveal that *sadah* have either remained in exile in Saudi Arabia or the United States after 1962.[20]

Among those who have remained in Yemen, some acknowledge that once the early cataclysmic years had passed, they have benefited from the revolution because a greater number of their kin have been able to achieve good positions. Some even judge the revolution as a liberation from the social constraints on certain types of work. Many *sadah* argue that as a result of greater proliferation in diverse occupational domains, houses can be influential in diverse spheres. Individuals had achieved greater independence from the patronage of older kin which however some still enjoy. Although the great 'Alid houses have lost much of their power and prestige, according to their own judgment the majority are economically better off than they were. Those who have suffered social drift have not pursued a modern education or were unable to adjust to the system for other reasons.[21]

The institutionalization of new types of knowledge has occurred within the frame of distinct political and economic structures that have been established in the process of republican state formation. A stratification system based partly on familiar criteria of mobility—education, wealth, and "being connected"—is emerging alongside a globalized consumer-oriented economy. The outcome is not just the reinvention of parts of the old elite as modern consumers, but a greater dependency of the new system on Western powers and Arab states whose monetary policies are designed to influence domestic politics. Chapter 9 continues to examine how the old elite situates itself within these novel structures of meaning.

Chapter 9

The Moral Economy of Taste

It emerged from chapter 8 that the majority of *sadah* reckon the choice of professions based on secular knowledge to be conducive to maintaining or gaining prosperity and to their accommodation to post-revolutionary Yemen. In San'a wealth accumulation and consumption contribute significantly to the definition of elite status. The cultural logic by which material and immaterial artifacts are understood is instructive about 'Alid self-identification in an era of rapid cultural and politico-economic transformation. Those *sadah*, who have successfully asserted their claim to elite status in republican Yemen, have employed their resources to construct a new identity which is, in important respects, one based on consumption. Consumption is also a means to reformulate their relations with others by adhering to the norms of the nascent "consumer society."[1]

Consumption styles objectify both distinctions of wealth and particular cultural values. It has been pointed out by various writers that the acquisition and enjoyment of material artifacts is an expression of a "process of social self-creation" which is always culturally specific (Miller 1987: 215; Appleby 1993: 172), and that goods are marked as a means of defining social relationships (Douglas and Isherwood 1978). Previously, the enjoyment of material artifacts was far less central to the moral economy of taste. By comparison with those states on the Peninsula which began to tap their oil wealth in the 1930s, in the Yemen consumer spending was both limited and discouraged. The most desirable commodity was land. Wealthy *buyut al-'ilm* would possess a well bred horse (very few had a car), a spacious house, land, and an expensive dagger and sword. They recruited private teachers, guards, *mazayinah*, and houseboys (sing. *duwaydar*). These types of consumption bolstered the status of those houses, but more prestige accrued to the ability to dispose wealth—for example in the form of religious endowments. Wealth was not considered inherently honorable and served mainly to corroborate a status which was defined through other criteria. However, in the recent boom years prior to the Gulf crisis, which has led to an economic down turn, and to an extent today, members of the old nobility have crafted their new identities within a rapidly expanding consumer economy enjoying lifestyles which previous generations might have reckoned to be vain and immoral. After briefly reviewing the verdicts of Zaydi authorities on the "life in this world," which ideally guide the behavior of the faithful, this chapter looks at elite consumption during the Imamate and the republican era.

Views of Zaydi Authorities

Shi'i sources demonstrate an inclination toward otherworldliness which in the case of the Zaydis does not conflict with the central doctrine of the Imamate according to which the exercise of political authority is the vocation of the Prophet's progeny. Imam 'Ali warned of the temptations of this world, saying that "it has decorated itself with deception and deceives with its decoration" (1981: 203–4). Shi'i 'ulama subscribe to values such as asceticism and modesty, and deviations are explained in terms of inevitable compromises with their impious contemporaries. For example, Imam al-Hadi Yahya (d. 911), apologizing to his cousin on his return from the market in Sa'dah wearing a fine new coat, bemoaned the impiety of his contemporaries:

> By God, were I among Believers, I would not wear a coat such as this. Indeed it is not my kind of dress. I would wear only rough clothing, but, if I wore such, people would think little of my position. So, I considered their ways and realized that they will obey only a leader who dresses in such clothing, but I feel as if a cloth of thorns lay against my skin (quoted in Mundy 1983: 531).

Despite the fact that Zaydi believers are not required to live the life of an ascetic (*zahid*), some 'ulama favored rigid life styles. In his treatise on *hisbah* (accountability), al-Hadi's contemporary Imam al-Utrush who ruled a Zaydi state in Tabaristan, insisted that mosques must not be decorated with pictures or gold, or hung with curtains. Musical instruments must be broken when found and singers are to be expelled (Serjeant 1953: 6–9). Imam Yahya b. Hamzah (d. 1346/47) went further by decreeing that if one uses one's speech organ for immoral purposes, for example by singing, one is no longer permitted to exercise the rights of an adult Muslim. The testimony of a person who sings is no longer permissible. Whoever sings is *fasiq*, godless and sinful (Sharh al-azhar, Vol. 4: 383). Thus the singer is placed into the same category as people who lie, drink, or are classified as mad, all of whom cannot be witnesses in court. The Zaydi writer Salih al-Maqbali, who took issue with Imam al-Mutawakkil Isma'il's verdict on the *kafa'ah* (chapter 6), said that "one of the best things he has done there [in the Yemeni mountains] is to prevent swinging and dancing . . . for their (the imams') doctrine entails the proscription of singing" (Madelung 1999: 143). The attitude toward singing was not different from that toward playing games. Imam al-Utrush ruled that carpenters are prohibited from making backgammon and chess pieces, idols and dolls (Serjeant 1953: 8). Many centuries later, I learnt from the son of one of Imam Ahmad's secretaries, who was a child during the last days of the Imamate, that these moral codes had currency among the elite.

> One day I went to a coffee shop and played domino with the men there. Suddenly everyone got up in silence. I had not noticed my father. When I looked up, he made a gesture asking me to come. I knew I would be beaten. We passed a government building that was guarded by two soldiers. "*Ya Hashidi* [assuming he was from the Hashid tribe]," I said to one, "this man wants to beat me." When the soldier came to talk to my father I ran away.

Edicts regarding consumption concern the maintenance of public morality and a person's moral disposition. Unlawful consumption threatens the social order and must be avoided at all costs. Debates among the 'ulama about whether and in what circumstances a man is permitted to wear silk, gold, silver, or pearls reveal their concern with the gendered integrity of their world. The Faqih Sharaf al-Din al-Hasan b. Abi Baqa, quoting the Prophet, prohibited men from wearing gold except that which is in pieces (Sharh al-azhar Vol. 1: 177; 4: 109).

Disreputable Pleasures

Among the elite of the twentieth-century Imamate, sophistication and refinement were communicated through a devotion to knowledge, the production and recitation of poetry, honorific language, and control over one's body and emotions. The austerity of the Imams' mansions reflected their lack of interest in profit-making and the accumulation of material goods; indeed, the houses of the merchants were often more lavish.[2] Although the Imams were aware of the necessity of economic development, this consideration was undermined by anxieties about foreign domination and limited state budgets. There were always foreign goods in the market, but during colonial rule in South Yemen and other Arab countries, those coming from Europe were stigmatized.

> I was given a watch by my mother which I wore on my upper arm because wearing a wristwatch was not respectable. It was *'ayb*. What made it worse was that it was a lady's watch. (One of Imam Ahmad's nephews)

During the first half of the twentieth century, the rigid view Zaydi 'ulama took toward aesthetic delights other than spiritual accomplishment was widely shared among the *buyut al-'ilm*. Smoking, dancing, and playing and listening to music were associated with moral laxity and said to draw attention away from pious contemplation. Imam Yahya's Prime Minister Husayn al-'Amri did not tolerate smoking in his house. His brother hid his water pipe in the wall of his room and smoked on the upper floor in order to prevent the minister from hearing the bubbling. One of Imam Yahya's grandsons, the late Husayn b. al-Qasim, recalled that he never smoked in front of his father, but when they were traveling together he would offer his son some tobacco. "My mother did not mind my smoking; a woman is more compassionate (*'andha rahmah*)." Young generation *buyut al-'ilm* who have not reached majority refrain from smoking in front of their parents, but some do so in the company of their friends. Women smoke at the *tafritah* but will not do so in front of senior men of the household.

When the radio was introduced during the time of Imam Yahya, many of his subjects regarded it as evil. The Imam himself only tolerated religious chants (*nashid*) and military music. Officials and men of letters were eager to listen to the radio in order to gain knowledge about daily events outside their country, but some even refused to keep a radio in their house. Among them was Imam 'Abdullah al-Wazir (d. 1948) whose soldiers had to listen to the news in the house of the governor of San'a and present him with a digest.

The Authority of Clothes

Chapter 5 analyzed how the 'Alid body is substantiated through a series of discrete performative acts. In the Imamate, specific items of clothing provided the visual cues of the mental dispositions shaped by these acts; they exteriorized the moralized body. With the exception of a few items, neither men's nor women's dress revealed their precise descent affiliation. The classical outfit of men of the *buyut al-'ilm* was a gown (*qamis*) whose sleeves were knotted out behind the shoulders.

The *qamis* was worn over a *zannah*. With the onset of puberty, men wore a headgear called *'imamah* (pl. *'ama'im*) which consists of a pill box hat and a rectangular piece of white cloth wound around it. Outside the house a *jukh* was worn over the *qamis*. The *jukh* is a dark-colored, almost ankle-long woolen coat ornamented with gold thread banding and buttons. It was highly improper for a man to walk the streets without his *jukh* and his *'imamah*, which was a symbol of manliness, erudition, dignity, and noble descent. One scholar told me that a *sayyid* would not have been accepted in court without his *'imamah*.[3] Men attached sentimental value to certain prestige objects such as their swords and daggers (sing. *thumah*), which were symbols of both masculinity and elite status (figure 9.1).[4]

Figure 9.1 Sayf al-Islam Muhammad al-Badr b. Imam Yahya (1898–1931)

These kinds of clothes cannot "be understood only as metaphors of power and authority, nor as symbols; in many contexts, [they] literally *are* authority" (Cohn 1996: 114; emphasis his). Several of the 'ulama who were captured and imprisoned in 1948 arrived at the prison of Hajjah wearing the *ma'raqah* (sweatcap) which is worn underneath the *'imamah*. When tribesmen loyal to Prince Ahmad ransacked San'a, they took the headgear and daggers of those men who were involved in the revolt, thus symbolically emasculating them (vom Bruck 1999b: 164–5). During the revolution the *'imamah* was identified with the "reactionary" forces. Radio broadcasts spoke of the polluted *'ama'im* (*al-'ama'im al-najisah*) (Mundy 1983: 538). In spite of having been worn by non-'Alid noblemen, the *'imamah* has become the emblem of the *sayyid*'s otherness, a metonym for reaction whose ideological referent is lacking commitment to the republic. The 'ulama and shari'ah-trained judges have continued to wear this kind of outfit. Like other professional San'ani men, the majority of *sadah* now wear suits the quality of which is the only indicator of the wearer's identity. Most young *sadah* wear the traditional outfit only at their weddings, but some have abandoned it completely. One of them told me after this wedding "if you wear the *'imamah*, they say 'he wants to be an Imam'."

Women used to wear the *sharshaf* or the *sitarah*, a colored printed cloak. The *sitarah* was more often worn by low status women but was not a clear marker of status. Status differentials were more apparent in women's visibility in the streets. High status women would be visited rather than visiting unless they attended life crisis rituals. A governor's wife told me that she had not left the house for seven years until her husband left the country in order to take up a diplomatic post abroad. In the hijrahs, these women had to be invisible to the extent that their bodies, even if entirely covered, would not be exposed to the gaze of men who in theory were entitled to marry them (non-*mahram*). When they attended rituals at the houses of relatives and friends, they left after dusk in the company of their husbands or guards. One of Imam Ahmad's nieces, who lived in al-Qaflah near Shaharah, recalled that when darkness fell, she went out in a group of three or four other women, a soldier walking ahead, holding a lamp.

The Road Toward Consumption

We shall see next that in the light of new consumption regimes, wearing the *sharshaf* has taken on new meanings. Since the end of the civil war in 1970, consumption has become an important factor of status competition and elite formation and transformation. The ability to acquire goods is one of the defining criteria of newly emerging elites who tend to be more inclined toward conspicuous consumption than the old nobility (P. Burke 1993: 158). In the Yemen, a specific type of consumption unites the well-to-do from among the old elite and the nouveaux riches made up of some merchants and technocrats, senior army and police officers, and shaykhs.[5] Traveling abroad for advanced education, hospital treatment and business, and the hiring of foreign domestic labor demonstrate cosmopolitan life styles and prosperity.

Once the old status hierarchy began to weaken, new consumption regimes have produced inegalitarian structures of their own kind. They have deepened divisions between the rich and the poor rather than produced cultural homogeneity. The new affluence of the late 1970s and 1980s was accompanied by the influx of a great number of luxury and consumer goods. In the mid-1970s, President al-Hamdi's ambitious projects marked the birth of the Yemeni consumer society.[6] The material consummation of the image of the good life during those years has changed both the consumption habits of the majority of Yemenis and the parameters that had defined their positions in the social hierarchy. Social prestige can now be derived from the accumulation and display of wealth more than ever before. Until the 1980s,

> it was possible to maintain a good life-style without even having a job. You would live off relatives or your land. This is no longer possible. Everyone needs a job now, but they are in short supply. Many people eat meat only once a week. People no longer celebrate the *yawm al-sabah* [part of the wedding celebrations following the wedding night. Relatives and friends are invited for lunch at the groom's house]. I haven't been invited to one for two years. (A university lecturer)

Since the late 1980s ushered in a period of economic decline and political instability in the region, the process of embourgeoisement has been jeopardized and cosmopolitan life styles have become the privilege of few. After the Gulf War of 1990/91 the Yemeni government lost a huge amount in yearly remittances and had an unemployment rate of over 25 percent (Carapico 1998b: 9–10).

Education and Other Kinds of Consumption

One of the key domains of investment by those who have nonetheless remained (or become) prosperous is education. Among the elite of the Imamate, knowledge was a major resource that provided power and self-definition. Even after the printing press had been introduced and books were no longer copied by hand, they were regarded as precious commodities—so precious that some young men had to beg their fathers to be allowed to read them. Imam Yahya, who had books printed in order to make them more widely available, ordered Bayt Zayd, a wealthy house which had produced only one son over several generations, to transfer their library to the Great Mosque so that others could take advantage of it. When more sons were born to the family, the library was returned.

The old nobility has maintained their dedication to knowledge. Among the nouveaux riches, many of whom come from rural families who never read a book other than the Qur'an, there is a great aspiration to turn economic into cultural capital. Education is seen as a road to refinement and urbane life styles among all sectors of the population. The kidnappings of tourists and oil company personnel in the 1990s, often intended to exchange hostages for schools and hospitals, testify to successful media campaigns presenting education as the magical key to prosperity.

The establishment of San'a university in the 1970s expressed the new regime's self-assertion as a modern nation-state. However, the elite considers university education abroad to be superior. This trend has increased since Kuwait ceased to sponsor San'a university following the Gulf crisis and the quality of education deteriorated. Degrees from local universities carry far less prestige and provide few employment opportunities outside Yemen. There is fierce competition for scholarships from foreign governments. Expert knowledge obtained at foreign universities forms part of consumption styles, which at once unify and lead to competition among the new and the old elite.

Claims to membership of the elite are partly based on knowledge of events in the world at large, whether acquired at the university, through the media, the internet, or the satellite television.[7] Literacy per se does not project respectability; one must be able to communicate ideas acquired from foreign sources. Professionals and well educated government officials are expected to be aware of specialists' analyses of current affairs. When abroad these men buy books rather than souvenirs and fine art for their friends and colleagues. Works such as Fred Halliday's *Islam and the myth of confrontation* (1995) are considered appropriate gifts for members of the enlightened political elite.

One of the indications of low status is the exclusion from the primary experience of things foreign: the poor observe the outside world through the ubiquitous television. They neither travel nor do they consume imported goods other than those that are cheaper than local produce. Some people rationalize the revolution in terms of consumer goods. Once at a *tafritah* the daughter of a Sunni professional dismissed Imamate rule by saying "there was no television (*ma kansh fi television*)!" Prestige goods, with the exception of daggers and qat, are usually foreign. Among these are luxury cars and all kinds of new technology ranging from computers to video cameras. Some travelers spend considerable time purchasing goods and cartoons for transporting them. Since the economic crisis began, ordinary foodstuff people had got accustomed to have been transformed from consumer goods into prestige goods. In the late 1990s, a friend who visited London bought cornflakes and croissants for her children as a special treat, explaining that she could no longer afford them in San'a.

One item of new technology that has been readily appropriated by the educated elite is the fax machine. Throughout Yemeni history, scholars have been preoccupied with the circulation of poems and pamphlets written either by themselves or others. It is common that on the occasion of a relative's death or the arrival of a new child, poems are sent via the fax machine by friends and family from all over the world. When abroad, men also fax love poetry to their wives. When a scholar underwent major surgery, his son set up a website asking people to pray for him and informing those who cared about him about his progress. Surat al-Shu'ara' (26:80) was quoted and it was said "we wish him the best of health and good fortune among his family, friends, admirers, and the countrymen for whose betterment he works constantly." Within a span of ten days, 824 people had visited the website, and well-wishers from Yemen, India, Britain and elsewhere had signed the "Guest Book" on the web.

Women's Gatherings

Women's daily gatherings (*tafritah*) are events where women of the old elite—who are by no means all wealthy—and the nouveaux riches mix and compete with each other.[8] Women of the old elite, who used to be visited by women of less distinguished houses, are now reciprocating the visits of those whose families have gained status after 1962. At a wedding the wife of a governor, on seeing the daughter of the Imam's deputy entering the room, brought her finger to her nose to indicate her speech, and said "they felt so superior; now we are all the same."

At the *tafritah* women wear long dresses made of polyester or silk imported from Japan, Korea, and India. These materials are desirable because they are foreign yet cheaper than French and Italian ones, and they cater for women's predilection for shiny and delicate fabrics. Most dresses are produced either by the women themselves or by tailors, and some are bought abroad. The tailors, many of whom are Somalis, display European fashion magazines in their stores from which women choose designs. Part of a woman's outfit is a headband (*al-masar al-tal'i*) consisting of cardboard wrapped with a piece of patterned silk worn over a scarf. Pret-à-porter from Europe is available at high prices. A woman who had returned from exile in Beirut in the 1970s told me that some girls who had lived there displayed their short skirts at the *tafritah*, which they had never worn before. This pretense at self-transformation was both true and false. They had imagined and appropriated Beirut where they attended school and made friends, but had lived much more secluded lifestyles than most of their Lebanese peers. Their families had been denounced by the new Yemeni regime, but they returned to their country feeling superior to their peers because they had lived in a glamorous metropolis few Yemeni women have ever visited. More recently, some young women have started to wear trousers at the *tafritah*. Most are still chewing qat, but cake, tea and biscuits are also served on small tables where mobile phones are placed.

The daily *tafritah* is always held in women's houses. Weddings, however, are often celebrated at international hotels. They represent an emulation of Cairene and Damascene lifestyles by the Yemeni elite.[9] At the weddings women's decorated exposure—outlandish make up and dress, heads adorned with artfully placed feathers— captures the other guests' imagination of their husbands' spending power. The wealthy never fail to let it be known that their daughters' wedding dress was bought in Paris. Those parents who allow the newly wed to spend the wedding night at a hotel demonstrate both their affluence and modern outlook by recognizing their right to privacy and intimacy which they never enjoyed. The older generation had to spend their first night in the patrilocal household.

Women welcome hotel weddings as a relief of the burden of home cooking, but these occasions expose the moral and material dilemmas faced by women of the old religious elite. Although female guests are taken to the hotel by their male kin and are guided to the function rooms, elderly women feel that hotels are improper places for women to enter. When they grew up, public guesthouses were avoided by high status men and women unless they were traveling. The groom's party celebrates in another function room or at his house, yet some women who attend hotel weddings fear exposure to non-*mahram* and are reluctant to take off their outdoor garments (*sharshaf* or *balto*) lest a waiter will enter or the groom will come to fetch the bride. The failure of women to remove their street clothing reveals their discomfort in an inherently ambiguous space; it also allows less affluent women to escape exposure to shame.

Weddings are but one phenomenon that exemplifies the tension experienced by the old elite between concepts of piety and the new parameters of respectability, which require them to engage in lavish consumption. Many *sadah* and *qudah* argue that the boastful display and waste of food at weddings are incompatible with Islamic principles.[10] This moral dilemma is reminiscent of the anecdote relating to Imam al-Hadi's reluctant purchase of a fancy coat, which I referred to earlier. The Prophet's dictum that one should not eat unless one is really hungry was often quoted to me. However, it was argued that a family of means which would arrange a simple wedding party would be regarded as misers rather than good Muslims.[11]

In some cases consumer passions are compensated for through charity. For example, on his return from the mosque to his house at midday, a wealthy *sayyid* picked up some wretched people and offered them lunch. He had money, clothes, and food taken to the houses of the poor and gave to all who came to him asking for alms. He was critical of the *sadah*'s seeking of personal gratification through material goods and attributed their loss of the Imamate to their acquisitiveness in the political domain.

> A *sayyid* should not aspire to accumulate mundane goods; he should sacrifice on behalf of the poor. If people ask for help he should be the first to respond and the last one looking for profit. The *sadah* have inherited this responsibility from al-Husayn and al-Hasan. When they first came to Yemen, the *sadah* were just mediators. They changed when they began to acquire land and to act like kings. Then people repudiated them. Today they are after money like the others, but not as much as the *qaba'il*.

To sum up, in spite of these predicaments, among well-to-do San'ani *sadah* consumption serves as a means of redefining themselves and their relations with others, and to maintain their former social position. The accumulation of prestige through success in the new (and some of the old) professions, and the ability to subscribe to new consumption regimes, to an extent compensates for their loss of political power. Through consumption the *sadah* demonstrate detachment from the ideals of the Imams which centered on asceticism. For them, consumption aims at their incorporation into the republic rather than the maintenance of difference which scholars such as Veblen (1970 [1898]) have ascribed to elite consumption. Their successful adaptation to the challenges of post—revolutionary Yemen provides one of the clues to the persistence of negative stereotyping of the *sadah* which will be explored in chapter 10.

Part IV

Engaging Difference

Chapter 10

Defining through Defaming

Late 1950s. A judge from a prominent qadi house entered a guesthouse on his way from San'a to Hajjah. Handing over a riyal to the owner, he asked him to cook a small lamb for him and his companions. After a while a waiter came and brought water for the guests to wash their hands. He asked "Who is the sayyid?", and when the guests pointed to him, the waiter handed the bowl of water to him. He returned to the kitchen and brought the shafut,[1] again serving the sayyid first. Then meat was brought by another waiter who also asked for the sayyid. Thereupon the qadi turned to the waiter and said "This time I am the sayyid."

This anecdote of a scene at a rural guesthouse, recalled by the *qadi* some thirty years later, reveals that the *sayyid*'s privilege to be served first was taken for granted by everybody except him. Business was as usual in this village guesthouse; few people had access to the media, and broadcasts from Cairo, which agitated against the monarchic regimes on the Arabian Peninsula still had little impact in rural Yemen. The *qadi*, however, was resentful of the *sayyid*'s treatment at the restaurant and boldly challenged his prerogative. One of the *qadi*'s sons narrated another story, which according to him demonstrated the *sadah*'s sense of superiority.

After the opening of a maternity hospital in the southern city of Dhamar in the 1970s, women were waiting outside the gate in order to be admitted. A *sharifah* jumped the queue, gave the receptionist her name and demanded to be let in. Shortly thereafter her name was called. One of the women waiting in the line exclaimed in Dhamari dialect: *'Inna limma mahi sharifah aw ma ahna la nitkhannath bi-shar'?'* (Why is she honorable (*sharifah*), are we from the street?)

There are many anecdotes like these, but they are rarely told in the presence of the *sadah*. Unflattering stories and jokes circulate in the *maqayl* of those who used to compete with the *sadah* for positions in the Imam's administration, the *qudah*, and others who have gained prestige and power since 1962.[2] Deprecatory stereotyping occurs in all central institutions: the family, the school, and the office. Stereotypes are held by men who are linked to the *sadah* through friendship, joint enterprises, and affinal relations, and who respect the achievements and propriety of individual *sadah*.[3] They focus on the *sadah* collectively but especially target those who belong to the administrative and technocratic elite, among them members of the *buyut al-'ilm* and formerly underprivileged *sadah*. Anti-'Alid commentary is employed without provocation. The *sadah* are readily blamed for any injustice inflicted by the Imams and the country's misery; domestic affairs are interpreted in view of whether they are to the *sadah*'s advantage or disadvantage. The discourse utilizes complex stereotyping elements drawing on alleged character attributes such as presumptuousness and self-righteousness. It also invokes images of stigmatized groups such as the Jews who have had an uneasy relationship with Muslim majorities throughout Islamic history. Irrespective of their individual political orientations or their role in the revolution, the *sadah* are collectively referred to as *malakiyyun* (royalists), a term that carries pejorative connotations because of its link to the Hamid al-Din dynasty.

The *sadah* are said to be "the Jews of the Middle East" and are compared to the Banu Qaynuqa', the Jews who lived in the town of Medina at the time of the Prophet. They had entered a contract which obliged them to refrain from uniting with his Qurayshi enemies, but defied it and have been portrayed as conspirators and traitors ever since.

> The *sadah* are the Banu Qaynuqa'. The Qur'an says that they destroy their houses with their own hands and the hands of the believers. (A former clerk in his sixties)

The image of "*the* sadah" is a trope of otherness. As noted in the introduction, representations of the *sadah* as "strangers in the house" center on their putative 'Adnani origin.[4] Those who claim descent from Qahtan see themselves as rooted in remotest antiquity and as the indigenous inhabitants of Yemen. But the young republic also claims Islamic credentials and thus cannot model itself on the pre-Islamic empires alone, just as Ba'thist Iraq sought to root itself in both the ancient Assyrian and Babylonian empires and the 'Abbasid era (Bengio 1995: 142). In Yemen, republican ideologues evoke a tradition which is both Qahtani and Islamic, and which is an alternative to the *taqlid ahl al-bayt* and Zaydi-Hadawi history.

There is refusal to recognize the *sadah* as *awlad al-balad* ("genuine Yemenis"), and statements made in favor of their expulsion do not provoke protest. On several occasions I was told that they should "go home"—a reference to the places from where the *sadah*'s ancestors emigrated to Yemen (the Hijaz, Iran, and Iraq). During the Iran–Iraq war (1980–88), the *sadah*'s sympathies for Iraq's enemy were taken for granted on the assumption that Iran's supreme leader, Ayatullah Khumayni, was a descendant of Imam Husayn.[5] Other factors were claims made by some 'Alid houses to descent from Tabaristani Zaydis (for example, Bayt al-Daylami).[6] The war served as a vehicle for sustaining images of the *sadah* as foreigners and as enemies of the Yemeni republic. Calls for their denaturalization are based on claims that Yemeni citizenship presupposes allegiance which some allege the *sadah* are lacking.

Views expressed toward the Jews demonstrate the historical contingency of "othering." The Jews, who in the past were subject to a politics of exclusion, are said to be "pure Yemenis" because they have been in the country for several millennia and because they have practiced endogamy (compare Halliday 2000: 65). This point is stressed by anti-'Alid ideologues who argue that unlike the *sadah*, the Jews are not foreigners. In these disputes over authenticity, the Jews' involvement in despised occupations such as peddling, cleaning of latrines, and removal of carrion from the streets (Lewis 1989: 19) is not given consideration.

A couple of years after the war, the YAR and the PDRY merged and South Yemeni politicians moved to San'a which became the capital of the unified republic. The stigmatization of the *sadah* took them by surprise.[7] Prior to the dissolution of the PDRY, the number of *sadah* in the government had not caused irritation. Some former PDRY politicians who are members of the government in San'a have attempted to combat the prejudices held against the *sadah*.

Against the backdrop of their history of rule, which is often interpreted as a subjugation of the "indigenous" population, it is assumed that the *sadah* have no right to feel insulted by the stereotypical language directed at them. It is even justified as a kind of protective shield against their return to power. This mode of reasoning is quite different from that pertaining to defamatory speech directed at members of subordinated groups who are perceived as inferior by those who victimize them (on this issue, see Delgado 1993: 94–5). Whereas this type of speech is rarely censored, there is fear that discrimination against the former servile classes may provoke criticism. Those who fall into this category are themselves disinclined to employ anti-'Alid speech. Few of those previously labeled *qalil asl* have climbed up the social ladder, and are therefore less likely than others to view the *sadah* as rivals. Some have remained loyal to their former employers. During the revolution they opened their houses to the female kin of men who were arrested, and hid their valuables when their houses were occupied. Among manual laborers in the old town, songs in praise of Imam 'Ali are chanted (see Marchand 2001: 114–7). However, a woman who had helped a family with domestic chores and continued to bake bread for them after the revolution, was asked by her neighbor to stop working for the *sadah*.

Positive stereotypes depict the *sadah* as being dedicated to learning and committed to Islam, industrious, disciplined, and willing to take responsibility. The anti-'Alid construes these virtues as vices, advocating to withhold influential state offices from them lest they attempt to reclaim power. An 'Alid accountant reported that the clerks in the ministry where he was employed assumed that he could surmount any task given to him.

Images of the *sadah* as strict adherents of religious rules co-exist with claims that they are lax in their religious observance and lack control over their women. The case of a *qadi* family who had accepted a *sayyid*'s marriage proposal to their daughter who thereupon took up the *sharshaf* because "the *sadah* like their women to be covered" exemplifies attitudes toward their abidance to religious principles (see chapter 7). However, the grandson of one of Imam Yahya's civil servants of *qadi* extraction held different views.

> The *sadah* are not as strict as we are. Men and women chew qat together. (The joint consumption of qat by men and women who are not *mahram* to each other has connotations of licentiousness.) They do things that are *haram* for the *qudah*. Bint al-Farisi and her mother are living on their own. They come from an important family (*usrah kabirah*). The *qudah* could not do this kind of thing.

The *sadah* are represented as a closed community whose members promote each other's interests.

> Since my aunt married a *sayyid* her links to our family weakened. She became entirely absorbed in the *sayyid* community and hardly mixed with anyone else. I was engaged to a doctor before I got married to my husband. On learning that he was a *sayyid* I broke the engagement. (A diplomat's daughter)

The Problem of Authenticity

Non-'Alids tend to downplay the religious component of 'Alid identity in favor of the political one. Some trace this representation to their early education. One man, a factory owner in his fifties, recalled how he came to understand the concept of "sayyid" as a child. "I went to school in [the northern town of] Hajjah where I was taught to revere and to respect the *sadah* because they descended from the Prophet. I created in my mind a link between them and the Imam. I thought of the Imam as a *jinn* who could fly and do miracles." Because of these identifications between the Imam and the *sadah*, republican rhetoric against the Imams has fostered anti-'Alid sentiments.

Men like him are aware that the ideology of descent from the Prophet provided a strong legitimating device among the population of the north not least because of the affinity between key Zaydi concepts and those held by many *qaba'il*. For centuries moral and political authority was formulated in the idiom of knowledge and pre-eminent heredity. Tribal leadership has been based on such notions, and they are the hallmark of the Zaydi theory of the Imamate. In spite of the widespread resentment of 'Alid rule, there is still regard for the *sadah*'s genealogies.

> A *sayyid* should be judged by his morals and by what he does for other people. I do not care about the Zaydiyyah, I do not adhere to any [religious] school. I have no special respect for the *sadah*, but every Muslim must acknowledge that their *nasab* ranks highest. (A doctor, grandson of one of Imam Yahya's senior officials)

Assigning the *sadah* to the highest place in a belief system creates a moral dilemma for those non-'Alids who wish to repudiate their former inferiority. They escape from this dilemma by denying the authenticity of 'Alid genealogies and by subscribing to anti-Shi'i ideologies. Polemical contestations of the *sadah*'s descent from the Prophet has precedents elsewhere (for example, Sebti 1986: 441–3). The Moroccan sultans made efforts to verify the authenticity of sharifian descent when some *ashraf* were accused of advancing false claims. On assuming power, Sultan Isma'il (1727–57) found that "the subjects would almost all become sharifs" (Sebti 1986: 446; compare - Combs-Schilling 1989: 318 n. 7). Like many contemporary Yemenis, Moroccans did not challenge the validity of the hereditary principle but were troubled by the problem of authenticity. Scholars who deny recognition of the *sadah*'s genealogies refer to them by their names rather than their titles. In Isma'il al-Akwa''s recent study of the Yemeni hijrahs (1995), even famous 'Alid 'ulama are not referred to by their titles, whereas most other scholars are introduced as *qudah*.

Two decades after the revolution, the Ministry of Information and Culture published a book about Muhammad al-Amir by a group of historians and politicians (Qasim Ghalib Ahmad et al. 1983).[8] The book's subtitle indicates that readers will be given a "picture of the struggle of the Yemeni people." In the first chapter we learn that al-Amir's forebears, among others, had been opposed by al-Qasim b. Muhammad who became Imam in 1597. The authors note that the Imam referred to himself as a descendant of Imam al-Hadi Yahya, and that he lived in Banu Mudaykhah in al-Sharaf near Hajjah. The first of his ancestors who lived there at the beginning of the fifteenth century was Muhammad b. 'Ali b. al-Rashid. According to the authors, elsewhere in Yemen the descendants of al-Hadi (al-Hadawiyyun) had carefully demonstrated their genealogy and there was little doubt about their background, but the people of Banu Mudaykhah had no knowledge of al-Hadi's social origin. Therefore they could not challenge al-Qasim's claim that he was one of the Imam's descendants. Indeed, so the authors insinuate, al-Hadi never had a descendant whose name was al-Rashid, a name which is uncommon in Yemen. The authors go on to explain that it has been argued that Muhammad b. 'Ali b. al-Rashid came to the area with the Turkish army, but that his origin remains unclear. His goal in Yemen was to exercise power. Having come to know Yemeni society, he realized that he could rule only if he claimed to be a *sayyid* and conducted himself in a virtuous manner.

These arguments are put forward by the authors in order to show that Imam al-Qasim, who traces his descent from Muhammad b. 'Ali b. al-Rashid, was not a descendant of Imam al-Hadi. They argue, furthermore, that the descendants of al-Hadi and al-Rashid looked different from each other. Unlike al-Hadi, al-Qasim had Turkish and Persian features, displaying a broad forehead, big eyes, and a wide beard. The authors also allege that before he became Imam, al-Qasim claimed that he had detected a false messiah who could only be seen by him. Therefore, al-Qasim tried to mislead the people. If someone makes such claims, so the authors conclude, he can claim anything (Qasim Ghalib Ahmad et al. 1983: 21-7).

The book's aim is to cast doubt on the claims of several important Imams to 'Alid descent. While the authors suggest that al-Hadi's offspring had convincingly demonstrated their genealogy, they still insist that it was uncertain. They wonder why Imam al-Qasim claimed descent from al-Hadi for there were even doubts about the latter's origin. Several writers had alleged that while Yahya b. al-Husayn was an authentic figure who belonged to the Prophet's progeny, the first Yemeni Imam had only claimed to be that man (Qasim Ghalib Ahmad et al. 1983: 21). During the present period, the authors' assertion that Imam al-Qasim and his progeny, among them the Hamid al-Din, were not genuine *sadah* is of crucial significance. It renders their rule invalid and raises questions as to why Yemenis ever accepted it. Because it is argued that even al-Hadi's *nasab* was uncertain, it is implied that the majority of the Yemeni *sadah* had no genuine claims to membership of the Prophet's House.

A question which was often raised in conversation with me was whether the Prophet could have had as many descendants as there are in the Muslim world today. The son of a schoolteacher who had become a diplomat argued that "Fatima was not a chicken farm." He denied that there were any *sadah* in the Yemen. Once, as we were talking about his neighbor, an 'Alid woman, he said "this *sharifah* will go to heaven and meet Fatima who will say 'this is not one of my children.' "

When in the mid-1980s a Ta'izzi family was said to have made false claims to 'Alid origin, an 'Alid friend of mine reacted with a sense of bitterness and irony. The daughter of this family had married a wealthy man in Sa'dah. One of his friends, who carried the same patronym as his new affines, failed to trace their line in his genealogical atlas. Rumors spread that their claims were false and that indeed they were *mazayinah*. Under pressure from his father, the groom divorced the girl but took her back shortly thereafter because she was pregnant. My friend commented, "If they want to be *sadah* in times like this, we should not stop them."

A minister's secretary, who often praised his teachers who claim descent from the Prophet, and argued that there was no reason to discriminate against the *sadah* merely because some had made mistakes, nevertheless reproduced common anti-'Alid jokes like the following:

> The *sadah* think of themselves as higher than others because they are of Fatima's progeny (*awlad Fatima*). Just imagine what they would say if they were descendants of the Prophet (*awlad al-nabi*)?[9]

When told at male gatherings where no *sayyid* is present, jokes like that, somewhat similar to Mediterranean nicknames, are used "to mortify, patronise, criticise or humiliate . . . The unavoidable impression is one of a sadistic joke secretly and lovingly savoured" (Gilmore 1982: 693–4). The jokes provoke smiles and even roars of laughter, and sometimes the occasion is used to ridicule a high-ranking *sayyid* whom most of those present know. The intention always is to insinuate the futility of making claims to status based on alleged blood ties with the Prophet. According to a shaykh whose sister was married to Imam Yahya,

> Yazid killed all descendants of 'Ali. Even al-Husayn's pregnant wife was harassed by Yazid's army. Those who claim to be al-Husayn's offspring

came to Yemen from all sorts of places. They were no more than a group (*fi'ah*) with its own political aspirations (*hizb*).

There are, however, tribal leaders who argue that during the Imamate respect for the *sadah* stemmed from religious considerations.

> Respect for the *sadah* was based on religion. A common *sayyid* ranked higher than a learned *qadi* (*qadi 'alim*). (A shaykh from the Jawf talking about the past)

In 1988 the Iraqi president Saddam Husayn visited Yemen. Several Yemeni dignitaries and officers accompanied him and his entourage to the airport. A member of the Iraqi delegation asked one of the Yemeni shaykhs whether the last rulers [the Hamid al-Din] came from a good family. "Yes," he told him, "in fact they are from the progeny of Muhammad who are very respected in our country." One of the Yemeni officers, slightly annoyed, remarked that there were no *sadah* in Yemen because they had all been slaughtered by Mu'awiyah. He was advised by the Iraqi official that some had been spared because his president was a *sayyid*.[10] After telling me about the incident the shaykh, who had fought the royalists in the civil war, raised the rhetorical question "Saddam is an infidel (*kafir*), so how can he be a descendant of the Prophet?" By explaining to the Iraqi official that the Imams "came from a good family" because they were descendants of the Prophet, he admitted to his conviction that claims to high status and moral authority are based on descent. In his eyes an infidel like the Iraqi president could not belong to Muhammad's progeny. A descendant of the Prophet must be a devotee.

Cases: Muhammad and Hamid

Unlike some of his counterparts, the shaykh who accompanied Saddam Husayn to the airport did not dispute the *sadah*'s descent from the Prophet. Nonetheless, the majority of shaykhs dispute the *sadah*'s right to rule.[11] Some depict them as colonizers who capitalized on the ignorance of the tribes. This discourse tends to be anti-Shi'i and is prevalent among people who were brought up in Zaydi areas but have turned to those legal schools which recognize no authority other than the Qur'an and the Sunna. By the time the United Reform Party (*al-tajammu' al-yamani li-'l-islah*) became one of the official political parties in 1991, Sunni Islamists had already propagated their message in several rural areas where Zaydi 'ulama had had strong influence.[12] In some parts of the country, the Wahhabi branch within the party has gained a considerable number of followers. In parochial local settings, Islah's role as a unified national party is not always reflected. As Weir (1997) has shown, in Jabal Razih Islah and the Wahhabiyyah (*muwahhidun*; Wahhabi school of thought) are virtually synonymous.[13] There is a notable correlation between ideological attachments and status affiliation. In the 1980s, those who became susceptible to Wahhabi ideas were young men of tribal affiliation and of lowly birth; some shaykhs also tacitly approved. In Razih, the *sadah* used to be the local governors and continued to hold positions in the administration during the first decades after the revolution. Since the late 1980s the influx of Wahhabi ideas has contributed to disputing the legitimacy of the former holders of power. In chapter 7 I referred to the song Razihi men sang on the occasion of the first wedding of an 'Alid woman to a non-'Alid, challenging the validity of the *taqlid ahl al-bayt*. By invoking a rival religious tradition, which devalues those traditions promulgated by the *sadah*, they produce counter-memories (Wachtel 1990: 12) aiming at the oblivion of others. In San'a, I never heard such bold and public challenges to the authority of the 'ulama. Displeasure with their ideas was not expressed in their presence. In one case, an elderly, well-respected Zaydi scholar explained at a *maqyal* that irrespective of a man's heredity, in the here-after the one most pious would be closest to the Prophet.

> According to a hadith, the Prophet cannot do a thing for his progeny unless they are good people. If the *ahl al-bayt* do as well as others, they will be better; that's what the Prophet had hoped for. Then others will take them as an example and do well in society. It is only for this reason that a *sayyid* will be rewarded in paradise. If he neglects his duties, it will not help him being from the Prophet's Family.

Later on, when I talked about the afternoon's discussion with an engineer with Sunni leanings, he brushed aside the scholar's speech and smiled bemusedly. "Where do they [the 'ulama] take it from?" he asked. "The Prophet's last sermon was addressed to all Muslims. He did not single out anyone as his successor." Such criticism of descent-based authority was also voiced among *qaba'il* who have recently adopted Wahhabi ideas. I met Muhammad, a young tribesman in his twenties, in the surroundings of the town of Sa'dah where Sunni Islamists have been particularly active. He comes from a family loyal to the Imams but has been exposed to propaganda against them since childhood. Since a Sunni-oriented religious institute was built in Sa'dah, he came under the sway of the Sunni reform movement and renounced the Zaydiyyah.

In our area, most *sadah* worked for the government. They always stressed their religious status and the religious nature of the regime. People thought that if they did not respect the *sadah* they would not enter Paradise and would be punished in the afterlife. The *qaba'il* insisted on kissing the *sadah*'s knees. They thought that by doing so they were pleasing the Prophet. I was told I should be obedient to the Imams because they, too, were the "sons" of the Prophet. The Imam was the legitimate ruler because he was from the *ahl al-bayt* and because he was religious. My parents and my grandmother still think that way. They are nostalgic about the time of the Imam because there was security. People were fighting less than today. There was always fighting but the Imams managed to contain it. My grandmother used to say "If there is rain and the land is fertile, there is an *imam 'adl* (a just Imam)." My father and my brothers think that the *sadah* must be respected because they descended from the Prophet. They say they are blessed people. Sometimes they quote a hadith—I cannot recall which one—that says that one is only safe if one treats the *sadah* well. They feel insecure since the Wahhabi movement has gained strength. I feel sorry for them because they are committed Zaydis. There will be difficult times ahead for them. They were taught that in the absence of the *imam 'adl* their prayers are invalid and that they cannot pay their zakah. Can I only be a dutiful Muslim if there is a *sayyid* who rules? People take soil from the graves of the great 'ulama and rub it against their bodies [in order to receive a blessing and to be cured].[14] This I find unacceptable.

Muhammad reckons his father's commitment to Zaydi ideas to be an anachronism, a kind of defensive reaction rather than a basis for a meaningful selfhood. As he was speaking, he suddenly turned his head as if looking for somebody, and asked with a slightly cynical undertone "Where's the *sayyid* so that I can kiss his knee?"[15] He continued to talk about his resentments toward the *sadah* and his encounters with 'Alid women.

> Once I passed a *sharifah* who returned to her house after she had just locked the door. Obviously she had forgotten something. I overheard her saying "*Allah salli 'ala jaddi Muhammad*" (May God bless my ancestor Muhammad). [She invoked God to help her concentrate better]. That moment I felt belittled.

He spoke to 'Alid women for the first time when he attended the university. One of these women, a San'ani, did not cover her face and was happy to have conversations with him. Sometimes he teased her by saying "We ask you *ahl al-bayt* for a blessing." When he found her approachable and unpretentious, he concluded that she was not a "real" *sharifah*. He wondered whether she would have accepted him as a husband had she not been married. He went on telling me about the reasons for the success of the Wahhabiyyah in spreading among some young tribesmen and their leaders.

Islah is now well represented in Sa'dah. The government supports the party because it promises to overcome the old social divisions. The government wants the country to be unified, and this is the only way forward. It is not surprising that many people convert to the Wahhabiyyah. Several shaykhs are Wahhabis. They have power and people follow them. Just remember that the Yemenis became Muslims overnight; they quickly adopt new ideas. The Wahhabis challenge Shi'i beliefs.

He paused in order to think of an example.

They ask how the unbeliever whose leg was cut off by 'Ali b. Abi Talib could fight on and kill seven people and three camels with his leg in his hand ['Amru b. 'Abd Wudd]. People are not really attracted to the Wahhabis, but it is good that they stress that all Muslims are equal. The Wahhabis do not accept anyone's elevated position based on his ancestry. They are only there to break the power of the *sadah*. The *sadah* do not have much power now, but people still have certain ideas about them. We want to get the *sadah* out of our minds. People no longer want to be obliged to have respect for certain kinds of people. Whenever I discuss these matters with the *sadah* and tell them that no one should claim a distinct status for himself, they say "we were born *sadah*, what can we do?"

Muhammad drew a picture of the *sadah* as imperious, ambitious, conscious of their status, and dedicated to religion. He thought of the *sadah* as being reserved, explaining that they resented physical contact in public even among men. He wondered how I proceeded with my research. "Isn't it difficult to meet them?" he would ask. He believed that the *sadah* would always publicly endorse Islam, and were easily recognizable because they would often refer to their identity. Politicians such as 'Ali Salim al-Bid confused him because he did not allude to his 'Alid descent. One of Muhammad's friends called Hamud, who had left his tribe, the Dhu Husayn [in the eastern provinces] in the 1970s in order to become a government employee in the capital, was not troubled by socialist inclinations among *sadah* like al-Bid. "Al-Bid is a Bedouin *sayyid* like the *ashraf*. The *ashraf* are like us, they are not from the city." He implied that al-Bid had grown up in an uncorrupted place, the Hadramawt, his home province, and therefore was different from other *sadah*.

For men like Muhammad the image of *sadah* who disavow the stereotypical symbols of 'Alid identity is an uncomfortable one because they can be less easily denounced. Those *sadah* are rendered "unauthorised others" (Gell 1998: 141), crypto-*sadah* who have renounced their forefathers' ambitions. *Sadah* who disapprove of the notion of 'Alid supremacy and have been loyal civil servants of the republican government are referred to as *sadah haqq al-mizan*. The *mizan* is the place where cloth is sold cheaply by the meter. The *sayyid haqq al-mizan* is the mirror image of the *sayyid* who is referred to as *hashimi sirf* ("pure Hashimite"), an uncompromising adherent of the ideology of Hashimiyyah.

Like Muhammad, Hamid, a shaykh in his mid-sixties (see chapter 7), was troubled by the idea that some *sadah* adhere to secular ideologies. Hamid's family had been loyal to the Hamid al-Din, and he spoke of Imam Ahmad with admiration. During the civil war, he had fought the Egyptian "infidels" in the name of the Zaydiyyah; now he disclaimed its legitimacy and subscribed to anti-Zaydi ideologies. He viewed the history of the Imamate in terms of the unjust subjugation of the indigenous population.

Since 'Ali b. Abi Talib's death, there has not been any *sayyid*. Those who came after 'Ali were all ordinary people. Just see what they are like—they are not stable people. Some of them are socialists and Ba'thists. The Imams monopolized everything, the people were only slaves. Under the Imams, people were not at ease (*mutadayiq*). You want to know about the 'ulama? The tribes always co-operated with them and defended them. But the tribes were always stronger than the 'ulama, there was never a real balance of power. The 'ulama needed the tribes for their protection. The *qaba'il* benefited religiously from the 'ulama. They [the *qaba'il*] had no understanding of Zaydi tenets; they followed the Imam because he was of good progeny. It did not matter to them whether the Imam was a Zaydi; what was important was that he was 'Alawi-Fatimi. Hamdan [one of the tribal confederations] were the first who adopted the Zaydiyyah because they were divided. Later the Zaydiyyah became politicized, it lost its religious cause. There was no need for mediators from the House of the Prophet. Any respected man can be a mediator; he might even belong to one of the parties at odds. Those who really benefited from Zaydi rule were the *qudah*. They became powerful under Imamate rule. They served the Zaydi Imams, and they always favored this or the other side [of rival Imams] by making tactical alliances.

I do not want to be told that I am only a good Muslim if a *sayyid* is in power, or that he must rule because he is a *sayyid*. Now we [the *qaba'il*] have the power (*quwwah*), the arms, and the elections (*intikhabat*) [i.e. democracy was introduced under the rule of the *qaba'il*]. In the final analysis, all is in the hands of the shaykhs—they have the authority. A scholar (*'alim*) might teach me about religion, but he should be in the mosque and not exercise political power. In the past, the *qaba'il* were ignorant (*jahilun*). They thought that a *sayyid* must rule over them because he is from the Prophet's House. The *sadah* say that they go to Paradise because they are *ahl al-bayt* and we are not, and that they are infallible (*ma'sum*) because the Prophet was infallible. They tell us that it is improper (*'ayb*) to marry foreigners, but they themselves do so. We sacrificed our blood [in fighting on their behalf] and our women [by giving them in marriage], but we never got anything in return. Tell me, how can a *sayyid* be a socialist and tell people there is no religion? [He was referring to Yahya al-Shami, a member of the Politburo of the Socialist Party]. All legal schools (sing. *madhhab*) are concerned solely with political objectives. The *sadah*'s religious mission [in the ninth century] was only a camouflage for their endeavor to subjugate us to their rule. We were nothing but tools in their power struggles. I no longer want to know about their *madhhab* [*madhhab ahl al-bayt*]. For me, there are only the teachings of Muhammad 'Abdullah. The Sunna is the first and the last thing.

His eldest son had been listening to the conversation. As if to confirm his father's words, he said "Yes, a *sayyid* who is a socialist cannot be a proper *sayyid* (*hashimi haqiqi*)." Referring to his father's original commitment to the Imam's cause, he explained: "My father went to war against the republicans because a *qarawi* [person of inferior status] was in power [President 'Abdullah al-Sallal].[16] I would rather be ruled by a tribesman or a *qarawi* than by a *sayyid*." Similar views were expressed by *qaba'il* like Mahmud (see chapter 7).

The *qaba'il* didn't want to be lower than anyone. Those *sadah* who lived among us and were poor farmers had little respect unless they became knowledgeable. The *sadah* taught us about the Prophet and his Family; they created a link between Allah and themselves. They exploited our ignorance. The Imam asked them to do that. There is nothing in the scriptures that says that the *sadah* should rule. [In the past] rival Imams were fighting each other, and the *qaba'il* fought on either side. The victorious ones settled among the wretched who were forced to sell their land at low prices.

According to Mahmud, the *qaba'il* have a legitimate claim to power and he credits them with introducing democracy to Yemen. The shaykh asserts the right of men like himself to rule by virtue of their descent from Himyarite leaders and long-established historical precedent. Like Mahmud, he feels he is second to none in dignity and prestige. The decline of the Zaydiyyah and the rise to prominence of rival religious movements has offered new perspectives for the *qaba'il*. Resurgent nativism invokes the legacy of pre-Islamic glory and is cast in anti-Shi'i rhetoric. Hamid and Mahmud contest the standard interpretations of Zaydi historiography as a mission to implement just rule. Hamid concedes that the early Zaydis were welcomed in Yemen because they were descendants of the Prophet and played a significant role as mediators between parties at odds. However, he claims that they had purely political objectives, and the establishment of the Imamate resulted in their domination over the indigenous population. In his eyes, the *qaba'il* soon realized that moral agency and power were divorced from each other. Zaydi historiographers had often depicted the *qaba'il* as infidels who neglect their religious duties. By arguing that the Zaydi rulers had not been guided by religious principles, the shaykh turned the tables on them.

Zaydi discourse is reworked in order to fashion a critique of the Imamate and to reject the attribution of moral inferiority. However, there is an unresolved tension between these men's challenge to birthright and their endorsement of the notion that a person's heredity must determine his conduct. For example, Shaykh Hamid challenges President Sallal's right to rule because he was born into a family of blacksmiths. Muhammad, the young *qabili* from Sa'dah, desires to marry a *sharifah* yet would be offended by a butcher's proposal to his sister. Hamid disputes the *sadah*'s special status but argues that their descent obliges them to devote themselves to their religion. Neither the *sadah* who in the past claimed the leadership in the name of religion nor contemporary ones, who are leaders of secular parties, are afforded legitimacy. In the shaykh's view, by adhering to ideologies which discourage religion, socialists of 'Alid descent fail to respect the moral imperatives associated with the status they acquired at birth. Thus, *sadah* who aspire to cross boundaries by adopting ideologies and practices incompatible with classical 'Alid stereotypes, are dismissed rather than approved of. They face a predicament: if they act in accordance with the ideology of Hashimiyyah, they are suspected of affinities to the ancien régime. Those who no longer identify with that tradition are depicted as mediocre people (*haqq al-mizan*). Hamid and Muhammad sympathize with a political party which discourages hereditary social divisions, but are themselves caught up within the very ideological system they dispute. Their anti-'Alid rhetoric reveals that what they cannot afford to admit about themselves they personify externally as the "sayyid." While paying lip service to the principles propagated by Islah, they still hold descent to be an important attribute of the person. Since the overthrow of the Imamate, descent has continued to be politicized, albeit under different premises. For people like Muhammad and Hamid, being Yemeni means being of pure South Arabian stock, yet being purer than those of lowly birth. Because they cannot claim that their descent is superior to 'Alid descent, they must either deny the *sadah*'s genealogical credentials or assert that they had failed to honor their ancestors.

Liaisons and Moral Authority

The stereotyping of the *sadah* which has become a catalyst for debates about history, legitimate rule, and moral authority, has a predominantly androcentric focus. However, through aspersions against their propriety, women have been made into tokens of male competition over the moral high ground. Their bodies have become political signs like in the rhetoric of motherhood (chapter 7).

> Many *shara'if* are married to their cousins. They are not given a choice. There is no erotic love in those marriages. This is why the *shara'if* are inclined to enter into liaisons with us. I would not mind having a relationship with them but would never marry them. (A businessman)
> The *sadah* are unwilling to punish their women regardless of what they do. (His cousin, a diplomat)

Chapter 6 argued that the Imamic social order was secured through forms of sexual control which discouraged marriages with non-'Alids. The latter, in turn, contest previous power relations by taking 'Alid women in marriage; and some defy images of those women as the paragons of moral purity. Those men who claim that they would deliberately neglect to obtain religious sanction for their relationships with them act impudently toward 'Alid men. Historically, concubinage demonstrated the strength of rulers who could afford to forego permanent marriage contracts (*nikah*). A man's sexual conquest of a woman without benefit of marriage is a severe assault on her family's honor; he also denies potential wife-givers the respect that affinity ideally entails. The men quoted above have obtained university degrees abroad and belong to the affluent new elite. They declare their newly acquired status by dismissing the importance of marriage alliances with the *sadah*, and by refusing to accept the obligations of affines who are expected to afford each other respect and affection.

The diplomat represents the *sadah* as holding values which are not shared by society at large. He insinuates that 'Alid men's failure to reprimand the Prophet's daughters is the reason for these women's alleged moral aberration claimed by his cousin. The image of the indecent *sharifah* becomes a metaphor for the decline of the *sadah* collectively. In anti-'Alid rhetoric her body is made a trophy of Qahtani supremacy. The virility which is asserted by men like the businessman becomes an idiom of national renewal in the wake of elite transformation. This masculine discourse, and the prominent display of the *janbiyyah* as an ubiquitous national symbol, is suggestive of the desire to emasculate and triumph over those who, rather than representing an effeminate class of religious specialists, claimed a monopoly on legitimate violence.[17] The diplomat's statement also insinuates that the *sadah*'s power had diminished to the extent that they had even lost control over their women. They are implicitly likened to men of lowly birth who have lost their honor and whose women are represented as unconstrained.[18] Some prosperous non-'Alid men project images of the *sadah* and those labeled *mazayinah* as effeminate and unable to exercise authority over those on whom much of the family honor depends. These stereotypes serve to lower both—the *sadah* and the social climbers among the *mazayinah* who claim their share of power. The *sadah*'s adversaries can represent 'Alid women in this way because they know that social mechanisms that might discourage these men's behavior no longer exist. The imaginary abolition of prohibition is the expression of the *sadah*'s defeat, a defeat which however has proved to be not quite as irremediable as these men would have hoped.

* * *

This chapter analyzed stereotypes of the *sadah* in the context of state transformation. It is noteworthy that anti-'Alid discourse has continued to be pervasive *after* the overthrow of the Imamate. This discourse constitutes a denial of former compliance with 'Alid prerogatives and voluntary subjection to their rule. Stereotyping of the *sadah* is but one dimension of the political mobilization against those accused of sympathies for the Imamate. Since the old elite has successfully established itself in various fields of secular knowledge, economic interests are also at stake. Among well-to-do and upwardly mobile non-'Alids, the category label "sayyid" has taken on predominantly negative connotations, but questions remain about the extent to which stereotyping conceals alternative views and is motivated by the expectations of their peers to join the defamatory chorus. The denial of the truth value of 'Alid genealogies points to a rationale of stereotyping fostered by the bothersome belief that others possess qualities they themselves cannot obtain. This preoccupation with genuineness reveals that descent ideology has remained meaningful to their lives, a fact which their flirtations with anti-Shi'i teachings cannot conceal. Given the reinvention and continuation of dynastic rule in Arab monarchies and republics, this is as significant as it is ironic. In spite of the doctrinal emphasis on descent-based authority, the politics of the Imamate was, for the most part, not determined by the dynastic principle which has come back into favor in several republics.

Chapter 11 examines how the *sadah* come to terms with the images others hold of them. Past experiences are reworked and channeled into new self-fashionings (Fischer 1991: 25) which to an extent are shaped by the expectations and prejudices of others.

Chapter 11

Memory, Trauma, Self-Identification

To begin with, there must be a will to remember.

Pierre Nora

'Alid identity which is in significant ways constituted through a genealogical history, is at the same time the site of uncertainty "where the dynamic of identification is at play" (J. Butler 1993: 143). Even while the majority of *sadah* have erased visible differences between themselves and others, and have abandoned notions of divinely sanctioned rule, prejudice against them is only gradually diminishing. The *sadah*'s pride in their descent from the founder of Islam is tempered by anti-'Alid stereotypes which portray them as people guided merely by political ambitions. Having considered themselves Yemeni for centuries, in 1962 they learnt that they were not, or at least less so than other Yemenis. The very spectrum of identifications, it would appear, has become severely limited.

Presenting fragments of life stories, this chapter examines how individual *sadah* who were brought up as a Muslim elite come to terms with the stigmatization they have experienced since the 1960s. Their stories speak of their struggle to cope with prejudice and of a deeply felt sense of insecurity. They have come to understand that escaping the constraints of their identity is illusory not least because, as I was once told by a friend, "as a *sayyid* you know that you are always looked at in a certain way." The narratives demonstrate these people's quest for full recognition as republican citizens and their desire to be accepted as *sadah*. They are caught between their allegiance to their distinct identity, both desired and scrutinized, and the premises of Yemeni citizenship, which discourage descent-based particularist identities.

The composition of life stories is shaped by the predicaments in which the narrators find themselves at the moment (Rosenwald and

Ochberg 1992: 6). Following on from the themes of chapter 4, life stories help us understand how the ideology of descent is individually appropriated in the construction of the self; and how the experience of being of 'Alid descent is molded by this ideology as well as by material conditions. Thus Qasim al-Taghi, a man of poor background whose father took part in the revolution and whose brother married into a once preeminent house, feels socially distant from both high ranking *sadah* and non-*sadah*. His deprived childhood and the stigmatization he suffered after the revolution have produced a refracted identity, constituted in ambivalent displacement and belonging. This condition, it must be kept in mind, is not entirely contradistinctive to the past. Qasim's case highlights the range of possible self-definitions within the category "sayyid" and the potential of social location to shape remembering. He was not brought up to conduct himself like his peers from more august backgrounds, nor does he believe that kinship with the Prophet obliges him to be pious. He repudiates religion as an essential affirmative signifier of self-identity and subverts idealtypical notions of "sayyid" as discussed in chapter 5. Beyond these considerations, Qasim's story also reveals a certain continuity between the pre-and post-revolutionary period. The demeaning experience as a low ranking *sayyid* in some ways echoes that of later days during which he was associated with a vilified regime.

However, socio-economic status does not correlate with religious orientations in any straightforward way. 'Abd al-Karim al-Hani, who comes from one of the most preeminent houses, does not have pretensions to piety either. He is self-employed and cultivates cosmopolitan lifestyles; he enjoys travels abroad and has sent his sons to foreign universities. He no longer wishes to adhere to the rulings of the Imams and is critical of his kin who oppose the government. This view is shared by his older relative Murtada, a government employee, who maintains that the *sadah* must seek people's esteem through exemplary moral conduct. Murtada, who approves of the Imamate as a concept, has been taught painful lessons by the rise and fall of his father's and mother's house. These houses both masterminded and endured the major upheavals of twentieth-century Yemen.

One of the strategies aimed at detachment from this historical location of the self is relinquishing first names and patronymics which conjure memories of ancestors who are linked with Imamate rule. The cultural production of names and naming has figured in formulating an ethical discourse for recentering subjects (Battaglia 1999: 124). This chapter documents several cases where this goal is pursued through either name dropping or name recovering. The *sadah* wish to distance themselves from what the name represents, whereas those *qudah* who recover names linked to pre-Islamic glory, reclaim a time and memory that speak of the rule of their forebears.

Bayt al-Hani: 'Abd al-Karim and Murtada

'Abd al-Karim's father worked for Imam Yahya's government. He and other relatives took part in the Constitutional Movement and were executed after Imam Yahya's son recaptured the capital. One of 'Abd al-Karim's lasting childhood memories is of passing the wall where the heads of the slain men were displayed. He and his older brothers were sent to prison, where 'Abd al-Karim became so emaciated that he was able to escape through a small toilet window. The following years were marked by misery and poverty until Imam Ahmad agreed to support his education at a high school in Cairo. After the 1962 revolution 'Abd al-Karim took up government employment for a brief period and then studied economics abroad; thereafter he established himself as a businessman. Most of his relatives, including his brothers, are employed by the government; others dispute its legitimacy and have joined opposition parties. 'Abd al-Karim takes a critical view of the rule of the Imams and has little sympathy for his kin in the opposition. "The Imamate was a rule which discriminated against others. We made others suffer for one thousand and two-hundred years." 'Abd al-Karim has concluded that his relatives' participation in the Constitutional Movement only harmed his family, and that the *sadah* should not therefore engage in politics. "These things belong to the past. Our fathers died for what we have now. Our relatives who are in the opposition only cause trouble for the *sadah*. They speak on our behalf but they only make us suffer. If they came to power and things would go wrong, I would be the first to be killed." His cousin was even more outspoken:

"they destroy the Hashimite families. People do not want Hashimite rule. They [his kin] provide those of our enemies who argue that we should not hold high positions with ammunition. If they took power I would kill them before the republicans would. It is better to mix with the people and to have good jobs. Those who still dream about the *imama* are nothing now, they remember only what they were before. What matters is that we [the *sadah*] offer something to the people."

'Abd al-Karim accepts the authority of the Qur'an but is not prepared to recognize specific Zaydi-Shi'i prohibitions. When during a lunch at an ambassador's house lobster was being served, he was reminded by a scholar that this food is *haram* for the followers of the *madhhab ahl al-bayt* (Shi'i doctrine). 'Abd al-Karim refused to take notice and enjoyed the lobster. He commented later on the incident. "The Qur'an says 'eat everything between the earth and the sky as long as it is not forbidden.' Only blood, pork and wine must not be consumed." He also dismissed the legal judgments of individual Shi'i authorities, arguing with a slightly cynical undertone that these rulings were motivated by pragmatic reasoning. When he bought himself a pair of trademark spectacles, I teased him by pointing out that according to several 'ulama of Sa'dah and Imami-Shi'i authorities, men must not wear golden spectacle frames.[1] He replied saying that "people who make rules like these are those who cannot afford nice things." On other occasions, he adhered to traditional views, asserting that heredity was a person's destiny. Once I complimented him on doing so well in his job without having enjoyed any patronage or practical training. "It's in the blood," he said. "But none of your forefathers was a merchant." "Who told you? The Prophet was a merchant. He took a commission from Khadija [his first wife] and invested her money. She was older than him, but he married her. He swapped love for commissions."

. . . and Murtada

Murtada has had regular encounters with the outside world since the 1950s when he studied in Egypt. He has been employed by the Yemeni Foreign Office since the reconciliation between royalists and republicans in 1970, and has been stationed abroad for most of his life. On one of his visits home, he told me about his views toward Zaydi history and what being a *sayyid* is all about.[2]

A *sayyid* is nothing more than any other human being, that is, unless he has certain qualifications. It is not a matter of "blue blood" at all. Salman Farisi who was a Persian slave [the loyal companion of the Prophet and 'Ali who refused to give allegiance to Abu Bakr] was more respected than Abu Lahab, the uncle of the Prophet. It is not a matter of being a *sayyid* as such, unless he follows certain principles. Then, together with his social status [deriving from descent], he will be liked and respected especially among the tribes. The influence of the *sadah* among the tribes is fast disappearing. There was a certain time when the *sadah* were granted privileges; now they themselves are giving up these privileges. They mix with the people and live like any non-*sayyid* citizen. Their privileges created jealousy and that is why I am not in favor of the continuation of that tradition. What the *sadah* have suffered in the last thirty to forty years is the result of that tradition because they are no longer like their ancestors. Their ancestors enjoyed these privileges because they stuck very strictly to the rules of Islam. The *sadah* violated the rules through tyranny and bad behavior. I am not saying that all Imams were tyrants, but there were some.

The condition that the Imam must be from the progeny of Muhammad [according to Zaydi-Hadawi doctrine] made others feel resentful. As a *sayyid* you know that a great number of non-*sadah* hold a certain view toward a *sayyid* just because he is a *sayyid*. That is because of the condition that the Imam must be a *sayyid*. So any *sayyid* is disliked by this extremist feeling of the anti-*sayyid*. We are lucky that this feeling is now diminishing. I was here [in San'a] when the revolution broke out, and I was thrown into prison. So being a member of a family who opposed the Hamid al-Din did not protect me. I was in prison for several years for nothing, and without being tried. After that, the Red Cross arranged an exchange of prisoners between the royalists and the Egyptians and I was set free.

I am concerned with the quality of a regime, the name itself means very little. As a *sayyid* I would say that I adhere to the republic, and this is because it relieves all *sadah* from bearing the responsibility [of holding power]. If the Imamate will ever be re-established—of course, if it is Zaydi, the Imam must be a *sayyid*—who will guarantee that this rule will be the best? No one, and we have experienced this. So whenever an Imam is acting unlawfully, all *sadah* will pay for it. Therefore it is better not to have an Imamate. For me the end of the Imamate is a great relief—and you know the history of my family. We have experienced bloodshed, confiscation, imprisonment, exile; we have had enough. So who is really foolish enough to believe in the restoration of the Imamate? Not because the Imamate is a bad system, but the system is one thing and the rulers are something else. It is not that I hate the rule of the Imamate, but I do not think that it benefits the *sadah* in Yemen. We have had enough. If you look at the history of it you find the good and the bad. What remains in the public mind are the bad things, and we have paid for it. So relieve yourself and your sons and grandsons, and do not expose them to any danger which may result in what I and my forefathers have endured.

Bayt Al-Taghi: Hasan and Qasim

Hasan was brought up in a learned family in a northern hijrah. He came to live in San'a in the 1970s and read law at the university. He speaks about the Zaydis' duty to believe faithfully in the fourteen principles but also acknowledges that they are discriminatory. He is nonetheless committed to the notion that a non-'Alid should rule only if no suitable *sayyid* can be found.

A real *Hashimi* [*hashimi qaqiqi*] is tenacious in his beliefs (*mutamassik*), he stands up for himself and declares "I am a Hashimi, and it honors me." He should not be afraid to admit that he is a Hashimi. It is important for a Hashimi to be a man of religion and to teach his children.

Today the Zaydi principles are only in books; they cannot be implemented. The fourteen principles are from the days of Zayd b. 'Ali. 'Ali
b. Abi Talib did not rule that his descendants be his successors. There
is some discrimination (*'unsuriyyah*) and partisanship (*'asabiyyah*) in
the fourteen principles.[3] It is the duty of a Zaydi to believe in them but
if there is no power (*quwwah*), he can do nothing. Al-Husayn b. 'Ali
followed his duty and was killed; al-Hasan [his brother] stayed at home.
They were much better 'ulama than Mu'awiyah who was a man of politics not of religion.

The problems the Hashimites face derive from the fourteen principles, throughout history from 'Ali until today. Because of these principles they have been persecuted and killed. 'Ali was the righteous leader
after the Prophet. If there is no suitable Hashimi for the leadership
[*khilafah*], someone else with good qualities should be chosen. Even
within the Zaydiyyah some people say anyone from Quraysh [the
Prophet's tribe] may be chosen for leadership. However, the
Hashimites are more suitable for guiding the Muslim community
(*ummah*) because they are devoted to their religion [*din Allah*]. They
have produced many 'ulama and Imams.

. . . and Qasim

Qasim's family belonged to the less fortunate of Bayt al-Taghi. They
trace descent to a famous Imam who ruled in the seventeenth century
but in the last two centuries, none of his family was eligible to compete for the supreme office because they no longer fulfilled the
requirements of excellent scholarship. The absence of wealth did not
exclude people from the higher ranks of the religious elite; but lack of
education and poverty reflected broad divisions in Imamate society,
underscoring the social heterogeneity of the category "sayyid."
People like Bayt al-Taghi experienced the cleavages of descent-based
dominance and internal social distinctions differently from other
sadah because of their ambiguous positions, as both subordinates
in the hierarchy and—by reason of their kinship with the ruler—as
representatives of the Imamate in their own right. Qasim's father's
perception of the *sadah* who held the reigns of power was not funda-
mentally different from that of their enemies who contested Imamate
rule. In spite of his participation in the revolution and his subsequent
assumption of a senior position in the republican government, he suffered as much discrimination as the Imam's officials. Those underprivileged *sadah*, who were somewhat less masculinized than their
kinsmen and had risked their lives for the republic,[4] have come to realize that in the aftermath of the revolution no distinctions were made

between them and the Imams' loyal servants. They both identify with and feel disillusioned with the revolution. During the 1950s, Qasim's father questioned his loyalty to a state which purported to represent the *sadah* but had little to offer to men like him. Collaboration with other revolutionary officers meant that he abandoned the ideology of Hashimiyyah and risked denunciation by other *sadah* without gaining full recognition as a republican in the newly created state. Therefore, he dissociates the memory of the revolution from the republican regime by which he feels betrayed. The experience of debasement as a low ranking *sayyid* indeed echoes that of later days during which he was identified with a vilified regime.

Unlike some boys from learned houses who thought about becoming Imams, Qasim never cultivated such dreams. When he was still young, he experienced prejudice by his schoolmates and still feels ambivalent about his identity even today. However, it would be wrong to assume that his identity resides in nothing more than *not being Qahtani*. Describing Qasim's self-perception, Martha Robert's verdict on Kafka who in her eyes "was Jewish even in the way of not being Jewish" comes to mind (cited in Bauman 1991: 182). Qasim held the view that the *sadah* were different from other people by virtue of their history and their exclusion from certain occupations, but not that they should be granted any privileges. Qasim spoke in favor of social advancement through education and dismissed the salience of heredity as a status credential. He argued that the *mazayinah* who had become educated and obtained good positions should not be denied social recognition.

> When I was a child, my parents never talked to me about our virtues as *sadah* or the history of the *ahl al-bayt*. We have a documented genealogy but it was never shown to me. I and my brothers learnt who we are through other children who called us "sayyid" in a derisive way. When I attended the elementary school in the 1960s, the other boys informed me that they belonged to *qabilat* so-and-so. This made me wonder who I belonged to. Our teachers told us about the Imams' struggle against the colonizers, but most of the time they portrayed the Imams as evil. They told us that the revolution would lead the nation into a bright future. The students were brought up to hate the Imams. I learned more about my background when I began reading about the history of the *ahl al-bayt* and the Yemen in my twenties.
>
> In those days [after the revolution], many *sadah* changed their names. Titles like 'Imad al-Din ("the pillar of religion") were no longer used.[5] For several years I did not even know that my cousin's [personal] name was Badr al-Din ("the full moon of religion"). I only heard him

being addressed as Badr. A few years after the revolution, my uncle suddenly called him Badr al-Din. I was confused and told him that his name was Badr. Then I understood.

My father supported the republicans. During Nasir's heyday, he listened to the radio programs from Cairo all night, and then talked to people about them when he went to the suq the next day. He stopped wearing the *'imamah* and dressed in a suit. He was seen as a traitor by other *sadah*. I grew up in a revolutionary milieu, and I almost hated the *sadah*. We were very poor; we did not benefit from the system at all. Yet all *sadah* were held responsible for what the Imams did. After the revolution my father was assigned a prestigious post [in one of the provinces where his more fortunate kin had held positions]. However, he became estranged from the new regime when it began to attack the *sadah* collectively. He felt he was living in a hostile environment, and he retired early. When my brother Mansur married the daughter of a royalist, he was delighted, though other members of the family were less approving. It took time until they got accustomed to each other. My brother was not used to addressing senior kin by a title [for example, *sidi*, *mawla*]. He referred to his wife's eighty-two year aunt by her first name. That caused indignation.

We are the *mazayinah* of the *sadah*. Unlike them [the 'Alid *buyut al-'ilm*], we are no longer like our forefathers who were Imams and had great knowledge. We do not know how to recite poetry; we do not address each other in the plural, we do not call each other *ukhti* ("my sister") and *akhi* ("my brother").[6] We just use first names. Our children are called 'Adil and Samir. The great 'Alid houses denounce us because we do not call our children after the *ahl al-bayt*.[7]

After Mansur's wedding, one of his wife's kin told me "they do not share more than blood." Qasim was conscious of the difference between his family and the high ranking *sadah* who had a classical education. Although he and his brother obtained university degrees and have succeeded in their careers, they felt that the 'Alid *buyut al-'ilm* do not recognize them as equals. Mansur, who often appeared careless to his wife, was told by her that "he was a *qabili*" and should not act as a role model for their child.

Qasim shared a sense of insecurity with other *sadah* and resented the prejudice held against them by those representing themselves as Qahtaniyyun.

The others think we should be expelled from the country, that we have no right being here. They think that none of us should be in a good position. My half-brothers are different from me and my brother because their mother is not a *sharifah*. They are more relaxed and happy, they do not worry about life like us. The mother of my sister-in-law is also a *qabiliyyah* [of tribal origin]. My brother always tells his wife

that he enjoys that side of her character because her father's family is
very rigid and she is a bit like that as well.

Because Mansur's son resembles his maternal grandmother, he
teases his wife by telling her that her son is a *qabili*. He thus embarks
on a phantasmatic and illusory self-deconstruction via his child—
partly motivated by his wife's derisive talk. Yet by telling his wife that
her son has become what she fears he might become were he to follow
his father's negligent ways, he also reminds her of her family's status
as nobles who have fallen from grace. However, he is quite aware that
adopting the label *qabili*, carried by the present governing elite with
pride but used by him and his wife as a negative stereotype in their
latently polemic exchanges, would be refused recognition.

Mansur and Qasim's discomfort and ambivalence about their identification with the normative expectations that "sayyid" entails might
also explain why Qasim does not feel obliged to express filial piety in
public.

> I no longer know how to pray. When friends ask me to lead the prayer
> I apologize.

Qasim's wife told me what happened when the couple lived in the
United States and the martyrdom of al-Husayn was commemorated at
the mosques. This ritual is an occasion for expressing personal allegiance
to the Prophet and his Family.

> Of course I was eager to attend and I encouraged my husband to join
> me. He asked me "Why mourn someone who died so long ago?"
> My mother then told me to stay home with my husband and I didn't
> go either.

Qasim's detachment provides a vehicle for reformulating 'Alid
identity, a concept which in important respects is predicated on symbolic activities which enact fundamental Shi'i values. His failure to
attend the emotionally charged rituals, which are chief agents of
"memory bonding" (T. Butler 1989: 20) and which interpret the past
on behalf of believers, helps Qasim to distance himself from a subjectivity which is structured around discourses of persecution and martyrdom. By declining to attend the ritual, Qasim repudiates the
melancholic turning inward which characterizes these occasions.
Against the background of Qasim's experience of marginality, poverty,
his family's revolutionary activity, and subsequent discrimination by
reason of his remote kinship with the ruling dynasty, his refusal to
attend the ritual is a poignant reminder of the potential social and

psychic cost of living under the sign "sayyid." Why should he identify with a personalized history of death and tragedy, a history that was enacted by the powerful "others" among his kin? The question of how Imamate history since the days of Imam 'Ali is to be remembered will occupy young *sadah* in the following decades. Qasim, for one, insists that remembrance was better directed to the living and to the future than to the past.

Bayt Al-Husaynat: Ahmad

Ahmad al-Husaynat has been searching for alternative ways of being a Zaydi and a *sayyid* which are compatible with both his self-image and official ideology. He has followed directions in his life which diverge significantly from those of his forebears. Rather than embarking on a career based on religious learning, he studied economics. He accepted his sister's marriage to a non-'Alid of modest means (see chapter 7). He has felt the bias against the *sadah* since his early childhood, and has tried to shape relations with others so that his children would not experience the same kind of prejudice. Ahmad was five years old when he, his brother, and his sister had a fight with the neighbors' children.

> We threw stones and swore at each other. The other children abused us, and one of the words they used was "sayyid." I did not understand the meaning of the word. When we returned home our mother explained it to us, telling us not to use it and keep this knowledge to ourselves. The incident left me very confused. How could I be a *sayyid* and be treated like that? When I later learnt that some people do not like Imam 'Ali, I felt betrayed and sad. When I attended the elementary school, I felt that the other students were reluctant to befriend me. Some boys had been told by their parents to beware of me because I was a *sayyid*. I was unable to resolve the tension between the teachings of Islam which call on believers to love the Prophet, and people's hatred of his progeny. I was so affected by those experiences that I could not concentrate well in school.[8]

Ahmad witnessed the humiliation of *sadah* in the streets and was belittled by his Egyptian teachers who arrived in Yemen after President Nasir had sent his troops to support the republic. At school, the certificates of children of 'Alid descent recorded their names with the specification *mulaqqab sayyid* ("*soi-disant sayyid*"). The honorific title and generic label "sayyid" was treated like a nickname and the genuineness of ascendancy from the Prophet disputed.

Once Ahmad became a father, he refrained from telling his son about his identity because "once he knows he will also feel the pain.

He will feel that people do not like him because he is a *sayyid*. I want to protect him from the harassment we have experienced ourselves." When his son 'Ali had reached the age of eleven, he told him that he had to start praying. 'Ali asked why he had to assume his religious duties earlier than Muhammad, his close friend whose father is one of Ahmad's colleagues. His father explained to him that he had to conduct himself in the best possible manner. Dissatisfied with the reply, 'Ali repeated his question. "Because you are a descendant of the Prophet (*ibn al-nabi*)" he was told. "Am I?" 'Ali was astonished and proud. In the end, perhaps spurred by his own pride, Ahmad told his son about their social location and drew his attention to the fact that as a *sayyid*, he was expected to perform especially well in society and had to be serious about his religious duties.

Reflecting on that encounter, Ahmad told me that his son had probably suspected that he was from the Prophet's progeny. Ahmad's children have asked questions that had troubled himself during his childhood. "Why do people dislike Imam 'Ali? Why don't they love his descendants (*awladuhu*)?" Their father explained to them that people are resentful toward those whom they cannot emulate. " 'Ali was very honest, he was too generous and kind; he was close to God, and immensely courageous. He put people on trial; they had to learn to love him. 'Ali taught us to be kind and humble." It was Imam 'Ali's heroism that most impressed the young 'Ali, and he asked his father to let him read his biography. On the eve of the twenty-first century, however, Ahmad felt that the more radical elements of the Sunni reform movement had strengthened to the extent that the Zaydis should no longer declare themselves. "If my son asks me any questions about the Zaydiyyah, I no longer answer him. I do not want him to be in trouble."

Ahmad's daughter, Zaynab, has a girlfriend who belongs to a family which holds decidedly anti-'Alid sentiments. Ahmad thinks that the girl has been trying to prove to his daughter that her family is better than hers.

> They give her more sweets than we give our daughter. That way they try to outperform us. My daughter is aware of this. When she visits her friend, she asks me to give her a bigger amount of sweets than she would normally receive. Once her friend told her "You are not a *sharifah*" to which my daughter replied, "How do you know?" The girl said "I know you are not." Later that day my daughter told me that it was wrong to challenge her friend. She felt she should have just kept quiet.

Ahmad was occasionally given homage when he traveled in the countryside.

When I was spending a few days in Khawlan, two *qaba'il* came up to me. They looked at me and asked whether I was a *sayyid*. When I told them that I was one, they said *ahlan wa sahlan* (welcome) and walked away.

In the workplace, it was a different story. He was adamant that his promotion had been hindered because he was a *sayyid*, a suspicion which was confirmed by his trainees.

If a *sayyid* studied abroad and reached a high position, they say "the *sadah* are powerful again, they will take over." It is better if the *sadah* do not occupy top positions. If they are guilty of misconduct, it will affect all of us. Some *sadah* prefer to be in lower positions so that they cannot be attacked so much. My cousin who holds a degree from an American university has a good job in the government. She works extremely hard and efficiently but has never been promoted. Some of those who passed her by are far less qualified. After so many years she has lowered her expectations. Her older relative 'Abd al-Rahman al-Nasir, who holds a senior government position, told her "we must wait, we must carry on doing good things for the people, then they will look at us differently." Sometimes she tells me with a sense of irony, "I am cursed, I am a Hashimi" (*Ana mal'unah, ana hashimiyyah*). When she was elected to a political office, many *qaba'il* told her that they voted for her. Alluding to her background, some of her colleagues commented upon their vote saying "They still love you [*the sadah*]." Her colleagues do not value her personal qualities. She tries to make them understand that she wants to be respected only for what she does for them. All her relatives are living abroad, so no one can say that she works on their behalf [i.e. using her position to offer them patronage]. Slowly people start to understand that.

For Ahmad one way of approaching his family history was his reading of Zaydi history. He took pride in his descent from these men, but was troubled by their ruthless pursuit of power. He noted that even though Zaydi doctrine required the believers and above all the 'ulama to criticize violations of religious principles, some of his ancestors were known for their cruelty and had imprisoned and killed those who had censured their style of governance.

Imam 'Abdullah b. al-Hamzah killed many 'ulama because they opposed him, and so did several others. One of my ancestors who was an Imam read the Qur'an each night and spent long hours praying. He bravely fought foreign invaders which is why many 'ulama praise him. They think that he was an Imam of both the pen and sword. But he treated his enemies with cruelty. After a revolt, the men captured by his soldiers had to carry the impaled heads of their slain comrades from one town to another. Some had their feet tied and were dragged across the rubble by horses.

Ever since I had known him, Ahmad had claimed that the notion of Hashimiyyah was obsolete and incompatible with social justice and equality. However, during a critical period, an historical figure which Zaydi writers hail as a savior entered his memory worlds more prominently than before. The war of 1994 caused despair; several *sadah* were arrested after being accused of links with the disgraced socialist leadership, and the economy declined even further. "Perhaps someone like al-Hadi should come," Ahmad reasoned. Earlier in his life, he thought that in order to prevent the *sadah* from further harm, forgetting, rather than remembering, would protect them and thus be a desirable goal (Dakhlia 1996). But when he again felt insecure, he returned to the early Zaydi history of salvation, hoping that its memory would have a liberating effect. On the other hand, his critical appraisal of Zaydi historiography might be an attempt at demythologizing the past, which is conducive to coming to terms with the present.[9]

Name-Calling

Ahmad developed his sense of self through negative identification after experiencing verbal abuse by his playmates. After the suppression of the Imamate's political culture that had drawn on symbolism associated with the *ahl al-bayt*, it was possible for children to grow up ignorant of their descent. When the young Ahmad was labeled "sayyid" by his peers, the term was intended to be demeaning, yet he did not understand it. When he finally found out he was asked to keep quiet about it. Experiences like these one have undermined both Ahmad's and Qasim's confidence. As J. Butler (1997a: 4) notes, speech which injures highlights "the volatility of one's 'place' within the community of speakers; one can be 'put in one's place' by such speech, but such a place may be no place." The force of a contemptuous name "depends not only on its iterability, but on a form of repetition that is linked to trauma, on what is, strictly speaking, not remembered, but relived, and relived in and through the linguistic substitution for the traumatic event . . . those injurious interpellations will constitute identity through injury" (J. Butler 1997a: 36; 1997b: 104–5). As demonstrated by Ahmad's case, a person's sense of distinctness can be potent, or perhaps more potent, when it is not deliberately taught. The circumstances in which he acquired a memory of the word his mother told him to forget, served to increase its mnemonic power. Halbwachs (quoted in Dakhlia 1996) argued that the individual's group acts as an interlocutor between their personal memories and the external space of history. In Ahmad's case this encounter was brought about by default through "outsiders." Halbwachs talked about the child's incorporating memories through semi-conscious, intuitive under-standings of events, such as her parents' reactions to strangers. For Ahmad, it was the sound of the unknown word shouted at him, the word which took the same trajectory as the stone thrown at him, which gave the event historical weight. In the 1960s, for some *sadah* the honorific title that had been so central to the Imamate's political culture became his private memory even while they suffered discrimi-nation in its name.

In my presence adults spoke so rarely to their children about the events of the 1960s that these could not be conceived as occasions for understanding how the memory of groups is conveyed and sustained (Connerton 1989: 4). Some of those families who had had allegiances to the Imamate and refrained from talking about the revolution in ways that contradicted the official version taught at schools, later real-ized that their offspring identified with those who had been responsi-ble for their fate. One man asked his eight-year-old nephew what he would like to do as a grown up. "I want to be like al-Fariq al-'Amri who was a great man." Al-Fariq (lieutenant colonel) was the title of Hasan al-'Amri, a leading revolutionary officer who belonged to the poorer branch of the renowned *qadi* house. He is held responsible for the execution of several people.[10]

People who are reluctant to pass the memory of being 'Alid to the next generation are forced to engage with the past even while attempting to shed it. One day I was talking to a scholar when his six-year-old daughter returned home in a confused state of mind. Nabila, one of her playmates, had told her "*anti hashimiyyah, anti shu'ah*" (you are a Hashimi, you are bad). "You had better stop play-ing with her," her father told her, declining to explain the meaning of "hashimiyyah" to her. After she had left the room, he explained to me "I do not want her to know, I do not want her to feel different from the other girls. If I tell her that she descended from the Prophet, I have to tell her about our family. I never told her that her uncle was exe-cuted [in 1962] and that we lost our property. She would start to feel resentful." However, when she had grown up, he agreed to her marriage to a member of the former ruling House. This marriage counteracted his earlier attempt to avoid raising her awareness of having a different status. Her husband is denied both Yemeni citizenship and compensation for his confiscated property. Like Qasim's father who had been an ardent revolutionary but welcomed his son's marriage to a girl of royalist background, this man was saddened by the ongoing discrimination against the *sadah*. "Even if you lived abroad and changed your ideas about the Imamate, it will not be forgotten that you come from a certain family. You will always remain a reactionary (*raj'i*)."

Name-calling may produce confusion and apprehension. However, uttering that name which is intended to be stigmatizing or left unpronounced may help in reasserting oneself. In one case, a seventeen-year-old girl who had an argument with a servant resisted the indignity she suffered by proclaiming her 'Alid descent.

> For some time Taqiyyah had suspected the Somali servant of stealing things; some underwear and even cooking oil and flour had disappeared from the house. She told the servant that she had not been cleaning well and that things had gone missing. The servant became angry, countering that although she was working for peanuts, she had done rather well in her job. As they were shouting at each other, the servant asked the girl "And who are you anyway?" Taqiyyah was deeply insulted; as she threw herself backwards onto the cushions, she whimpered tearfully "*Ana bint al-nabi*! (I am a descendant of the Prophet!)"

Taqiyyah's sentiment is reminiscent of Hasan al-Taghi's. By proclaiming "I am a Hashimi and it honors me," Hasan carried the name "Hashimi" like a badge of honor. Taqiyyah's behavior speaks of both vulnerability and self-possession. Not only had republican youths insulted her father by throwing his *'imamah* into the dirt while he was walking in the street in the early 1960s, but now even her servant tried to put her in her place. J. Butler (1997a: 19) raises the question whether hate speech and even subtle disparagement always work and whether linguistic injury indeed produces social subordination. The *sadah* resist identity formation through injury by parodying the speech of the adversaries. For example, polemic speech which is intended to question 'Alid descent ("Fatima was not a chicken farm") is countered by phrases like "Imam al-Hadi's member was solid" which are rhythmical in Arabic (*zubb al-Hadi sulb*).[11] One *sayyid* told me that whenever he heard speech that stigmatized the *sadah*, he would challenge the speaker by asking him "I am a *sayyid*, what do you mean?" Thus, he at once exposes and counters the impudent exercise of this type of speech, leaving the meaning of the term unspecified and open to interpretation. This talking back, a kind of counter-appropriation of offensive speech, screens the person from injury or at least mitigates it. Like Hasan al-Taghi, this man appropriates the very name by which the *sadah* are abused in order to deplete it of its degradation, revaluing affirmatively the category label "sayyid."[12]

Involuntary Memories

One response to anti-'Alid prejudice is "self-willed amnesia" (Kirmayer 1996: 193) enacted through name-dropping and renaming. This is significant because patronymics and certain first names possess mnemonic agency, thus motivating people to think about or to remember certain events or people. I was sitting in the office of a high government official when his secretary brought him a note. "Ah!" he exclaimed, "Yahya Muhammad al-Nasir! Why didn't he tell us straight away?" He had had two messages from a man called Yahya Muhammad, asking him to return his calls. Turning to me, interrupting our conversation, he said with a smile, "this man tries to be more republican than the republicans. When the army called upon professionals like him to undergo a few months of training, he signed up enthusiastically in order to demonstrate his patriotism. He wants it to be known that he no longer cares about his background. But he only makes things difficult for people like me. See how long it took me to find out who he is?"

Yahya al-Nasir grew up in a northern province where his father was governor. Yahya's father died before the revolution, while his brother was executed. In his youth, Yahya had been influenced by the ideas of the 1948 reformers. He was dissatisfied with the education system, and in his teens he left the country in order to study abroad. His brother's execution motivated him to join the royalist resistance to the republic, but he soon changed his mind and left the country again for university training abroad. On returning to Yemen, he no longer used his famous patronymic, which had been carried by several Imams and Muslim rulers before him. These acts of "'active', purposeful forgetting" (Battaglia 1992: 14) are notable because "the patrimony of a sharif is, in many ways, his name" (Waterbury 1970: 97). Patronymics like Yahya's do not just confer recognizability and legitimacy over time. They bring the whole family history, the history of the Imamate, and even transnational histories to bear on the newborn who is named thus. During the Imamate, the legal framework around political membership, ancestry, law, and names were constitutive of identity. Names such as these instituted a political reality and invoked relations of power and command which gave the names their force. Prominent patronymics, then, are narrative fragments which commemorate people, historical events, and relations of power—a connection that did not escape opponents of the Imamate such as Muhsin al-'Ayni, who became prime minister in 1972. Several years before the revolution, he wrote: "We want a ruler stripped of his holiness—we want a ruler called Salih, Sa'id, 'Ali, just like you, me and other people. We have tried these (Imamic titles) al-Mutawakkil 'ala 'llah, for 1,000 years, and the result is plain to see" (quoted in Serjeant 1979: 96).

Patronymics like al-Nasir as well as the category label "sayyid" mark the body as 'Alid and may become known through trauma as is demonstrated by the cases of Qasim and Ahmad. Those who carry the patronymic know that people's response to their names is politically tainted. Yahya's decision to abandon his name is a kind of self-disciplining aiming to foreclose "othering."[13] Failing to exercise their right to name themselves is tantamount to acknowledging defeat, but its primary goal is to establish their republican credentials. Those *sadah* who omit their patronymics induce forgetting in themselves and others, thus suspending memory and dehistoricizing the body. This produces rupture but opens up avenues toward recreating the self and reconstructing relationships with others. It is the untold, "dropped" name that provides a stimulus for the redirection of life; as the object of genealogical and political history, it contrasts with names on gravestones which inscribe the dead person's kinship with the Prophet (see figure 3.2). Name-dropping here is an "act of strategic ambiguity" (Battaglia 1999: 126), producing the effect of an identity at play, a situational disengagement from stereotypical notions of "sayyid." By repositioning himself as a person without a patronymic, Yahya obliterates a crucial aspect of selfhood, seeking to enjoy unequivocal recognition as a Yemeni like the *qaba'il* the majority of whom do not carry patronymics. Battaglia (1999: 124) stresses the detachability of names from both the physical person and political value, which indicates the importance of the contextual and relational nature of acts of naming. The relationship between Yahya and past bearers of the name is marked by the tension of the historically critical move between signifying distance and recording loss and inscribing presence (Bammer 1994: xiv).

Like patronymics, first names, too, are means of conjuring the memory of powerful personalities which "project a hoped-for similarity" (Herzfeld 1982: 289). These names signify both the persons them-selves and their pious prototypes; they are not meant to be unique (Lambek in press). Herzfeld (1982: 289, 292), writing about Greece, says that a child named after an ancestor will eventually displace the previous bearers in living memory so that this identification is actually a weak form of commemoration. Where there are recorded genealo-gies, no such amnesia occurs. For example, the name of boys who are called al-Qasim evokes the memory of al-Qasim al-Rassi, the grandfather of Imam al-Hadi, Imam al-Qasim the Great, and al-Mansur Qasim b. 'Ali al-'Iyani. A friend who is called al-Qasim recalled that when he was raised during the time of Imam Yahya, his brother Yahya told him

"I am better than you because my name is Yahya b. al-Husayn (Imam al-Hadi)." Their father, who had listened to the conversation, intervened: "You are both named after Imams, you (Yahya) after Imam al-Hadi and you (al-Qasim) after Imam Qasim al-Rassi." In some cases, both the famous forebear and the child's father's name are taken into account. Thus, a boy whose father's name is 'Ali may be called Zayd, thereby evoking the memory of Imam Zayd b. 'Ali. In one case, a governor had named his son after one of Imam Yahya's sons who had passed through his town during the year the boy was born. As few children grow up with the knowledge about the lives of the Imams, those names are losing their mnemonic effervescence, and being named after them instills little pride in them.

Names have always reflected people's ideological proclivities, but have more poignantly been understood as political statements after the revolution.

> When my father called me Asma' and and my sister 'A'isha, my uncle was angry because Asma' bt. Abu Bakr was 'A'isha's sister.[14] Later a friend told him that there was another Asma', namely Asma' bt. Umays [who had Hashimite leanings and was married to 'Ali's brother Ja'far], and he was reconciled. After the revolution it was a different thing altogether. The whole family was in exile in Beirut and my other uncle gave his daughter the Christian name Naïme. (The daughter of a former civil servant)

So far we have seen that names may embody the memory of the historical and biographical past, or exteriorize their bearers' feelings of detachment and loss. Some of the names recovered by the *qudah* also express detachment from the Imamic past, yet for different reasons. As the Imamate's "secondary elite," their relationship to the state was often uneasy and ambivalent. After the revolution, some have reappropriated their "south-Arabian" ancestors. Consider the example of Bayt al-Akwa' who were attested membership of the *ahl al-bayt* by Imam al-Mutawakkil Isma'il because of their descent from one of the Prophet's Companions named Salama al-Akwa'. Bayt al-Akwa' were thus of Qurayshi origin and entitled to "protection, honour and deference" from the people among whom they worked as tax collectors on the behest of the Imam. These documents were not only prestigious but, moreover, entitled Bayt al-Akwa' to grain and money (Meissner 1987: 356–7; Coussonnet 1993: 30). According to some of their contemporaries, they took pride in the fact that they were genealogically closer to the Prophet than those who traced descent to local ancestors. Since the revolution several *qadi* houses have reclaimed their "indigenous" ancestors, putting those who linked them to the rulers through both government service and descent to oblivion. Qurayshi extraction which was once confirmed by the ruler would now make these houses vulnerable to accusations of being "foreigners" like the *sadah*. Thus, Bayt al-Akwa' have called themselves al-Akwa' al-Hawali, tracing descent to Ibrahim b. Muhammad who adopted the patronymic al-Akwa'. According to Muhammad al-Akwa' (1980: 178–9), Bayt al-Akwa' are descendants of 'Amir Dhu Hiwal al-asghar, a Himyarite leader with impeccable Qahtani credentials. Al-Wadhah b. Ibrahim, a descendant of 'Amir Dhu Hiwal in the sixth generation, had three sons, Kurayb, al-Khattab, and al-Sabbah, the forebear of the Yu'firids (847–997).[15] Bayt al-Akwa' trace descent to Kurayb's second son Mazhar. The author establishes the genealogy of his family from his contemporaries up to Mazhar b. Kurayb. The historian Muhammad Zabarah (1940: 4) confirms this interpretation, saying that Bayt al-Akwa' trace their descent to Qahtan and Himyar and that the Faqih Lutf Allah Jahaf who had written about this house had confused them with Salama Ibn al-Akwa'.

Thus like the republican state, these *qadi* houses reinvent themselves by recovering a pre-Islamic, "Qahtani" past. After 1962 they have called their children after tribes, tribal heroes, and ancient places such as Qahtan, Ma'in (the capital of the Minaean kingdom), or figures of the pre-Zaydi period such as Dhi Yazan, a Himyari of the fifth century. A street and school in San'a is named after him. Muslim names like Kutaybah (who conquered land further to the East in India and China in the name of Islam), which are devoid of Shi'i connotations, are also fashionable. In one case, a boy of a *qadi* family was named Yazid (see chapter 7), a name which reveals their anti-'Alid sentiments and testifies to their desire to shed their own past as the Zaydi Imams' loyal civil servants. They declare themselves to be Sunni without committing themselves to any legal school.

The *sadah*, on their part, repudiate accusations of their "foreign" origin by pointing at the 'Adnani descent of some *qadi* houses. For example, the historian Ahmad al-Shami maintains that the reformer Muhammad al-Zubayri was 'Adnani because the Zubayris claim descent from al-Zubayr b. al-'Awwam, a Companion of the Prophet (Serjeant 1979: 95). Al-Shami also points out that Bayt al-Iryani, in spite of coming from Iraq in the fifteenth century (M. Zabarah 1984: 256), were now Yemeni, an attribute which has been denied to the *sadah* by republican hardliners.[16]

*   *   *

Based on the *sadah*'s personal reminiscences, this chapter analyzed the fluctuating self-understandings and moral dispositions of individual *sadah* in the post-revolutionary epoch. Acquiescence to the status quo and the desire to pursue a life free of harassment, along with self-criticism and the acknowledgment of moral failure, appeared as dominant themes in their personal accounts. Situated at the interface of the personal and the transpersonal, they demonstrate the extent to which the memory of personally experienced events is informed and shaped by the knowledge of an historical past. Indeed, in the process of remembering these events, knowledge of an historical past is evoked and interpreted. Returning to Tulving's "two system" theory of memory, one semantic, the other episodic, it would appear that the data bear out Baddeley's criticism of the distinction according to which memory at one level evokes that at another.[17]

As Malkki (1995: 17) has pointed out, life stories testify to the historically specific processes of making and unmaking categorical identities. When we examine how people like Qasim al-Taghi make sense of their world, what emerges most powerfully is a sense of rupture rather than an unproblematic identification with idealtypical notions of "sayyid." These notions must be reconciled with these men's self-estrangement, with their provisional selves. Their narratives reveal that remembering is contingent on an historical experience which is shaped by a (however diffuse) notion of shared descent, social position, and other idiosyncratic dispositions and interests. Just how biographical memory is interpreted in terms of the historical past is dependent on the relative power available to subjects to shape their own world. In this regard, such acts of interpretation are also a diagnosis of power relations. It emerged, furthermore, that people who grew up in similar milieux and were exposed to much the same experience may memorize the past differently: just how and why it impinges on the present as it does is not reducible to one or two factors and may ultimately be inexplicable.

While several chapters show how remembering plays a pivotal role in processes of self-identification, the way in which the *sadah* induce forgetting in order to dislocate the history they have been taught also deserves attention. Willful forgetting is about detachment from and decentering the ancestors so as to visualize a future and allow for the production of new histories. Renaming, name-dropping, *not* naming after the ancestors, and *not* commemorating them have generated an historically situated moral discourse for relocating the self. Yet the very act of refraining from doing, of forgetting, is a kind of remembrance of what the *sadah* once were and how they have made their contemporary lives.[18] Many *sadah* are caught up by the conflict between denial and remembrance (Slyomovics 1998: xiii) as when children are no longer brought up as "the Prophet's children" (*awlad al-nabi*).

However, as the case of Ahmad al-Husaynat illustrates, attempts at disconnecting themselves from this kinship-focused historical knowledge often are constrained by the imperatives of the past, with its psychological sedimentation of specific memory regimes. This raises questions about the extent to which the self comes to fully believe what the body is made to perform under conditions of refused identification.

Several cases demonstrate that traumatic memories, which generally tend to resist integration, have not lost their sharpness not least because official denial of their victimization and defamation have precluded the legitimation of these memories. Like others who feel the need to perform a new identity, Yahya al-Nasir publicly complies with the work of erasure initiated by nationalist and post-revolutionary projects. The majority of *sadah* are not eager to recuperate an identity which is linked to exclusive claims to power, but ask for the right to remember their past without fear of being declared "foreigners."

Many of the themes dealt with in the previous pages; the impossibility of detachment from a biography formed at the unstable point where stories of subjectivity meet the narratives about the House of the Prophet, the mnemonic efficacy of religious symbols, and generational time, are conveyed in the words of a young *sayyid*.

> In my heart I feel that I am a *sayyid*. When I read about the Prophet and the sufferings of his kin, I cannot help thinking that I am from that family (*usrah*). It arouses sentiments in me. This is only a personal issue; it does not make me feel superior to anyone. My parents are much more conscious of being *sadah*, and they spend more time among them than I do because they feel more accepted and respected by them.

Chapter 12 ties in with themes dealt with in chapters 10 and 11. An issue which has emerged in those chapters is the shifting political salience of the category "sayyid" which is also discussed, implicitly or explicitly, in the recent explorations of 'Alid writers into Yemeni and Islamic history. Their contestation of Hadawi notions of authority, which had provided the ideological underpinning of the Imamate, is an ethical reflection on the new political order and the *sayyid*'s place in that order. In spite of the revival of some Shi'i rituals and renewed production of Zaydi literature, this reflection provides further evidence that the politicized memory of the *ahl al-bayt* has been suspended.

Chapter 12

History through the Looking Glass

After the revolution, the second major challenge to the Zaydi *sadah* has been the Sunni reform movement which arose in the 1970s. The Zaydis, for their part, have begun to interrogate both Yemeni and Muslim history, and there has been a modest re-assertion of Zaydi Islam. However, Ahmad al-Husaynat's narrative (chapter 11) suggests that some *sadah* feel that to an extent Zaydi knowledge has become a kind of "unofficial" memory. Several chapters demonstrate how Zaydi thought and history shape but also haunt people like him. Some have abandoned the Zaydiyyah and have turned to Sunni schools of thought. Others like Husayn b. Badr al-Din al-Huthi, a former Member of Parliament, challenged the government on current political issues and paid with his life. Generally, the main carriers of Zaydi Islam (but by no means exclusively) are members of the old elite who have suffered prejudice in the republic and experienced hostility from Sunni radicals who consider the Shi'a to be heretic.[1] The Shi'a, above all its Zaydi branch, has always engaged in discourses of righteousness that often resulted in rebellious activities. The 1979 revolution in Iran, which ended centuries of quietist anticipation of the mahdi's arrival, became strongly associated with that tradition. Meanwhile, with few exceptions, this pattern has been reversed. As Nakash observed, "since the 1991 Gulf War, radical Islam in the Middle East has been largely shaped by Sunnis of the Hanbali-Wahhabi legal school, whose hatred of America is rivaled only by their distaste for Shiism."[2] Some Zaydi scholars have officially advocated forsaking the militant forms of "rising," arguing that in a democracy opposition can be expressed through elections and the parliament. They have even renounced the central Hadawi dogma according to which the Imam must be of 'Alid descent, invoking the rulings of Imams who deviated from the mainstream. The earlier chapters provided commentaries on these issues by *qaba'il* and *sadah* of diverse socio-economic background. This chapter deals with the doctrinal revisions of historians that indicate a form of accommodation to the new political reality generated by the revolution and the Sunni reform movement.

The Sunni Islamists and their Zaydi Critics and Followers

The notion of descent-based authority, which is enshrined in the Hadawi doctrine conflicts with the republican constitution and has come under attack by Sunni reformers. As already noted, followers of the movement root moral and political authority in a literal interpretation of the Qur'an and Sunna rather than in birthright, and reject the interpretation of religious texts through *ijtihad*. This controversy is, of course, a characteristic feature of many Muslim countries, and inextricably linked to various social, economic, and political factors salient in these countries. In Yemen, one of the key factors in the debate is the attempt by adherents of the movement to delegitimize the former holders of power and to eliminate Shi'i Islam, a pursuit which is supported by influential Sunni bodies elsewhere in the Arabian Peninsula. Since the mid-1970s, the sponsorship of Sunni political organizations and schools by affluent donors has been vital for the creation of an "indigenous" Sunni Islamist movement in Yemen. Prior to unification, the government perceived the movement as a bulwark against the PDRY and as a counterweight to Zaydis whose commitment to the republic was suspect.[3] Since Yemen embarked on two significant historical experiments in 1990, namely unification and democratization (Hudson 1995:19), the Sunni Islamists have attempted their own balancing act. For example, the major opposition party, Islah, has occasionally participated in government as a coalition partner. As noted by Rouleau (2001), the Islamists have been successfully co-opted in countries such as Jordan and Yemen, where they have representation in parliament and even in government.[4] In Yemen, the integration of Sunni Islamists and tribal leaders (some of whom are prominent members of Islamist parties) into the government has been an important feature of republican state consolidation which, however, continuously exposes its precariousness.

To a certain extent, Sunni Islamists have filled the religious vacuum, which was left after the last Imam went into exile after the Civil War. Only a few years after the demise of the Imamate, political groups demanding an "Islamic state" have grown more confident. Yemen has a profound Sunni-Shafi'i tradition which is not entirely in harmony with the recent past.[5] As in previous centuries, the patronage and promulgation of strands of knowledge is contingent on the interests of local and regional powers.[6] Haykel (1999:196) points out that

Saudi Arabia has recently welcomed the ideological use Yemeni republicans have made of scholars like al-Shawkani who enjoys great respect among Wahhabis. "This identity of vision in religious matters played an important part in rehabilitating Saudi–Yemeni relations after a civil war, in which the Saudis had backed the Hamid al-Dins." Consequently, unlike some Zaydi scholars who reckon that Saudi promotion of the "scientific institutes" (*al-maʿahid al-ʿilmiyyah*) was an attempt to efface Zaydi knowledge, the Saudis argue that they supported indigenous strands of knowledge.[7]

The Islah party and the *maʿahid* have been the backbone of the Sunni reform movement. Founded during al-Hamdi's rule, the *maʿahid* gave primacy to religious instruction and offered an alternative to badly funded government-run schools.[8] The *maʿahid* enjoyed the patronage of Shaykh al-Ahmar who embodies the twin pillars of republican nationalism, the *qabaliyyah* and anti-Hadawi Islam. After many attempts to bring the *maʿahid* under government control, in 2000 it was ratified by the Parliament to integrate them with government schools. The main neo-Salafi college, *al-Iman*, was closed too and only permitted to reopen on condition that it adopted al-Azhar principles of teaching. In the eyes of Yemenis who had opposed these institutions, the decision to close them represented an admission by the government that they had promoted radical ideology. However, Salafi teaching has continued in private houses.

While the *maʿahid* were blossoming, among the old Zaydi elite reactions to these institutions reflected diverse ideological orientations. Many of them, among them Sayyid Muhammad al-Mansur, one of the few remaining *mujtahids*, were concerned about the impact of the *maʿahid* on Yemeni society and dismayed at the intolerance of the Sunni Islamists towards rival schools. Al-Mansur has called for the re-opening of the *Madrasah al-ʿilmiyyah* which was closed in 1963.

> In education, there should be more concentration on the ethical and religious aspects of life. If there is good faith and belief in a society, that is the best safeguard against social divisions and crime. The *maʿahid al-ʿilmiyyah* were built in order to achieve those goals. There has been opposition to the *maʿahid* because their curricula have failed to take account of the social fabric of the areas in which they were established. This can only lead to conflict. Sometimes difference in thoughts can create more problems than ignorance. The government accepted the establishment of the *maʿahid* but disapproved of the continuation of teaching at the *Madrasah al-ʿilmiyyah*. It objected to a dual system of teaching religion, but if this is needed why should we not have it? Since the *maʿahid* have been accepted, we should have the *Madrasah al-ʿilmiyyah* as well because it was successful.

No school of jurisprudence (*madhhab, madhahib*) should deny any other its right to exist because all have the same source, the Qur'an and the Sunna. There is no harm in co-existing *madhahib*; each *madhhab* should respect the other. When we teach the Zaydi *madhhab*, one of our basic principles is to teach the students to respect all other *madhahib*, and to accept that none of them is wrong. When a student has reached a high level of understanding, he should not follow any *madhhab*. He should neither follow the Zaydi Imams nor any other authority.[9] The Zaydiyyah is very liberal. If, for example, I find a student following the Hanafi school, I encourage him in this. Every *mujtahid* is correct (*kull mujtahid musib*). As long as the Zaydiyyah accepts all other *madhahib*, there should be no conflict among different religious groups. There are some extremist groups who want to impose their own beliefs on their followers and to create problems among people of different schools.

Unlike these 'ulama, some members of renowned Zaydi houses have attempted a break with the values that guided their childhood and youth, and have become attracted to the Sunni Islamist movement. The Islah party and the institutions affiliated with it offer them economic security and the prospect of embarking on a new mission the broad features of which have been familiar to them for centuries. For example, the late Hamud b. Muhammad Sharaf al-Din (d.1993) left the declining town of Kawkaban and took up the post of Under-Secretary in the *ma'had* administration in the capital. He believed that the institutes had a valid mission and made light of the disparities between the various schools of Islam.

There is no important difference between the *madhahib*, all are based on the Qur'an and the Sunna. There is a difference only in [the style of] prayer. The *ma'had* helps to overcome the old divisions [between the *madhahib*] and provides new perspectives. In Kawkaban, the *ma'had* students study alongside other students at the mosque. What matters to me is that Islam is being taught. There is no break with tradition; religion has always been taught. The difference between the Zaydiyyah and what we teach is of minor importance. The Zaydiyyah focuses on the principle of the *imama*. The President is accepted as the Imam. Anyone who can do the job alright is accepted, whether he is black or white.

One *sayyid* explained these Sunni inclinations among Zaydi *sadah* with reference to the bias held against them.

The *sadah* must find a solution to their predicament. They leave the country, study abroad, or join a party. It is understood that if a *sayyid* joins Islah, he no longer feels superior [by virtue of birth]. That way Islah avoids speaking about the *imama*. It wants to inform other *sadah* that there are *sadah* in the party, that they have come to the right path, and that they have moved away from the [notion of] Hashimiyyah and discrimation (*'unsuriyyah*) against others. Those *sadah* [who have joined Islah] can be used as propaganda for the party. It is like declaring that they do not believe in 'Ali and his successors. One of the leaders of Islah, 'Abd al-Rahman 'Imad, who is a *sayyid*, demonstrates that the *sadah* can obtain leadership positions—albeit according to Sunni principles. The party knows that the *qaba'il* respect the *sadah*, that's why they think it is good to have them in the leadership.

Another *sayyid*, who was critical of the rule of the Hamid al-Din, interpreted the popularity of the Sunni movement in terms of the violations of Islamic and Zaydi principles during the last century. "People were oppressed in the name of the Zaydiyyah. It is only natural that they hate it now."

Shifts in *madhhab* affiliation indicate a determination among men like Hamud Sharaf al-Din to maintain or regain influence especially over religious affairs. As yet this career strategy carries little prestige among the old Zaydi houses, but nostalgia for their *madhhab* is likely to wane among the new generation who seek to obtain positions according to their qualifications. There is a parallel between members of prominent Zaydi houses who have turned their back on the Zaydiyyah and the established religious families in Ottoman Syria. After Istanbul had gained full control over the appointments to religious posts in the eighteenth century, a significant number of Damascene families who wanted to compete effectively for positions switched from the Shafi'i to the Hanafi school (Vol. 1 1975:56–7).[10] Some Yemeni families have shifted several times between different schools. In one case, a Zaydi *sayyid* from San'a had taken up employment with the Turks in Ta'izz. His sons remained there and worked in the administration of the Imams until 1962. By the time they went into exile in Lebanon, they had been Sunnis for half a century. In Beirut some men of the family became involved with Imami Shi'a leaders. Two married Lebanese Shi'i women. One of the men contracted a temporary marriage (*mut'ah*) and called his son Ja'far after the sixth Imam.[11] His niece commented "they became fanatical. They were in exile, and when they returned they found themselves in much lower positions than their fathers."

Old generation *buyut al-'ilm* argued that the Sunni Islamists had especially targeted young *sadah* whose fathers had enjoyed considerable prestige in the Imamate and to whom the government has had little to offer. Some of these young men responded to requests for volunteers to fight in Bosnia and Afghanistan. Whilst visiting a

Yemeni friend in the early 1990s, I noticed a photo in the reception room which I had not seen before. "This is Amin, my cousin," she explained, "he recently died in Bosnia. He was twenty-three. Nobody knew he was going there. He left the country under a pretext. He left a wife and two young children. Next time al-Zindani [the chairman of al-Islah] wants to send fighters to Bosnia, he should send his own children."

Those members of the Prophet's House who desire to maintain their identity as both Zaydis and *sadah* see the Sunni Islamist movement as a threat. Whether or not they approve of the Hadawiyyah, they value the "open-mindedness" of the Zaydiyyah embodied in the *ijtihad* tradition. They sum up their views by saying *taqlid al-hayy ahsan min taqlid al-mayyit* (following those who are alive is better than following those who are dead). They defend the Zaydi rule rendering unlawful any *mujtahid*'s judgment which is informed by that of his predecessors by stressing that Zaydi 'ulama are concerned to find solutions to the problems of the present day. In their view the survival of the Zaydiyyah depends on the ability of the 'ulama to offer valid advice to their constituents while not seeking power for themselves.

The strengthening of the Sunni movement and the more relaxed political environment following unification inspired a moderate Zaydi revival. A Zaydi-based political party, the Hizb al-Haqq, was established but attracted few followers.[12] There was an increase in Zaydi instruction and publications. Some Shi'i rituals, most prominently that commemorating Ghadir Khumm which stopped after 1963, were once again performed. The Ghadir ritual has been conducted at San'ani mosques since 1994, and in Sharafayn, Razih and other places in Khawlan b. Amir even earlier.[13] The occasion of Imam 'Ali's arrival in Yemen on the first Friday of the month of Rajab, which used to be celebrated like *'id*, is also given attention. However, following the clashes between the army and Husayn al-Huthi and his followers, in 2005 the Ghadir ritual was prohibited and several bookstores closed.

The ritual marking the birthday of Imam Zayd b. 'Ali, an apparent innovation, is significant by virtue of the ambivalence inherent in its contemporary performance.[14] On the one hand, by commemorating a martyr who founded the Zaydi school, the participants declare their adherence to Zaydi Islam. On the other, Zayd's teachings have recently been promoted by Zaydi thinkers wishing to highlight those aspects of Zaydi orthodoxy which do not conflict with official republican ideology. They argue that Zayd did not declare 'Alid descent to be a prerequisite for legitimate rule. Zayd's verdict, according to which 'Ali b. Abi Talib had a more legitimate claim to the caliphate than

Abu Bakr and 'Umar, was based on considerations of 'Ali's superior capacity to interpret God's revelation rather than his kinship with the Prophet. Abu Zahra (1959:188–9) relates that in spite of Zayd's conviction that 'Ali was the most expedient to succeed the Prophet, he insisted that Abu Bakr had been chosen in the interest of the Muslim community and Islam. Claimants from the House of 'Ali were the most eligible, but their election was not a foregone conclusion. Based on such considerations, Sayyid al-Mansur argued that if the nation voted for the president, the result should be respected. Reference is also made to schools within the Zaydiyyah. Deviant views by scholars such as Hasan b. Salih (d. 784) had only a moderate impact during the Imamate, but have recently been revived by Zaydi writers who are opposed to Hadawi teachings. In Hasan b. Salih's view, the principal confirmation of the leader's legitimacy is the loyalty demonstrated by the Muslim community. He argued his point by referring to the caliphs who ruled after Muhammad even though 'Ali was the most eligible candidate for the caliphate. He distinguished between the most excellent (*al-afdal*) and the less excellent (*al-mafdul*) leader of the Muslim community, contending that the latter was acceptable, an idea however rejected by al-Qasim b. Ibrahim (Madelung 1965:142; Subhi 1980: 101–2).

Rethinking Issues of Authority

Thus, liberal Zaydi thinkers establish continuities with non-Hadawi traditions within the Zaydiyyah which, unlike al-Shawkani's teachings, are not acknowledged by the government. Among those attempting to reconcile the doctrine of the Imamate with the current constitution are Ahmad b. Muhammad al-Shami and Zayd b. 'Ali al-Wazir who come from eminent scholarly families. In his recent work, Ahmad al-Shami (d. 2005) draws his conclusions from both his personal history and from religious injunctions regarding the *sadah*.[15] He argues that the *sadah*, who were singled out by the Prophet as the only people who are forbidden the zakah as a form of alms, should not be entrusted with sovereign power (*al-wilayat al-'ammah*). He rejects the idea that God entrusted a certain group of people (the *ahl al-bayt*) with the task of "rising" against oppressors. The holding of power, particularly its monopolization, were to the detriment of the *sadah*. The author maintains that the Imamate had been all too often a travesty of justice. Al-Shami is aware of the corrupting influence of power, suggesting that the exercise of authority in uncompromising obedience to God bears elements of Sisyphus' endeavors. Because an Imam's assumption of office was all too often accompanied by the use of violence, he stresses in particular that the blood of the faithful should not be spilled to achieve political power. As pointed out by Madelung (1987:176), the non-hereditability of the Imamate caused political problems. Al-Shami notes that from its very inception, Zaydi history witnessed bitter struggles among brothers and relatives, beginning with the family of Imam al-Hadi until the recent conflict between the al-Wazir and Hamid al-Din in 1948 in which the author was personally involved. According to al-Shami, the *sadah* always fought and killed each other, until they were finally killed by those who had elected them as their rulers.[16] He stresses that 'Ali gave his allegiance to 'Umar and Abu Bakr, and that Zayd b. 'Ali had also accepted them.

Al-Wazir's doctrinal revisions also focus upon the issue of sovereignty.[17] He takes a critical stance toward both the principles of election as laid down by various Shi'i doctrines and Sunni governments. He rejects the view of the orthodox Shi'a—including the Hadawiyyah—that God determined the Prophet's successor. In his view, the hereditary succession of the Imams revered by the Imami Shi'a conflicts with the principle of *shurah* (consultation) which was also violated by Sunni rulers. Following the death of Imam 'Ali, the rule could have been democratic. Such hopes were dashed by Mu'awiyah who introduced an autocratic style of rule which then came to characterize the course of most of Islamic history.

Al-Wazir's argument in favor of the ruler's nomination by the Muslim community leads him to comment favorably on the early period of Muslim rule when the people of Madina paid their allegiance to both the Prophet and 'Ali. This system of election, he suggests, is preferable to the Zaydi system according to which a new Imam is confirmed in his office by the *ahl al-hall wa-'l-'aqd* (people who make loose and bind). This "committee" comprises mainly the 'ulama, the *dawi al-shawkah* (people who have the power and means to lead the *khuruj*), and tribal leaders, but it excludes "the people." In another respect al-Wazir considers the Hadawiyyah to be exemplary among the various doctrinal strands for it defines special conditions which must be met by any candidate for the highest political office. The one condition he disapproves of is the stipulation that the Imam must be a descendant of 'Ali and Fatima, a requirement which had caused the self-destruction of the 'Alid community. It led to disunity among the *sadah* and rivalry between them and the shaykhs.[18] The author approves of the Hadawi principle according to which obedience to an Imam is contingent on his adherence to the law. He also refers to the problematic issue of autocracy in the Middle East, raising the question whether it is possible to avoid the killing of oppressors and the violent contests for power which have been a common feature of Zaydi history. He suggests that the period of an Imam's rule be restricted in the same way as the term of office of the republican president. Al-Wazir attaches special significance to this proposal because it has never been implemented in Islamic history. He asserts that according to both Zaydi doctrine and the republican constitution, the ruler's power is subject to supervision by either parliament or an equivalent consultative body. He points out that such a theory of the Imamate does not contravene the republican constitution. Rather, we might perceive the republic as a progression from the Imamate. By referring back to the Prophet's dominion and by revising Shi'i political doctrine, the author advocates a theory of Islamic rule which is compatible with democratic systems of government.

Al-Shami's study raises important questions about the legitimacy of power and the place of the *sadah* in a society which no longer recognizes them as chosen people. He interweaves both subjects, attempting to set moral guidelines for the Prophet's descendants. On a doctrinal level, it is instructive to note that Zaydi thinking reveals an inherent dogmatic connection between a preoccupation with justice and an acute awareness of the potentially corrupting nature of power. This is indicated by the emphasis Zaydi doctrine places on the *khuruj* and the injunction that holders of power such as Imams and governors must refrain from commercial activities. They are to be deterred from manipulating political institutions for their own economic ends. In order to prevent political rule from becoming an instrument of oppression and injustice, the doctrine stipulates that only the most erudite and pious of Ali's progeny may hold the leadership. Based on their analysis of a millennium of Zaydi rule, contemporary Zaydi scholars such as al-Shami have grown disillusioned with this fundamental premise. In al-Shami's writings there is an admission that so grave was the failure of Imamic rule that the author advocates the *sadah*'s abstention from political authority. His writings also reflect the trauma of the revolution and the subsequent search for alternatives which would enable the *sadah* simultaneously to maintain their identity whilst renouncing claims to political authority.

Al-Shami proposes that instead of sullying their purity by holding political authority, the *sadah* ought to dedicate themselves to religious learning. By acting in a pious manner and by abstaining from power they may earn people's respect rather than their hatred. The 'ulama from among the *ahl al-bayt* are to act as moral critics of the holders of power. The author's position diverges significantly from the Hadawiyyah, but he retains the notion of the *khuruj* which he believes is compatible with the responsibilities of the 'ulama. Thus, the renunciation of the 'ulama claim to sovereign power which he urges does not entail uncritical compliance with authority. In his view it is paramount that the *sadah*'s concern with social justice rather than temporal power is to be made explicit. This may lead non-*sadah* to accord moral legitimacy to them by virtue of their descent and force of character.[19] Two factors are crucial here. During the revolution it was mainly the *sadah* who had been associated with political leadership and those who were persecuted and killed. Those who lived in their hijrah and were politically aloof were not harassed, and have continued to enjoy respect. Furthermore, Zaydi *sadah* take into consideration that their Sunni peers in the South have not been discriminated against by virtue of belonging to the Prophet's House. Their holding of political offices has not caused resentment because they had never claimed that rightful rule was the prerogative of 'Ali's progeny alone.

* * *

The current scrutiny of central features of their history and doctrine by Zaydi writers indicates both the marginalization of the Zaydi *madhhab* since the revolution and their endeavor to ensure its survival. Most Zaydi 'ulama are aware that calls for an implementation of Hadawi doctrine would jeopardize this survival. The cultural response to the Sunni reform movement has been formulated in such a way as to simultaneously reconcile the Zaydiyyah with republican ideology and to present a reformed Zaydiyyah as a viable alternative to certain Sunni schools.

The doctrinal revisions regarding political authority made by scholars such as al-Shami and al-Wazir demonstrate the mutability of Zaydi orthodoxy. By pronouncing the ascendancy of piety and erudition over descent in relation to the leadership question, they avail themselves of the intellectual weapons of their Sunni rivals. The repudiation of the Hadawi doctrine by the two authors conveys several political messages. They are aware that the images other people hold of the *sadah* tend to be associated with the former rulers. By disavowing the claim that the *sadah* are best placed to interpret Muhammad's teachings and to guide the Muslim community, they hope that they gain moral authority based solely upon their learning and piety. Other statements implicit in the authors' expositions are that Zaydi ideology does not present a threat to the republic, and that those *sadah* who hold political authority are not to be regarded as wolves in sheep's clothing with a hidden agenda to restore the Imamate. Doctrinally based pronouncements about politically correct practice serve at once to define the *sadah*'s place in republican Yemen and as vehicles for debating relations of power.

Some *sadah* of the older generation have grown so disenchanted with Zaydi ideas that they question their present-day viability. They reason that the Imamate belongs to the past and that in a pluralistic democracy, the principle of *khuruj* has become obsolete. In 1948, provisions for a "constitutional Imamate" marked a significant shift away from earlier styles of government, but the principle of 'Alid leadership was upheld. The abandonment of the very notion of the Imamate, the establishment of which is obligatory on the Zaydi community (Madelung 1971:1160), shows a remarkable independence from central tenets of the doctrine. There is a readiness among many Zaydi leaders and members of the House of the Imam to contemplate alternative political realities.[20] In the past Zaydi 'Alid scholars deviated from the mainstream by claiming that honor and dignity stemmed from pious asceticism rather than the exercise of authority, or that any Muslim could be Imam,[21] but it is only now that a considerable number of them are committed to giving up claims to what has been their hereditary right for over a millennium.

Conclusion

Frontiers of Memory

The object of this study has been a Yemeni hereditary elite who ruled until the Imamate was dismantled, in a revolution that introduced republican rule. It analyzes how in the process of state transformation, members of the old elite, the *sadah*, locate themselves within shifting contexts of kinship and marriage, education and occupation, and moral and political authority. The book's overall theme is an exploration of how remembering is implicated in concepts of personhood and morality. 'Alid personhood is idealtypically defined through the memory of the *ahl al-bayt* and the holy scriptures, such that kinship with the venerated ancestors becomes enacted. It is argued that patrilineal descent is predicated on religious learning in the performance of kinship. Furthermore, in distinction to post-ghaybah Imami Shi'ism according to which the twelve Imams are infused with everything that can be learnt, and need only to remember, the Zaydis merely claim that the Prophet's descendants are endowed with the potential to better understand the scriptures. A number of practices, above all prolonged study and memorization, at once serve to gain morally sound knowledge and to commemorate learned forebears. Orthodox texts, some of which were channeled through or produced by the ancestors, through these practices become objects for validating and remembering relationships. During the Imamate these relationships were politicized such that the descent metaphor provided the state's raison d'être. Once glorified, now stigmatized, the knowledge that afforded it legitimacy has become an "unofficial," unauthorized memory.

The book examines quotidian recourse to the corpus of verdicts and actions of eminent religious authorities, referred to as *taqlid ahl al-bayt*, at a crucial historical juncture. The hallmark of this complex body of knowledge is its genealogy of oppositional discourses that can be mobilized in unexpected times and places—for example, to refuse or affirm compliance with dominant political agendas, and to build new, republican subjects. Utilizing Lambek's concept of memory as "moral practice," it is argued that for those who embrace this knowledge, invoking the memory of specific verdicts is a moral engagement, which is linked to claims about the past and present and ultimate value. The selective exhumation of dead ancestors, so to speak, gives these processes of remembering not just a special moral edge but also a historicity.[1] This, of course, has a political dimension because both the government and Sunni Islamists wish to put the knowledge on which they center into oblivion.

The *taqlid*, an orthodox tradition that was central to the culture of the Imamate, has become a discourse from the margins, which does not seek to fracture national grand narratives. The book traces how it is employed in such a way as to make the *sadah*'s practices commensurate with them. Because novel phenomena are interpreted within traditional schemes of analysis, these practices are attuned to their past. The *taqlid* thus provides a personal and interpersonal sphere for articulating value judgments outside the officially sanctioned narratives without contradicting the ideology of the state; and for repairing relationships with others without necessarily acknowledging defeat or explicitly engaging in appeasement. Within this sphere, the *sadah* have found their own voice in remaking their world, and in putting a moral spin on what is occasionally little more than acceptance of the inevitable, like marrying one's daughter to a non-'Alid. Recourse to the *taqlid* creates continuity with the past but also offers opportunities for critical disengagement from it. Invoking particular rulings, to the exclusion of others that encode different values, achieves detachment and constitutes an interpretation of the present. The citation of verdicts of 'ulama who approved of non-'Alid claimants to the Imamate is one such example (chapter 12).

There are other areas of remembering where continuity with the past is being established. For instance, a boy's experience of victimization (chapter 4) is grafted onto the memory of the tragedy of Karbala, which is central to Shi'i collective memory. In similar vein, some *sadah* interpret the memory of the revolution as but one critical moment in the history of the persecution of the *ahl al-bayt*. The book's broader concern is the impact of the revolution on individual subjectivity. Several chapters indicate that the *sadah*'s memory of the revolution is by no means uniform, and partly dependent on relative rank. While for the majority of previously privileged *sadah* it is remembered as a catastrophe, those who participated in it remember their involvement with some pride. (They take less pleasure in their achievements than they did a few decades ago, but nonetheless do not cast doubts on its justification. The image of executions without trial and of highly respected officials has to some extent displaced official memories of heroic achievement).[2] The case of Qasim al-Taghi, who comes from a revolutionary family, demonstrates that for men like him the memory of the revolution does not have the same magnitude it has for others. His family was positioned in a social space apart from that occupied by the eminent *sadah* houses. For him the burden of memory is not the history of persecution invoked by other *sadah*, but that all were treated alike after the revolution. He also refuses to carry the burden of memory by declining to attend Shi'i rituals which activate the memory of suffering. However, this longing for closure, which I analyze as willful amnesia, constitutes a profound working of the past rather than mere forgetting. This is particularly so in cases of name-dropping where people explain themselves differently to others and cease to perform 'Alidness in culturally expected ways. Along similar lines, some parents do not furnish their children with category labels even while they were scornfully used by other children. Others like Taqiyyah, whose family has with distress registered attacks on the symbols of 'Alidness in the wake of the revolution, reassures herself of her noble birth when the servant's derision conjures up memories of those attacks (chapter 11).

Official discouragement of particularist identities is congenial to acts of memory suspension. Chapter 5 dealt with family histories, which are silent about descendants of scholars who were no longer learned. Nowadays some *sadah* break the continuum of genealogical memory by refusing to inform family annalists about the names and occupations of their sons. Anxious to comply with official injunctions to forget, they determine how the future is to be memorized. With the imposition of self-censorship, the memory that authorized the exercise of power and afforded high status is no longer immortalized in prosopographical works. Then there are those who criticize 'Alid writers for documenting acts of violence by thirteenth-century Imams against their opponents and their women (chapter 11). Disapproval is expressed in phrases like "this is not the time"—neither to give one's adversaries more ammunition nor to write books like this *in times like these*.

Category Labels and their Ethical-Political Implications

The book set out to criticize the salience of reified collective identities, arguing that a monolithic understanding of *sayyid* as a "vessel of charisma" and "paragon of piety" fails to comprehend the complex imbrications the concept entails. Its aim has been to show how being born a *sayyid* assumes meaning in the lives of those labelled as such while acknowledging the historical and biographical contingency of this notion. Considering that the new republican project has inspired debates over who was really Yemeni and who was not, my interest has been how the *content* of cultural difference (Thomas 1999: 271) is being constructed as well as the ways relational definitions of personal identity are engaged by the actors. I have argued that recognizing how difference is "real" in the lives of people like the *sadah* necessarily requires us to draw attention to intersubjective realms where domains of difference are displaced. While in certain respects agreeing with Battaglia's (1995: 2) dictum of the self as a "a reification continually defeated by mutable entanglements with other subjects' histories, experiences, self-representations," my data has led me to conclude that it is precisely this entanglement which makes difference appear "real" to the subjects. At the same time, of course, in many respects this apparent difference is one of unacknowledged or disguised like-ness. Harrison (2003) stresses this point by arguing that representations of difference are always bound to perceptions of similarity, or rather that cultural difference ought to be conceived of as muted or broken resemblance. Negative valorizations of the "other" are often predi-cated on representations of censored and disclaimed attributes of the Self (2003: 357) such as attaching significance to one's ascendancy (see chapter 10). As several chapters have pointed out, drawing on Yemeni history both *sadah* and non-*sadah*, who wish to espouse republican ideas, accuse each other of an exclusive preoccupation with their own kind. The conflict-ridden nature of these perceptions tends to disguise aspirations toward social parity on both sides. The book draws the conclusion that in republican Yemen the category *sayyid* can only be rendered comprehensible through the tension experienced by those who carry the label, aspiring to be recognized as "different" whilst enjoying the status of an equal partner in social interaction. The study has sought to account for unpredictable mobilities within and across social categories, but has expressed caution about Derridan-style "incalculable choreographies" which aim to capture a continu-ously changing subjectivity (Derrida cited in Bordo 1993: 267). One of its concerns has been to analyze the contradictions experienced by the actors between the normative assignments of descent, their own interests and shifting self-perceptions, and their position within specific configurations of power relations. Several chapters have revealed that insofar as far as *sayyid* is an assignment, it is one which is never entirely carried out in accordance with the normative expectations of that term. In short, the actors' self-experience would appear to be one of "constrained contingency" (J. Butler 1997a: 156). This perspective allows for an analytic focus on both the fragmentation and strengthening of identity which tend to occur in the aftermath of cataclysmic events. The experience of revolution and persecution has certainly reinforced self-awareness of being 'Alid, but it is nonetheless impossible to speak of a unitary conception of what it means. The book sought to explore its various dimensions and to explain why and how it matters.

The study has also intended to show that generalizing statements about the *sadah*'s "situation" at any one time are misplaced. Instead of smoothing over what might appear as inconsistencies—a bewildering array of conjectures and statements by themselves and others about occurrences that have served either to "weaken" or "strengthen" them—I have come to conceive of them as a vital part of the real-life drama anthropologists struggle to describe. What has caught my attention over several years of study is how *sayyid* is situationally contingent and variously emphasized. What one cannot fail to remember are childhood stories recounting the first awareness of the label during an encounter with stone-throwing neighbors; a man explaining his marriage to a foreigner with reference to a remote past when Imam Zayd b. 'Ali's father decided to wed a woman from Sind. Would one come across a non-'Alid making similar statements?

Issues of categorical (in-)coherence aside, should identity labels be maintained at all? Even though many non-'Alid Yemenis conceive of *sayyid* as a "hegemonic" identity label, I have refrained from discarding or placing the label in quotation marks as has become fashionable in analyses inspired by postcolonial theorizing.[3] Doing so would only mimic rather than tackle the problematic of identity that categories such as *sayyid* (or Qahtani) represent; I have therefore relied on the ethnographic data to signify *sayyid* as a critical and political problematic. Moreover, abandoning terms such as these prevents us from analyzing how they act as political signifiers, establishing a set of connections or a given object as a political reality (Žižek 1989: 99). In any case, their deconstruction does not dissolve the political realities of which they are part. Navaro-Yashin (1996) is right to lament that category labels are "distressingly real" in the lives of their bearers, but it remains the task of the analyst to establish the histories of their truth-claims. Beyond these considerations, Appadurai (1998: 226) makes the important point that labels such as "Kurd" and "Muslim," which seem to be the same as long-standing ethnic names and terms, are often transformations of existing ones and serve substantially new frameworks of identity, entitlement, and spatial sovereignty. Chapter 11 provides examples of how the actors themselves historicize and even seek to suspend these labels in order to accommodate such new frameworks.

Scholars challenging the "politics of identity" approach have rightly taken issue with the analytical essentialisms deployed by colonial and nationalist governments, but have drawn far less attention to the use of category labels by post-colonial and post-revolutionary regimes. As pointed out in the introduction, in contemporary Yemen public endorsements of the *qabilah* paradoxically converge with pronouncements of "citizen" as a universalizing concept. Emphasis on equal citizen rights is important, but the assignment of cultural priority to the *qabilah* may not have integratory effects and merely serve to reproduce terms of exclusion. For example, to those traditionally associated with servile labor (for example, *akhdam*, *mazayinah*), official discouragement of the use of their social labels is certainly desirable, but their placement *outside* the *qabilah* and claims to their "foreign" origins continue to form part of everyday social knowledge. So far the new nationalist project has not succeeded in transcending conventional political identities not least because it has a double face, now universalist, now exclusionist.

Whether or not democratic politics should validate the distinctive status of diverse groups remains a sensitive issue. The question of how to deal with differences in (self)-ascribed origin, gender, and religion[4] is particularly delicate because in the Imamate they were linked to institutionalized status differences.[5] Therefore, an official endorsement of cultural identities and communities as they have flourished in, say, postwar America might be difficult to sustain because it would appear to serve as an impediment to rather than the promotion of participatory parity in social and political life.[6] Critics on the "left" of the political spectrum, who even think of the various tribal groupings as "quasi-ethnic" collectivities, argue that such recognition is likely to paralyze the formation of a national civic culture and to generate social strife. In this context it must also be borne in mind that constructs of national identity must eventually confront contradictions between emphasis on Qahtani heritage and discouragement of descent ideologies. As chapter 10 has argued, these ideologies are cultivated as much among those representing themselves as *qaba'il* (or Qahtaniyyun) as among the majority of the *sadah*. With so few people living up to the ideal of the republican citizen, anyone is a potential "other."

Against this backdrop the *sadah*'s location within republican society remains fraught with tension. The question arises whether the affirmation of the *qabilah* as a unified national category has the effect of effacing a social category whose influence in the administration and wider society remains a vital concern.[7] The case of the *sadah* reveals that efforts to displace category labels by that of "citizen" have failed to produce their depoliticization. Chapter 10 provides many instances of the deployment of linguistic power which, in its insistence on historical alterity, seeks to offend and deride. Stereotyping of the *sadah* is unlikely to subside as long as an identification with the new republican state is forged through a struggle against the Imams who are inevitably identified with the *sadah*. Seen in this light, *sayyid* remains an inextirpable sign of negative difference. Yemen is different from Algeria and Syria where "sayyid" is no longer associated with a formerly privileged stratum. As Schatkowski Schilcher (1985:124) explains, "one of the most telling indications that we have that the Syrian *ashraf* have declined is the now socially-accepted practice of using the form of address 'al-sayyid' as 'mister' as widely as this is used in western societies. The formerly precise connection of this form of address to the *ashraf* has been lost."[8]

The problem of labeling is intimately connected with the question whether subjects will be permitted to remember the past in specific ways without suffering the pernicious moral and political effects of "othering." As the flagship of democracy on the Arabian Peninsula, postwar Yemen may have the capacity to forge an inclusive national identity that is coupled with impartiality in the public realm. It certainly has the cultural resources to create a democracy that sustains a social vision whose ultimate value lies in promoting tolerance of difference. This may eliminate the pitfalls of identity politics.

Appendices

Appendix I: Wives of Imams Yahya and Ahmad

Imam Yahya

1. Amatullah 'Ali al-Wushali (*sayyid*)
2. Huriyyah Sayf al-Islam Muhammad b. Imam Muhsin al-Mutawakkil (*sayyid*)
3. Fatima 'Ali al-Madani (*sayyid*)
4. Safiyyah al-'Ansi (*qadi*)
5. Amat al-Rahman 'Ali b. Muhammad Ghamdan (*sayyid*)
6. Fayqah al-Aghrubi (shaykh)
7. 'Atika Abu Nayb (*sayyid*)
8. Amat al-Hafiz 'Abdullah Shirhan (shaykh)
9. Amat al-Razaq Muhammad al-Dirwish (*sayyid* and shaykh)
10. Amatullah Muhammad Ishaq (*sayyid*)

Imam Ahmad

1. Tuqqah Muhammad al-Mutawakkil (sister of Huriyyah, wife of Imam Yahya)
2. Huriyyah Muhammad al-Kuhlani (*sayyid*)
3. Safiyyah Muhammad al-'Azzi (*qadi*)
4. Amat al-Rahman 'Abd al-Ilah al-Nassar (*qadi*)
5. Shams al-Dhuhur Khurshid (of Turkish origin)
6. Aniza Muhyi 'l-Din Raghib (former Foreign Minister of Imam Yahya of Turkish origin)
7. Amat al-Latif Muhammad Hizam (*qabili*)
8. Samiha (slave; gift of King Sa'ud)
9. Muti'ah Muhammad Bashir, merchant of Syrian origin (marriage not consummated)
10. Layla (slave)
11. Amat al-Karim Ahmad Ishaq (*sayyid*)
12. Amat al-Rahim Muhammad al-Mutawakkil (marriage not consummated)

Appendix II: Professional Histories (1850s–1990s)[1]

1. Bayt Zabarah

Jamal al-Din b. ʿAli b. ʿAli
Born 1888. Trader; treasurer in al-Sudah during time of Imam Yahya.

Yahya b. Muhammad b. Yahya b. ʿAli
Born 1908. He was brought up in Sanʿa where he pursued a military career. When Imam Yahya's son Sayf al-Islam Muhammad was governor of Hudaydah, he asked him to stay with him. After the governor's death, he returned to the military. The reason for his assassination in 1933 remains unclear.

Muhammad b. Yahya b. ʿAli b. Husayn
Born 1895 in Jahana/Khawlan. Studied grammar and Zaydi *fiqh* in Shaharah; advanced his studies in Sanʿa.

The House of Muhammad b. Yahya b. ʿAbdullah

Muhammad b. Muhammad b. Yahya b. ʿAbdullah
1883–1961. During the time of Imam Yahya he led campaigns against the Ottomans (*raʾis al-mujahidin*); *ʿalim*, historian, politician. Author of several biographical dictionaries.

Yahya b.Muhammad b.Yahya b. ʿAbdullah
Known as a pious and austere man; killed in 1904 by the Turks under the leadership of Zakry Pasha during their assault on Sanʿa.

The House of ʿAli b. Muhammad b. Yahya

Zayd b. ʿAli b. Muhammad
Before 1962 treasurer in Dhamar; died 2001.

Muhammad b. ʿAli b. Muhammad
Before 1962 employed in Ministry of Finance; died late 1990s.

ʿAbdullah b. ʿAli b. Muhammad
Before 1962 in charge of supplies for the Court.

ʿAbbas b. ʿAli b. Muhammad
Studied with Ahmad b. Muhammad, the late Mufti; before 1962 belt maker; assistant of ʿAli b. ʿAli and Muhammad b. ʿAli.

ʿAli b. ʿAli b. Muhammad
1888–1976. He studied with ʿAbd al-Karim b. Ahmad Mutahhar, Muhammad b. Yahya al-Kuhlani, Ahmad b. Hasan al-Wazir and ʿAli b. Husayn al-ʿAmri. During Imam Yahya's rule treasurer in al-Sudah and thereafter in Sanʿa; later Imam Ahmad's deputy in Sanʿa from 1958–61; imprisoned in 1962.

The House of Muhammad b. Muhammad b. Yahya

Ahmad b. Muhammad b. Muhammad
ʿAlim, Minister of Justice and member of the Royal Court during the reign of Imam Ahmad; Mufti of the Yemen 1970–2000.

'Abdullah b. Muhammad b. Muhammad
Before 1962 studied Agriculture in Prague; 1980s Director General in the
Ministry of Agriculture.

'Ali b. Muhammad b. Muhammad
Before 1962 educated in Egypt; study of civil aviation in Russia and Italy; died
in 1962.

Amin b. Muhammad b. Muhammad
Before 1962 studied at the Military College in Cairo; first parachutist in
North Yemen. During 1980s Member of the Arab League in Cairo; retired
in Cairo.

Ibrahim b. Muhammad b. Muhammad
Before 1962 studied in Egypt and Italy, Ph. D. in Political Science from the
United States; by the 1980s worked at the Foreign Office; died 2001.

Hasan b. Ahmad b. Muhammad
No employment; living in Britain.

Mutahhar b. Ahmad b. Muhammad
Educated in Yemen; in 1980s employed by Ministry of Youth and Sport.

Husayn b. Ahmad b. Muhammad
Before 1962 education in Lebanon, Egypt, and Italy. Used to work for Yemen
Airways. Currently no employment.

Muhammad b. Ahmad b. Muhammad
Before 1962 education in Cairo; study of Political Science in Rome. After
1970 businessman; living in Cairo.

Yahya b. Muhammad b. Ahmad b. Muhammad
Born early 1950s. Educated in Saudi Arabia; study of Civil Engineering in the
United States, M.A.; runs an engineering office in San'a.

The House of 'Ali b. 'Ali b. Muhammad

Zayd b. 'Ali b. 'Ali
Attended secondary school in Cairo; obtained a Ph.D. in Economics and
Political Science in the United States; 1980s employed in the Prime Minister's
Office.

Mutahhar b. 'Ali b. 'Ali
Obtained a Ph.D. in Economics and Information Technology in the United
States; Professor of Economics and Head of Department at San`a University.

'Abd al-Khaliq b. 'Ali b. 'Ali
After 1962 attended a secondary school in San'a; 1980s wheat grinder and
grocer in the old suq.

Hasan b. 'Ali b. 'Ali
Born late 1930s. Before 1962 at Foreign Office, sent by Imam Ahmad to
Egypt to attend a course in Administration; after the reconciliation between
royalists and republicans, he was employed at the Yemeni embassies in the
United States and Geneva.

Muhammad b. 'Ali b. 'Ali
Born 1921. Studied at the Madrasah al-'ilmiyyah; during the time of Imam Ahmad he was his father's acting deputy; executed in 1962.

'Abd al-Rahman b. 'Ali b. 'Ali
Born 1941. Poet; before 1962 studied at Great Mosque; after 1962 he was an employee at the Ministry of Finance; during the 1980s worked as a librarian at the National Library (*al-maktabah al-wataniyyah*); retired.

Ahmad b. 'Ali b. 'Ali
Born 1922. Poet; sent by Imam Ahmad to Cairo for a diplomatic course; before 1962 representative of Yemen at the United Nations and the Yemen embassy in Washington which he established jointly with 'Abd al-Rahman Abu Talib; after the United States recognized the Yemen Arab Republic, he worked as an advisor of the Saudi embassy; in the 1970s he again worked at the Yemen embassy and retired in the 1980s; died in exile 20 January 1995.

'Abd al-Malik b. 'Ali b. 'Ali
Born 1948. After 1962 study at the Police College (colonel); later studied Business Administration in the United States; General Director of Yemen Computer Company; Honorary consul of Canada to Yemen.

'Abdullah b. 'Ali b. 'Ali
Faqih; before 1962 he worked as an assistant of his father, Imam Ahmad's deputy.

Yahya b. 'Ali b. 'Ali
Poet; before 1962 studied in Cairo; since 1958 worked at the Ministry of Agriculture; during the 1980s he held the post of deputy minister; now advisor to the Minister of Fishing.

'Abbas b. 'Ali b. 'Ali
Born 1942. Elementary school in Damascus; secondary school in Egypt, study of Medicine, Ph.D. in Paediatrics, Catholic University of Rome; Director of primary health care; General Director, Health sector development (IDA/World Bank); Advisor to the Minister of Health; since 2003 General Secretary of the Red Crescent in Yemen; teaches in Public Health at San'a University; General Secretary of the Yemeni–Italian friendship association.

Tariq b. 'Abbas b. 'Ali
Born 1974. Bank employee in Munich, Germany; Studies Slavic languages.

Raja bt. 'Abbas b. 'Ali
Born 1979. Works for a satellite company in Munich.

Hamza b. Mutahhar b. 'Ali
Study of Information Technology in New York.

Akram b. Zayd b. 'Ali
Study of Dentistry in San'a.

A'man b. 'Abd al-Rahman b. 'Ali
Study of Petrochemistry in the United States.

Muhammad b. Yahya b. 'Ali
Study of Law in San'a. Works at a Yemeni ministry and in his father's poultry company.

Majid b. ʿAbdullah b. ʿAli
Works as a tax officer for the government.

ʿAli b. Muhammad b. ʿAli
Educated in the United States; works for Citybank in Saudi Arabia and Texas.

Safiyyah bt. Ahmad b. ʿAli
Works as an office manager in the United States.

Sayyidah bt. Ahmad b. ʿAli
Worked as a nurse at Washington's Columbia Hospital; retired.

Safiyyah bt. Muhammad b. ʿAli
1980s employee in the Ministry of Education.

Muhammad b. Ahmad b. ʿAli
Born 1940s. Educated in the United States; Ph.D. in Political Science (Harvard); professor at Sanʿa University; emigrated to the United States in the 1990s.

Ahmad b. Muhammad b. Ahmad
Study of Computer Sciences in the United States

ʿAtikah bt. ʿAbdullah b. ʿAli b. ʿAli
Study of Literature in Cairo; teacher at a secondary school.

Muhammad b. ʿAbdullah b. ʿAli
Studied at the *Maʿhad al-ʿilmi*; 1980s Director of Finance at the *Maʿhad*. Used to be an active member of the Islah party but left recently.

Muhammad b. Hasan b. ʿAli
Teacher of English at computer company of his uncle ʿAbd al-Malik b. ʿAli.

Khadijah bt. Hasan b. ʿAli
Study of Dentistry in the United States; runs a surgery in Sanʿa.

Sayf b. ʿAbd al-Malik b. ʿAli
Study of Economics in the United States.

The House of Zayd b. ʿAli b. Muhammad

Ahmad b. Zayd b. ʿAli
Before 1962 worked at Ministry of Finance; 1980s customs officer in Saʿdah province.

ʿAbdullah b. Zayd b. ʿAli
During the 1980s worked as a customs officer.

The House of ʿAbdullah b. ʿAli b. Muhammad

ʿAbd al-Rahman b. ʿAbdullah b. ʿAli
Studied in Germany; after 1962 in Ministry of Transport; Captain in the Marine Forces; during 1980s worked at the Ministry of Transport.

ʿAli b. ʿAbdullah b. ʿAli
Study of Economics in Germany, Ph.D. Before 1962 advisor in the Ministry of Economics; 1990 Minister of Economics.

Muhammad b. ʿAbdullah b. ʿAli
Before and after 1962 in Ministry of Finance; 1980s worked in the tourism industry.

Ahmad b. Muhammad b. ʿAbdullah b. ʿAli
Studied in the United States; works for computer company belonging to ʿAbd al-Malik b. ʿAli b. ʿAli.

Muhammad b. Muhammad b. ʿAbdullah b. ʿAli
1980s worked in the Ministry of Water and Sewerage.

Bushrah bt. Muhammad b. ʿAbdullah b. ʿAli
Teacher in elementary school.

Salim b. ʿAbd al-Rahman b. ʿAbdullah
Trained as a dental technician.

Kamal b. ʿAli b. ʿAbdullah
Study of information technology in Moscow. Works as a computer specialist in Moscow.

The House of ʿAbbas b. ʿAli b. Muhammad

Muhammad b. ʿAbbas b. ʿAli
Studied at the *Madrasah al-ʿilmiyyah* and the Great Mosque; during the 1980s he was employed by the Ministry of Justice.

ʿAli b. ʿAbbas b. ʿAli
After 1962 Advisor at the Ministry of Civil Service and Administration.

ʿAbdullah b. ʿAbbas b. ʿAli
After 1962 he attended secondary school in Sanʿa; during 1980s employee at the Ministry of Finance.

Hasan b. ʿAbbas b. ʿAli
After 1962 study of Engineering in Algiers; during 1980s he was an employee of the Ministry of Public Works.

2. Bayt al-Wazir

Branch of Muhsin

Muhsin b. al-Hadi IV b. Salah
born late seventeenth/early eighteenth century (no data about his profession available)

Muhummad b. Muhsin b. al-Hadi
no data available

Muhammad b. Muhummad b. Muhsin
ʿAlim; secretary of Imam Yahya.

ʿAli b. Muhammad b. Muhammad
Born in 1850s. *ʿAlim.*

Ahmad b. Muhammad b. Muhummad
Born 1856. *'Alim*, supreme judge in Hajjah province during time of Imam al-Mansur and Imam Yahya; father of Imam 'Abdullah; died 1915.

Muhummad b. 'Ali b. Muhammad
Born 1894. *'Alim* of no great stature, had military leanings; supported by 'Abdullah b. Ahmad; helped recover Hudaydah from Idrisi forces; opposed the rule of Imam Yahya and spent years in prison; exile in Aden; executed in 1948 after the failure of the Constitutional Movement.

'Abdullah b. 'Ali b. Muhammad
Born 1913. Judge during the time of Imam Ahmad. After 1962 judge in Ministry of Justice. Died 1985.

Muhummad b. Ahmad b. Muhammad
Born 1887. Father's assistant in San'a. During time of Imam Yahya he acted as governor of Wasab near Dhamar when his brother was absent. In 1948 he was in charge of the *Qasr al-silah* (armoury). Died in the 1970s.

'Abdullah b. Ahmad b. Muhammad
Born 1889. Studied in al-Sir (Hijrat Bayt al-Wazir) and San'a (at the Great Mosque and al-Wushali Mosque under Ahmad al-Kuhlani and 'Abd al-Wahhab al-Shamahi). Governor of Dhamar at the age of 26. Pursued appeasement policies among various tribes such as Arhab and Hashid on behalf of Imam Yahya. Representative of Yemen during the peace negotiations between Saudi Arabia and Yemen after the war of 1934. Since 1940 no official position. Became Imam in February 1948; executed by Imam Ahmad later that year.

'Abd al-Quddus b. Ahmad b. Muhammad
Born 1904. Studied at the Great Mosque and the Madrasah al-'ilmiyyah; Deputy Governor of Dhamar and later Governor of Yarim; married to Imam Yahya's daughter Amat al-Rahman. Died of illness 1943.

Ahmad b. Muhummad b. 'Ali
Born 1919. *'Alim, hakim* (judge) of Jabal Sabir near Ta'izz; imprisoned in 1948 after the failure of the uprising; after 1962 vice minister at the Ministry of Justice; after 1962 Head of the Court of Appeal in Hudaydah; senior judge at the Court of Appeal in Ta'izz province; died June 2003 (father of Isma'il who held the post of Minister of Justice in 1997).

'Abd al-Samad b. Muhummad b. 'Ali
Born 1925. Member of the Royal Court (*al-diwan al-malaki*); worked at the Ministry of Endowments. After 1962 he worked as a judge at the Ministry of Justice and in various provinces. Died 1987.

'Abd al-Malik b. Muhammad b. 'Ali
Date of birth unknown. Member of the Royal Court in Ta'izz during the time of Imam Ahmad. After 1962 judge in Wasab; died same year.

'Abbas b. Muhummad b. 'Ali
Born 1936. Traditional education in San'a; employee of the Foreign Office before and after 1962; Consul in Syria, Saudi Arabia, and the United Arab Emirates. Presently at Foreign Office in San'a.

Yahya b. ʿAbdullah b. ʿAli
Born 1941. Studied in Prague and Cairo. Poet.

Nabil b. ʿAbdullah b. ʿAli
Born 1970. Civil engineer; heads the Yemen Heritage and Research Center in Sanʿa and works with the *Ittihad* party (see below, Ibrahim b. ʿAli b. ʿAbdullah).

Muhammad b. ʿAbdullah b. ʿAli
Born 1976. Poet and government employee.

Yahya b. Muhammad b. Ahmad
Died at the age of 22.

ʿAbdullah b. Muhammad b. Ahmad
Born 1929. Served as secretary of Imam ʿAbdullah al-Wazir in 1948. Executed.

Ibrahim b. Muhammad b. Ahmad
After 1962 Member of Parliament; at present president of the *Hizb al-ʿamal al-islami* (Islamic Worker's Party) and owner of *al-Balagh* newspaper; publishes on political issues; built a mosque in the quarter of Bustan al-Sultan of Sanʿa where he teaches.

Muhammad b. Muhammad b. Ahmad
After 1962 employee of the Foreign Office; Counsellor at the Yemeni Embassy in Oman; at present businessman (owner of *Dar al-Hikmah al-Yamaniyyah* publishing house; representative of Italian companies).

ʿAbd al-Rahman b. ʿAbdullah b. Ahmad
After 1948 in prion in Hajjah and Sanʿa. Employee of the Ministry of Health and the Ministry of Finance before and after 1962. Later employee at the Department of Antiquities (died 1987).

ʿAli b. ʿAbdullah b. Ahmad
Born 1947. After 1962 study of History and Philology in Yugoslavia; M.A. At present general manager of the Ministry of Oil and Mineral Resources.

Yahya b. ʿAbdullah b. Ahmad
Born mid-1940s. Before 1962 study of Medicine in Moscow; obtained a Ph.D. from a British university. Director of a private clinic in Sanʿa. Lecturer in Medicine at Sanʿa University.

Ahmad b. ʿAbdullah b. Ahmad
Born 1940s. Attended secondary school in Cairo. Study of archives in Hungary. Under President al-Hamdi archivist in Presidential Office. Worked as an archivist in the Department of Antiquities; currently runs a private company trading in gem stones.

Mutahhar b. ʿAbdullah b. Ahmad
Born 1941. After 1948 in prison in Hajjah and Sanʿa. Until 1960 secondary school in Cairo; study of Political Science and Economics at the University of Prague. 1966 Advisor of President Sallal for military and tribal affairs. After 1967 deputy minister without portfolio. At present businessman. Owner of al-Saʿida export/import company.

Muhammad b. ʿAbdullah b. Ahmad
Born late 1930s. After 1948 in Hajjah prison. Before 1962 trainee at the *Hayʿah al-sharʿiyyah* (Sharʿiah Tribunal), and for a short period at the Court of Appeal. Secondary school in Egypt; study of Political Science in Italy. Consul at Yemeni embassies in Czechoslovakia and Ethiopia. Chief secretary of Department of African Affairs at Foreign Office until 1996. Until July 2000 ambassador in Rome. Retired.

ʿAbd al-Karim b. ʿAbd al-Quddus b. Ahmad
Born 1932. During the time of Imam Ahmad Director General at the Foreign Office; Head of Public Security in Taʿizz. After 1970 Vice Minister of Ministry of Local Administration; then employee at the Foreign Office; Minister of Plenipotentiary in Iraq, Britain, Holland, Morocco, Tunisia; died 1999.

Muhammad b. ʿAbd al-Quddus b. Ahmad
Born 1933. Secondary education in Cairo; study of Economics and Political Science in Beirut; 1959 Chargé d'affaires at the Yemeni embassy in Rome; 1960 Minister of Plenipotentiary; 1970 ambassador to Lebanon; 1974 Minister of Agriculture; 1975 ambassador to Jordan; 1981 ambassador to Pakistan; 1986 ambassador to Japan. After the war of 1994 ambassador to the Arab League in Cairo; retired in Beirut.

ʿAbd al-Bari b. Ahmad b. Muhammad
Study of statistics in Cairo; M.A. from the United States. Works as an accountant at the Ministry of Oil and Resources.

ʿAbd al-Munʿam b. Ahmad b. Muhammad
Assistant at the Legal Office of the State; at present works as a lawyer at his private office.

Ibrahim b. Ahmad b. Muhammad
Study at Military College in Amman in 1960s. Deputy of the general manager of the Yemen Petroleum Company in Hudaydah.

Ismaʿil b. Ahmad b. Muhammad
Born 1940s. Study of Law in Cairo and Damascus. In 1970 Legal Advisor to the Parliament (*Majlis al-Shurah*); 1979 Minister of Justice; 1987 Advisor in the Legal Office of the State; Minister of State for Legal Affairs and Legal Advisor of the President; 1997–2001 Minister of Justice. Since 2001 member of the Consultative Council; works as a private legal consultant.

Muhammad b. ʿAbd al-Samad b. Muhammad
Study of Business Administration at the University of Sanʿa, M.A. from an American university; employee of the Ministry of Local Administration; since 1997 he has worked at the Ministry of Finance.

Ahmad b. ʿAbd al-Samad b. Muhammad
Works for the Canadian Oil Company Nexen.

ʿAbdullah b. ʿAbd al-Malik b. Muhammad
General Manager at the Ministry of Trade and Supply. Representative of the ministry in Saʿdah.

ʿUsam b. ʿAbd al-Malik b. Muhammad
Study of Civil Engineering in Cairo and Sanʿa. Works for the United Nations in Sanʿa.

Fadl b. ʿAbbas b. Muhammad
Study of Business Administration in the United States; since 1980 employee of the Ministry of Planning (CPO).

Muhammad b. Ibrahim b. Muhammad
Born late 1950s. Director General of *Dar al-Hikmah al-Yamaniyyah* publishing house; bookseller.

Ismaʿil b. Ibrahim b. Muhammad
Born 1950s. Lecturer at the Department of Law, Sanʿa University.

ʿAbdullah b. Ibrahim b. Muhammad
Journalist with *al-Balagh*.

Huda bt. Ibrahim b. Muhammad
Teacher in an elementary school.

Tariq b. Muhammad b. Muhammad
Born in the late 1960s; educated in Germany and Yemen; Member of the German Parliament, representative of the Green Party.

Faris b. Ahmad b. ʿAbdullah
Works as a physician in Budapest.

ʿAbdullah b. Mutahhar b. ʿAbdullah
Born 1972. Finished high school in 1991, studied Business Administration and Management at the American University in Washington; works in his father's company.

Nazar b. Muhammad b. ʿAbdullah
Born 1970s. Study of Engineering in the United States. Works for an engineering company in Atlanta, Georgia.

Ismaʿil b. Muhammad b. ʿAbdullah
Study of Medicine in the United States, MD. Works at a hospital in Dallas, Texas.

Nidal b. ʿAbd al-Rahman b. ʿAbdullah
Study of Computer Science in the United States; worked for a foreign oil company in Yemen, at present employed at the Ministry of Justice.

ʿAziza bt. ʿAbd al-Rahman b. ʿAbdullah
Born mid-1950s. Studied Law at Sanʿa University, B.A.

Usamah b. ʿAbd al-Karim b. ʿAbd al-Quddus
Born 1959. Study of Business Administration at the American University in Washington. Promotion Manager for the World Gold Council in Saudi Arabia.

Amal bt. ʿAbd al-Karim b. ʿAbd al-Quddus
Born 1953. Studied Home Economics in Saudi Arabia. Runs a financial investment office for women in Jiddah.

Khalid b. ʿAbd al-Karim b. ʿAbd al-Quddus
Born 1965. Study of Computer Science in Washington. Works for a telecommunication company in Washington.

Ahmad b. Muhammad b. 'Abd al-Quddus
Born 1953, San'a. Study of Economics in Washington, M.Sc.; ran an import/export business in Washington before he moved to Riyadh.

Buthaynah bt. Muhammad b. 'Abd al-Quddus
Born 1958, San'a. Studied International Relations and English literature at the American University in Washington, B.A.; M.Sc. in International Studies at the University of Islamabad where her father served as an ambassador. Until 2003, she worked at the United States Information Services in Riyadh. She now lives in Washington.

'Abd al-Quddus b. Muhammad b. 'Abd al-Quddus
Born 1967. Studied Humanities in Washington. He was employed as an image setter in his cousin's company in Cairo, now living in Beirut.

Branch of Muhammad

Muhammad b. al-Hadi IV b. Salah
'Alim.

'Abdillah b. Muhammad b. al-Hadi
Born late eighteenth/early nineteenth century; *'alim.*

Muhammad b. 'Abdillah b. Muhammad b. al-Hadi
Imam in 1854.

'Abdullah b. Muhammad b. 'Abdillah
'Alim; taught in al-Sir; died during *hajj* (pilgrimage) in Mecca in 1893.

Ahmad b. 'Abdullah b. Muhammad b. 'Abdillah
Born 1868. Was taught by his grandfather Imam Muhammad b. 'Abdullah and his father; worked as a judge; died young.

'Ali b. 'Abdullah b. Muhammad
Born 1885. Governor of Ta'izz, later Prime Minister in 1948; executed by Imam Ahmad in 1948.

Muhammad b. 'Abdullah b. Muhammad b. 'Abdullah
Born 1869. *'Alim*, judge during the Ottoman occupation and under Imam al-Mansur Muhammad; no employment under Imam Yahya.

'Abdullah b. 'Ali b. 'Abdullah
Born 1907. During the time of Imam Yahya governor of Dhi Sufal near Ta'izz. In 1939 he formed an opposition group in Cairo. Involved in the 1948 uprising. Died in exile from the effects of poisoning in the same year.

Ahmad b. 'Ali b. 'Abdullah
Born 1911. Supervisor of family property. Died 1993.

'Abbas b. 'Ali b. 'Abdullah
Born 1928. During the reign of Imam Yahya he studied with his father and at the Madrasah al-'ilmiyyah. After 1948 imprisoned in Hajjah; on release went to Cairo where he worked as vice president of the opposition party led by al-Zubayri and Nu'man; one of the founders of the opposition party *Hizb al-Shurah* (est. in 1958). During the civil war (1962–70) he was active in the

"Third Force" which allied itself neither with the republicans nor the royalists. After 1962 he settled in Khawlan among the tribes as an advisor, organising self-help groups to build schools and water wells.

Muhammad b. ʿAli b. ʿAbdullah
Born 1938. Study of Economics in Prague, M.A. Until late 1990s publisher in Jiddah; at present import/export business in Jiddah.

Ibrahim b. ʿAli b. ʿAbdullah
Born 1933. Since 1946 study in al-Sir under Ahmad b. Muhammad b. ʿAli al-Wazir. After 1948 student at the Madrasah al-ʿilmiyyah. Founder of the opposition group ʿUsbat al-haqq wa-ʾl-ʿadalah (Group of Truth and Justice); after imprisonment he was active in the opposition against Imam Ahmad in Cairo. After 1962 short stay in Sanʿa; in 1964 he established the "Third Force." Chairman of the Ittihad al-quwwah al-shaʿabiyyah al-yamaniyyah (Yemeni Union of Popular Forces), founded in 1960. Writer on political and religious issues. Owner of al-Shurah newspaper; lives in Maryland and Jiddah (on his childhood see chapter 4).

Zayd b. ʿAli b. ʿAbdullah
Born 1935. During the time of Imam Ahmad he studied at the mosques of al-Sir, Sanʿa, and in Hajjah prison. Member of the Hizb al-Shurah and al-Ittihad. Before 1962 he worked at the Yemen embassy in Beirut and Bonn. At present historian and writer on political issues; lives in Maryland and London.

Qasim b. ʿAli b. Abdullah
Born 1937. Study in Hajjah prison. Writer, poet, politician; until 1982 advisor to President ʿAli ʿAbdullah Salih; leading member of al-Ittihad; business interests; lives in Maryland.

Muhammad b. Muhammad b. ʿAbdullah
Born 1900. Court Judge (hakim al-maqam) during time of Imam Yahya. Involved in the 1948 uprising. Executed by Imam Ahmad.

ʿAbdullah b. Muhammad b. ʿAbdullah
Born 1926. Studied at the Madrasah al-ʿilmiyyah and in Hijrat al-Sir; no employment before 1962. After 1962 worked in the Ministry of Education; retired.

Yahya b. Muhammad b. ʿAbdullah
Born 1923. Studied in al-Sir and Taʿizz; ʿalim. No employment before 1962. After 1962 worked at the Ministry of Justice.

ʿAli b. Muhammad b. ʿAbdullah
Born 1913. Before 1962 assistant of the governor of Hujariyyah province; after 1962 judge in Ministry of Justice and at the Court of Appeal. Retired in 1992.

Ahmad b. Muhammad b. ʿAbdullah b. Muhammad
Born 1916. During the time of Imam Yahya secretary and deputy of ʿAli b. ʿAbdullah b. Muhammad; governor of Shibam. After 1948 imprisoned for eight years. After 1962 Director of the Madrasah al-ʿilmiyyah; after its closure he worked as a judge at the Court of Appeal in Sanʿa; hakim al-awwal in Anis. Author of books on Yemeni history (Hayat al-amir ʿAli b. ʿAbdullah al-Wazir, 1987). Died 2002.

Ahmad b. Muhammad b. Muhammad b. ʿAbdullah
Born 1922. Judge at Court of Appeal in Sanʿa, *hakim al-awwal* (supreme judge) Sanʿa; president of the Court of Appeal in Ibb.

ʿAbd al-Jalil b. Ahmad b. Muhammad b. ʿAbdullah
Born late 1940s. Study of Medicine in Cairo (Ph.D.); works at the military hospital in Sanʿa.

Salah b. Ahmad b. Muhammad
Studied at Sanʿa University; bookseller.

Muhammad b. Yahya b. Muhammad
Born in late 1940s. Study of Accountancy in Egypt. Works for *al-Ittihad* in Sanʿa.

ʿAbd al-Rahman b. Yahya b. Muhammad
Study of Business Administration in Egypt and the United States. Works for *al-Ittihad* in the United States.

Al-Baraʾ b. Qasim b. ʿAli
Born 1980. Study of Business Administration/Management information systems (MIS) at George Mason University in the United States.

Ruqayyah bt. Qasim b. ʿAli
Born 1977. Study of Psychology at Montgomery College in Maryland, United States.

Najiyyah bt. Qasim b. ʿAli
Born 1982. She attended the Islamic Saudi Academy in Washington; studied International Relations of Development in Brussels.

ʿAliyyah bt. Qasim b. ʿAli
Born 1983. Studying at the Islamic Saudi Academy in Washington.

ʿUmar b. Qasim b. ʿAli
Born 1985. At high school in Washington.

Tariq b. Zayd b. ʿAli
Born 1959. Attended the elementary school in Jordan in the 1960s; finished his junior high school in Lebanon in 1975; completed his high school at Fettes College, Edinburgh. Studied Civil Engineering in the United States; Ph.D. Holds three master degrees in civil engineering, structural engineering, and public work administration. Conducted post-doctoral research. At present works at the Washington branch of the Yemeni Heritage and Research Center.

Afaq bt. Zayd b. ʿAli
Born 1964. Study of Chemistry in the United States, B.S. and M.S. Part-time teacher at the Islamic Saudi Academy in Washington; worked as an accountant at her brother's real estate company; lives in London.

Luʾay b. Zayd b. ʿAli
Born 1964 (twin brother of Afaq). B.S. in Fine Arts at Rhode Islands School of Design and Bachelor of Architecture; M.S. in Architecture at Columbia University; worked as an architect and real estate developer in Washington, currently working in Sanʿa.

Atyaf bt. Zayd b. ʿAli
Born 1979. Study of Communication, Law, Economics & Government (CLEG) at the American University of Washington, DC; works for the National Endowments for Democracy in Washington.

Dhilal bt. Zayd b. ʿAli
Born 1982. Final year of Islamic Saudi Academy; study of Special Education at George Mason University in Virginia.

Al-Hadi b. Ibrahim b. ʿAli
Born 1968. Study of Aerospace Engineering in Maryland University, B.A.; M.Sc. in Engineering from Georgetown University. Worked for NASA; later worked as a banker in New York, now private business.

Al-Mahdi b. Ibrahim b. ʿAli
Born 1971. Study of Marketing at the American University in Washington, M.A. Worked in Bahrain in advertisement company; representative of Coca Cola for the Gulf States, lives in Manama.

Al-Rida b. Ibrahim b. ʿAli
Born 1977. Study of Economics and Law at Maryland University, B.A. Works for Unilever in Dubai.

Rahmah bt. Ibrahim b. ʿAli
Studies Psychology at George Washington University.

Najwa bt. Muhammad b. ʿAli
Born 1963. Relationship branch officer at First Virginia Bank in Virginia.

Mawaddah bt. Muhammad b. ʿAli
Born 1970s. Studied English literature in Jiddah. Works at the American embassy in Jiddah.

Ibrahim b. Muhammad b. ʿAli
Born 1979. Study of Economics at the University of Arizona.

ʿAli b. Muhammad b. ʿAli
Born 1965. Studied Computer Science at the American University in Washington, M.A degree. Works for IBM in Jiddah.

ʿAbdullah b. ʿAbbas b. ʿAli
Born 1948. Accountant of *al-Ittihad al-quwwah al-shaʿabiyyah* in Washington.

Suhayb b. ʿAbbas b. ʿAli
Born 1974. Study of Political Science at Goldsmiths College, University of London; (M.Sc.).

ʿAbbas b. ʿAbbas b. ʿAli
Born 1979. Studied Politics and Law at the University of Hartford in Conneticut.

Fausia bt. ʿAbbas b. ʿAli b. ʿAbdullah
Born 1952. During 1970s and 1980s teacher in an elementary school; homemaker. Now living in the United States.

Husayn b. Ahmad b. ʿAli
Born 1934. Successor to his father who supervised the family property.

Salim b. Ahmad b. ʿAli
Born in 1960s. Lives in Sanʿa; looks after the interests of his uncles who live abroad.

Appendix III: Genealogical Histories

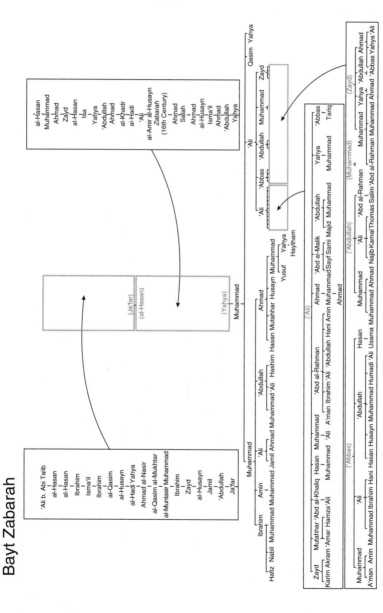

Figure A III.1 Genealogy: Bayt Zabarah

Bayt al-Wazir

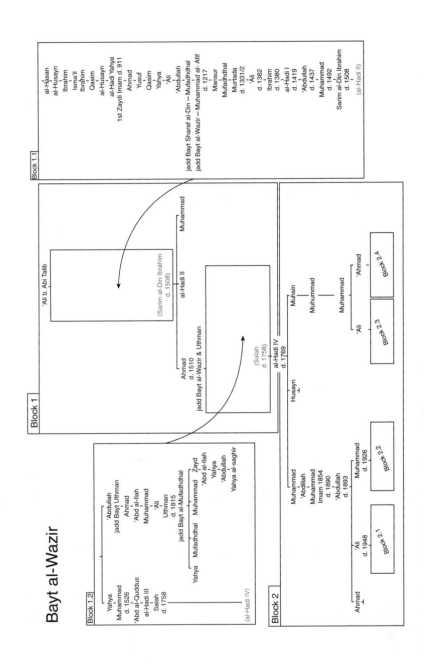

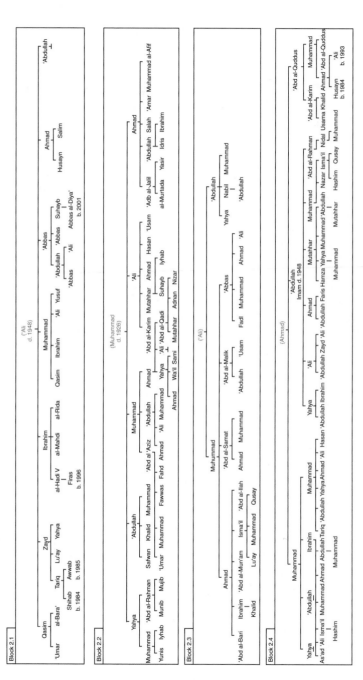

Figure A III.2 Genealogy: Bayt al-Wazir

272

Figure A III.3 Genealogy: Bayt Ibrahim

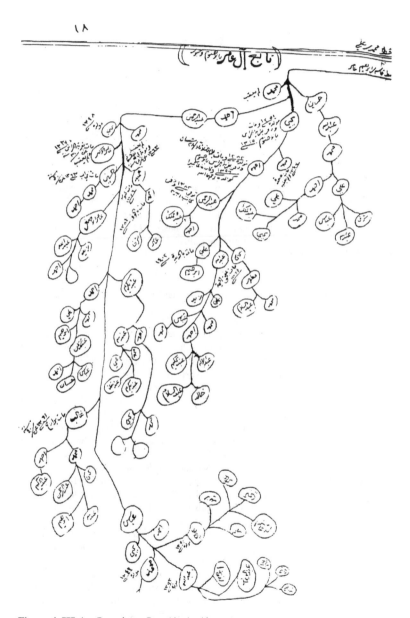

Figure A III.4   Genealogy: Bayt 'Amir, Ahnum

Notes

Introduction

1. Serjeant (1957: 3).
2. J. Butler (1997b: 96).
3. The term Imamate is a translation of *imama*, the "supreme leadership" of the Muslim community after the Prophet's death (Madelung 1971: 1163). The Zaydi school of Islam (Zaydiyyah) distinguishes itself from the mainstream Shi'a (Imamiyyah or Twelver-Shi'a) by recognizing Zayd b. 'Ali, the great-grandson of 'Ali b. Abi Talib, as its founder (see chapter 1).
4. Madelung (1965; 1988b: 88–91; 2002); Daum (1987b: 169). The most critical element of this communication was Maimonides's prediction that the messiah was to arrive in the year 1210 (Maimonides 2002).
5. I should note that references to the "Imamate," or "the past" relate to the twentieth century unless specified.
6. The ascendancy of the Hashimite houses of the Hijaz (1908–25), Iraq (1921–58), and Jordan (1921–) began with al-Husayn b. 'Ali (b. 1852) and his sons, 'Ali, 'Abdullah, and Faysal who were brought to Istanbul by the Ottomans in 1894 (Morris 1959; Dawn 1971: 263; Susser 1995). At the end of the thirteenth century, the 'Alawis came from Arabia to North Africa where they founded a dynasty which consolidated itself in the seventeenth/eighteenth century. The Moroccan Sultan Muhammad VI, head of the oldest monarchy still to wield actual power, belongs to that dynasty. The monarchy was founded in 789; the contemporary ruling dynasty came to power in the mid-seventeenth century (Terrace 1960: 355–8). The first Shi'i (Fatimid) caliphate in North Africa was founded by the Fatimids (909–1171) (Canard 1965: 852).
7. Layne (1994: 22); Susser (1995: 1); Kostiner (1995: 62); Teitelbaum (2001). The Sharifian state of Mecca in the Hijaz, whose last ruler was the Sharif Husayn b. 'Ali, was founded about 968 and lasted until 1925. It was governed locally by descendants of the Prophet through his grandson al-Hasan (Teitelbaum 2001: 11).
8. Since the abolition of the Imamate, Ayatullah Ruhallah Khumayni (1978/79–89) and Hujjat al-Islam Muhammad Khatami (1998–2005) remained the only Shi'i leaders who descended from the House of the Prophet. They have themselves refrained from referring to their descent, but their supporters have stressed this issue in order to buttress their legitimacy.

Literature which was published on Middle Eastern political elites in the 1970s and 1980s reflects the rise of new political forces. In Lenczowski's volume on political elites in the Middle East (1975), Hashimite leadership is mentioned only in the context of past regimes. Tachau's introduction to his edited volume on the same subject (1975) makes no reference to the Hashimites; nor does

Wenner's article on the survival of traditional elites in Saudi Arabia (same volume). In his seminal work on Middle Eastern elites, Zartman (1980) discusses the monarchies of Egypt, Iraq, Libya, Jordan, Iran, Morocco, and Saudi Arabia, but draws attention neither to the Hashimites nor to the Yemen. Susser and Shmuelevitz's *The Hashemites in the modern Arab world* (1995) does not even contain a reference to the Yemeni Hashimites despite the remarkable longevity of their rule. In spite of the book's focus on the Hashimite kingdoms of the Hijaz, Jordan, and Iraq, Hashimites are defined as the "descendants of Muhammad the Prophet" (Susser 1995: 1). Yemenis refer to the Prophet's descendants as either *sadah* or Hashimites (see chapter 1).

9. See Dresch (2000). I should also note that this study does not primarily aim at analyzing the *sadah* within the framework of either urban or national elite politics (e.g., Waterbury 1970; Bujra 1971; Schatkowski Schilcher 1985; Denoeux 1993: 99–133). Although my concern is to understand the lived experience of the ruling 'Alid families after the state they had founded was dismantled, their role in the state is given only minor consideration. By drawing out the personal and social characteristics of this elite, in certain respects the study follows a tradition initiated by the members of the M.I.T. School of Elite Studies (among them most notably Frey 1965 and Zonis 1971), without pursuing their objectives.

10. I use the term 'Alid for want of an appropriate adjective of *sayyid* and the term Hashimite when indicating descent from the Prophet irrespective of whether or not it is traced to 'Ali and Fatima.

11. Compare Zartman (1980: 1–2); on Morocco, Brown (1976: 71). The establishment of protectorates by colonial powers also served to destabilize the elites (Eickelman 1976). Writing about Morocco, Brown (1976: 74) says that "by the turn of the century it had become progressively more difficult and less useful for a man to forge an identity and create a livelihood on the basis of descent from the Prophet." In the Ottoman Empire, the power of the 'ulama had gradually declined toward the end of the eighteenth century (Heyd 1993: 40). In Egypt the effect of the reforms carried out by Muhammad 'Ali in the early nineteenth century was the destruction of the old leadership. The new administrators often came from religious minorities or from a humble background; the *iltizams* (tax farming) were abolished, new legal codes were developed and the old system of religious education undermined (Hourani 1968: 55–7). On Syria see Khoury (1984: 519); Schatkowski Schilcher (1985); Schroeter (1988: 186). In Syria, especially the 'ulama lost their control over the religious, judicial, and educational institutions (Ma'oz 1971).

12. The Yemen had previously been occupied by the Ottomans from 1539–1635. The descendants of the Prophet enjoyed prestige throughout the Ottoman Empire (Voll 1975: 55).

13. However, as Wedeen (2003: 682) notes, "there is also little evidence to suggest that the incumbent regime has succeeded in constructing a sense of membership that is coherent and powerful enough to tie people's political allegiances to the nation-state."

14. An exception is the Civil Code (*al-qanun al-madani*) which guarantees the Jews special rights in accordance with their religion.

15. In the Yemen those referred to as "tribes" (*qaba'il*) are people in pursuit of settled agriculture in the mountainous areas who sustain a memory of Arab tribal descent as specified next (Yapp 1987: 4). The term "tribe" is a conventional rendering of the Yemeni term *qabilah*, which refers to a distinctive social and political unit

within a given territory. On the role of the tribes in Yemeni history see Adra (1982); Gochenour (1984); Meissner (1987); Dresch (1989, 1991); Freitag (2000). On the debate whether northern Yemeni cultivators should be depicted as tribesmen or peasants see Mundy (1995). Recently, Dresch (2000: 24) argued that "in many areas the tribesmen, so called, did not belong to really much of a tribe. Their relation to their local shaykhs was largely that of peasants to greater landowners." For a scathing critique of the analytical category "tribe," see Blumi (2004). The impact of fashionable paradigms on recent scholarship on the Yemen is apparent in Willis's (2004: 124) conception of "tribe" in the former Protectorate as an imagined social space and "a form of colonial knowledge."

16. Government officials are aware that the exaltation of the *qabilah* may be a double-edged sword because the *qaba'il* are both incorporated into the government and a potential threat to its stability. Government-sponsored textbooks on national education ("*Al-tarbiyyah al-wataniyyah*") do not make any reference to the tribes. This does not justify inferences like those made by Shryock in regard of Jordanian textbooks. "The conclusion the student is meant to draw from this text is rather obvious: the *modern* history of the Arabs has nothing to do with tribes, nothing to do with subnational loyalties, nothing to do with political identities that are not inclusively Arab or Muslim" (1997: 305; emphasis his). For a similar observation regarding Saudi Arabia, see al-Rasheed (1999: 31).

17. These images are held predominantly by rivals of the *sadah* and are by no means endemic in the society at large. In the present political context, their significance is different from more generic classifications of the *sadah* as foreigners by people such as the Muslim Meos in northern India (Jamous 1996).

18. Glaser (1885: 202) boldly attributed the decline of the southern part of the Peninsula to this age-old conflict. "Since the penetration of Islam with which they entered the country, [the *sadah*] have endeavored to take power away from the descendants of the old Himyarite mountain princedom. In this continuous feud of the new aristocracy against the old lies the main cause for the decline of South Arabia." According to Halliday (2000: 63), Yemeni nationalists set "the sons of Qahtan (the Yemenis) against those of Adnan (the Bedouin, Saudis, etc.)." It would appear, however, that in the first instance, the claims to supremacy entailed in this discourse aim at the main component of the former indigenous political elite, the *sadah*. The Saudis trace descent either to Qahtan or 'Adnan. However, the prevalence of these identity labels should not invite analysis of political controversy in terms of an archaicized ethnic conflict (see, e.g., Wenner 1991: 22).

19. Yemeni nationalism nourishes itself on these ancient images. It must be stressed, however, that this is only one facet of Yemeni nationalism that co-exists with Islam (see chapter 10). I use the term "other" with some hesitation for it reproduces what it professes to deconstruct. Analytical constructs of the "other" are no less essentializing than those held by people who pursue a politics of "othering" and can therefore only serve as a heuristic device. As discussions of Yemeni low status categories in subsequent chapters make clear, the category of the "other" is also problematic because it assumes a false homogeneity of the majority against whom it is pitted. On this issue, see Bammer (1994). As noted by Derrida (cited in Moraru 2000: 52), othering is never absolute; each category, self and other, "is already 'infected' by the elements they purport to exclude: their conceptual 'other.' " What is crucial is that here the term specifies relational social configurations within the same national entity without excluding cross-cultural reference (see Feldman 1994: 416, n. 6).

On the problematic use of generic terms in academic writings, see conclusion, esp. n. 3.

20. The *Mithaq al-watani* (National Charter) was issued by the ruling party, the General People's Congress (al-Mawtamar al-sha'abi al-'amm), in August 1982. On the background and nature of the charter, see al-'Amri (2000: 4–5). The ancient temple of Marib, a symbol of Sabean state power, became the emblem of the Yemen Arab Republic (and the unified Yemen), and the university of San'a. Another potent national symbol is the *janbiyyah*, a dagger that has traditionally been worn by the tribesmen. The oil refineries represent prosperity and progress.

21. Al-Ayni (1973; emphasis mine). Five years before the revolution, when he and other reformers placed their hope on the tribes as national liberators, he wrote "all the tribes agreed on the rule of law: that things must be in the future as they were in the days of Ma'in, Qataban, Saba' and Himyar" (quoted in Dresch 1989: 242).

22. I borrow a term from Boyarin (1992: 77).

23. Anonymous personal communication; see Zubayri in Serjeant (1979). Since the mid-1950s, Muhammad b. Ahmad Nu'man described Zaydi rule in the Sunni parts of the country as an occupation (*ihtilal*) (Ahmad M. Nu'man, personal communication).

24. An extension of this concept is Shaykh al-Ahmar's reformulation of pan-Arabism in tribal terms. As stated, he conceives of the state as being constituted by tribal units. Yemeni nationalism—and by implication Arab nationalism—is defined here through notions of kinship (which are central to constructions of *qabilah*) and is sanctioned by the Qur'an. The fusion of kinship, regimes of rule, and religion, and an (albeit weak) concept of nationalism that entailed universalist claims also characterized Imamic constructions of the state.

25. In a poem eulogizing the Yemeni tribe Hamdan which had expressed loyalty to Imam 'Ali and was called upon to support the claim to the Imamate of the first Zaydi Imam in the Yemen, the latter says: "I hold them in the highest esteem since they are all, the sons of women and men of proud birth. *They share in the glorious rank of Muhammad's progeny*, surpassing all others of exalted line and lineage" (quoted in Eagle 1994: 113; emphasis added).

26. Although before 1990 genealogical charisma had quite different social and political manifestations in North and South Yemen, the proclamation of independent republics in both countries (1962 and 1967 respectively) signified the end of the *sadah*'s preeminence (Mermier 1999: 14).

27. As part of the religious elite, the *sadah* have been analyzed as leaders of local groups and religious communities (Mousavi 1998); as legal experts, mediators in dispute settlements and patrons of local shrines in predominantly rural settings (Evans-Pritchard 1949; Berque 1955; Barth 1959; Peters 1963; Gellner 1969; Crapanzano 1973; El-Zein 1974; Rabinow 1975; Brown 1976: 69; Eickelman 1976; Rassam 1977; Christelow 1980; Jamous 1981; M. Marcus 1985; Schroeter 1988: 34,97; Ensel 1998). Sanadjian (1996) focused on the *sadah* as mediators between local and national legal representations. Other writers explored their role in urban elite politics (Waterbury 1970; Bujra 1971; Brown 1976), their self-perception as a distinct group (Wachowski 2004), and as rulers (Hammoudi 1997; Cunningham 1999; Tripp 2000: 108–47; Milton-Edwards and Hinchcliffe 2001; Teitelbaum 2001).

28. A typical example of Zabarah's (1984: 124) description of famous Yemeni personalities until the mid-twentieth century is that of Sayyid Muhammad al-Bulugh of San'a. He introduces him by listing the names of some of his

forebears: "Sayyid Muhammad (called al-Bulugh) b. 'Abdullah b. 'Abd al-Rahman b. al-Imam al-Mahdi 'Abbas b. al-Mansur al-Husayn b. al-Mutawakkil Qasim b. al-Husayn b. al-Mahdi Ahmad b. al-Hasan b. al-Imam al-Qasim b. Muhammad al-Hasani al-San'ani. From then in our time [1950s] is the Director of Industry who is also the imam [prayer leader] of the Mutawakkil mosque, the scholar (*'alim*) Ahmad b. Qasim b. Muhammad b. 'Abd al-Rahman b. al-Mahdi 'Abbas."

29. Tulving (1983: 21, 28) characterizes semantic memory as "a mental thesaurus" and "knowledge of the world" (see also Neisser 1988: 553). For a critique of Tulving's distinction, see Baddeley (1989); Fentress and Wickham (1992: 20–2); Brewer (1996: 54–6); Bloch (1998: 114–27).

30. The widely varied cultural modalities of remembering and forgetting are indicative of how the person is conceived and vice versa. Humans do not inevitably remember the past (Bloch 1996: 229), and some writers have argued that the past does not determine what a person "is" at any one point (e.g., Astuti 1995a: 153–4; Bird-David 2000). These examples constitute one of the extremes on the continuum of memory worlds. Graeber's (1999) study of the descendants of Malagasy slaves reveals that memories matter but are never channeled into historical consciousness and oral history. At the other end of the continuum we find "anamnestic generational traditions" (Boyarin and Boyarin 1995: 17) which have fostered an obsession with memory by victims of survivors of Eastern European pogroms who have produced monumental memorial books (e.g., Wachtel 1990; Slyomovics 1998: xiii). Boyarin (1994: 23) argues that memory constitutes the person to the extent that identity and memory are virtually the same concept. On the "command to remember" in Jewish thought, see Yerushalmi (1982); Nora (1989: 16). Asserting that "only in Israel and nowhere else is the injunction to remember felt as a religious imperative to an entire people," Yerushalmi (1982: 9) fails to take Shi'i teachings into account.

31. Alternatively "imitation," the principle of following the doctrines of the *mujtahids* who descended from 'Ali and Fatima; the opposite of *ijtihad* (interpretive judgment of religious tenets) (see Madelung 1982: 169). As de Groot (1983: 22) explains, the views of religious authority as embodied in the *ijtihad* tradition stems from the emphasis on personal religious leadership among Shi'is. A component of the collective memory of traditionally educated *sadah* and Zaydis generally are works such as the *Sharh al-azhar*, based on the *Kitab al-azhar* (The Book of Flowers) by Ahmad b. Yahya al-Murtada (d. 1432), a compilation of the verdicts of Zaydi 'ulama combined with a commentary by 'Abdullah b. Miftah (d. 1472). (Note, however, that the verdicts of Sunni 'ulama, especially of the Hanafi school, are also referred to.)

32. I am not concerned with the nature of history writing nor with the authenticity of texts. On this subject see, e.g., P. Burke (1989); Dresch (1991); Jalal (1995); al-Rasheed (1998: chap. 5). On reconstructions of history which challenge official versions, see Shryock (1997); Karakasidou (1997); Sutton (1998: 3); al-Rasheed (1999). The notion that "history is kinship" has been elaborated by other scholars working in such diverse cultures as Amazonia (Gow 1991) and Langkawi (Carsten 1997). However, whereas 'Alid genealogies are conceived of as relatively stable over time, in Amazonia kinship relations reflect historical change. History consists in the ongoing transformation of kinds of people (Gow 1991: 270, 288).

33. They are particularly characteristic of the Shi'a which places greater emphasis on the *ijtihad* than the Sunnis. However, in Sunni circles processes of legal

interpretation never came to a halt (Eickelman and Anderson 1999: 12; for impressive examples, see Bowen 1998; Bälz 1999).

I should stress that I do not wish to give credibility to the storehouse conception of memory (for critiques see Trouillot 1995 and Bloch 1998: 118–19). I argue that the *taqlid* can work like personal memory precisely because of the simultaneous continuance and malleability that characterizes it. Conscious and purposeful mental operations alternate with incidental and involuntary remembering. For example, memories of statements, events in the lives of Imams and so forth may be triggered by emotional states (see, e.g., chapter 4). Here I am most interested in the ways memory is partial, fragmentary, and piecemeal.

34. On ritual practices, see Thaiss (1972, 1973); Chelkowski (1979); Schumacher (1987); Pinault (1992); Schubel (1993); Hegland (1995, 1997, 1998a, b); Torab (1996, 2002).

35. Likewise, Sunni ritual in the Yemen is much more elaborate. It is noteworthy too that the *taqlid*-based modes of reasoning this book focuses on are not tied to a *marja'iyyah* (lit. reference point for emulation, religious leadership) known to Iranian and Iraqi Shi'is.

36. There are few in India and Pakistan. However, those who adhere to Zaydi principles do not necessarily consider themselves to be followers of a "school" (Muhammad Kalisch, personal communication). Most Iranian Zaydis became adherents of the Twelver-Shi'a under the Safavids in the sixteenth century (Madelung 1988b: 92). According to Heine (1984: 146), there are some Zaydis in southern Saudi Arabia. In the Yemen the Zaydis inhabit most of the northern parts of the country. The Sunnis, who mainly adhere to the Shafi'i school, are most prevalent in the southern and the coastal areas where they make up about 60–65% of the population. Conflicts between Sunnis and Zaydis have centered on political rather than doctrinal divisions.

37. Compare Neisser (1988: 554) and Fentress and Wickham (1992: 25).

38. On this issue, see also Neisser (1988); P. Burke (1989: 98, 100); Rosenwald and Ochberg (1992: 1, 6); Peacock and Holland (1993: 368); Robinson (1996); Stoler and Strassler (2000); Garro (2001: 107); Neyzi (2002); Wilce (2002). Fascinating examples of this type of analysis are provided by Zerubavel (1991, 1995) and Zur (1997). They illuminate psychologists' assumptions that the loss of memory of personal experiences becomes transformed due to interference with the temporal coding of stored events (Tulving 1972: 391).

39. In recent times, Yemeni authors have provided fascinating accounts of their childhood (e.g., M.al-Akwa' 1980; al-Shami 1984).

40. They have been taken up again more recently by writers such as Bruner and Fleisher Feldman (1996).

41. For an exemplary study of the reproduction of domestic economies in the context of wider political processes in rural Yemen, see Mundy (1995).

42. However, there are Marxist analyses of the traditional order as a class society by Yemeni authors (see Mundy 1995: 7; Halliday 2000: 68).

43. Cohen cited in Stoler and Strassler (2000: 17).

44. However, since the turn of the century dismissive remarks about the entire Imamate period as the age of oppression occasionally have provoked criticism.

45. People of "lowly birth" are not necessarily more accessible than those of high status, and there are remarkably few ethnographies about them. They resented being questioned about their forefathers' occupations; some would turn their faces away when asked. Nor could my visits to their houses be explained to my high status friends (see Walters 1987: 30).

46. This might even involve a broken limb or an illness. As A. Marcus (1986: 174) notes regarding eighteenth-century Aleppo, information is an effective means of social control which people tried to conceal "precisely because it could be used to their disadvantage."

47. Compare Mir-Hosseini (1993: 29) on Iran and Morocco.

1 The House of the Prophet

1. Literally "People of the House." The term occurs in Qur'an 33:33 "God will remove the stains from you, O people of the House, and purify you completely." It was interpreted especially among Shi'is as referring to the Prophet's cousin 'Ali, his daughter Fatima, and their sons al-Hasan and al-Husayn who are also mentioned in the famous "Mantle hadith." Muhammad is reported as having covered the four with the a mantle (*kisa'*) which he used at night. He pronounced a prayer which gave rise to the revelation of Surah 33:33 and called them *ahl al-bayt*. In the Qur'an the term *ahl al-bayt* occurs twice, regarding the House of Abraham (11:73), and the House of the Prophet (33:33) (Goldziher et al. 1960; Huart 1974: 31; Sharon 1986: 172; Madelung 1992: 24; van Arendonk 1996: 331). On hadith accepted by both Shi'is and Sunnis which glorify the status of the *ahl al-bayt*, see Hoffman-Ladd (1992: 623).

2. Serjeant (1979: 114); Madelung (1996: 420; 1997: 15). With the exception of the Hanafi school, all legal schools recognize these rights (Madelung 1992: 24–5).

3. Abu Bakr rejected Fatima's claim to her father's land (*fadak*) arguing that he had heard the Prophet saying that "Prophets have no heirs" (Gibb 1960: 382). During the days of Abu Bakr the *ahl al-bayt* were denied their share in the *khums* and the *fay'*. In protest the Prophet was buried privately by his family in his house and they refused to pledge allegiance to the new ruler for several months (Madelung 1992: 16; 1996: 422; 1997: 50–1).

4. 'Ali himself informed the Muslims of the hadith of Ghadir Khumm, asking those of the Prophet's companions who had witnessed it to testify on his behalf (Madelung 1996: 420). The Prophet is also reported to have said "He, whose master I am, has also 'Ali for his master"; "He who has loved Hasan and Husayn has loved me and he who has hated them has hated me." Surahs which Shi'is interpret as being in support of the *ahl al-bayt* are 4:59 "Oh you who have attained the faith! Pay heed unto God, and pay heed unto the Apostle and unto those from among you who have been entrusted with authority"; 42:23 "Say [O Prophet]: "No reward do I ask you for this [message] other than [that you should show] love for [my?] relatives (*al-mawaddah fi-'l-qurbah*)." On the issue of succession, see Momen (1985: 11–22; Madelung 1997).

5. Accession to the Imamate by inheritance is illegal though several Yemeni Imams were nominated by their fathers (see e.g., Haykel 1997: 84–5).

6. The Zaydis also distinguish themselves from the Imamis by claiming that 'Ali's designation as Muhammad's successor was obscure (*nass khafi*) rather than explicit. According to the Zaydis, only 'Ali, al-Hasan, and al-Husayn had been designated by God and the Prophet (Kohlberg 1976b: 91). In the sphere of law, the Zaydis agree with aspects of both Sunni and other Shi'i legal traditions. Messianic tendencies exist only rudimentarily and mysticism is discouraged. Notions of the Imam as mahdi and his temporary concealment, and of *taqiyyah*—dissimulation of faith under circumstances where display would be dangerous—are absent from Zaydi doctrine (Serjeant 1969: 287; Ende 1984: 89).

The Zaydis have in common with the Twelver-Shi'a the call to prayer "come to the best of works," the *hay'alah*, which was prescribed by al-Hadi (Strothmann 1974: 651; Madelung 2002).

7. Of the seventy-three Yemeni Imams sixty were al-Hadi's direct descendants, and six were offspring of his brother 'Abdullah or his uncle Muhammad (al-Shamahi quoted in Eagle 1994: 114).

8. Subsequent references to the Zaydiyyah focus upon the Hadawiyyah unless specified otherwise. Following his grandfather Qasim b. Ibrahim, Yahya b. al-Husayn contended that it was valid without the consent of the community or the *bay'ah* of two or more Muslims. He issued his *da'wah* shortly before his arrival in the northern town of Sa'dah in 897 and thus claimed the Imamate (Madelung 1965: 142; personal communication).

9. This principle, which is a basic Zaydi tenet, has three dimensions: (1) *qawl*; opposition to injustice by means of speech and writing; (2) *qalb*; the feeling that injustice prevails; (3) *'amal*; fighting injustice by force (personal communication Ahmad al-Shami; compare Strothmann 1912: 2–6; Landau-Tasseron 1990: 255).

10. Madelung (1992: 5–26). Al-Kumayt describes the Banu Hashim as "the highest of creatures" and "the peaks of splendid nobility," who are granted "a preeminence among all humankind" (van Arendonk 1996: 331).

11. See, for example, Crapanzano (1973: 22); Brown (1976: 66–75); Sebti (1986: 440).

12. Unlike the *sharif*, the *da'if* did not bear arms (van Arendonk 1996: 330). Even by the mid-1980s, in parts of the Yemen persons reckoned to be "weak" were not entitled to carry rifles which is the mark of a tribesman's standing (Gingrich and Heiss 1986: 24; Dresch 1989: 38).

13. The plural *shurafa'* is not used in the Yemen. Gochenour (1984: 88 n. 15; 218 n. 29) claims that during the earlier medieval period the term *sharif* was applied to virtually any member of the Prophet's House, while later the *sadah* differentiated themselves from other members of the House who traced descent to al-Husayn rather than al-Hasan by calling them *ashraf*. However, contemporary *ashraf* such as Bayt al-Dumayn, a prominent family of the Jawf, claim descent from al-Hasan. Their view is shared by the 'ulama I consulted on this matter (compare Serjeant 1982: 39).

14. For an historical analysis of the term, see Dietrich (1982); Bosworth (1995); Eshkevari (1999).

15. In San'a the term is no longer heard since critics of the old regime have argued that it implied that other women were not honorable. For example, Würth (2000: 125 n. 38) notes that judges at a San'ani family courts never use the term.

16. Anyone called *hashimi* may or may not trace descent through 'Ali and Fatima. During the Imamate those who failed to do so, among them Bayt al-Mudwahi, Bayt al-'Azzi, and Bayt Muta', were excluded from candidacy for the office of the Imam. Imam 'Abdullah b. Hamza (d. 1215), an orthodox Hadawi, insisted that the office of the Imam must be reserved for a descendant of al-Hasan or al-Husayn. Anyone else who claimed the Imamate, however just, erudite and pious, was to be killed. Alluding to the difference between 'Ali's descendants and others, he proclaimed that "pearls are not like cameldung," and "a precious stone is not like mud" (al-Shami 1987, Vol. 3: 152–3).

17. Wenner (1991: 121) suggests that 'Ali, who had close ties with the region, influenced Muhammad, who made several favorable statements about the Yemen (e.g., "faith is of the Yemen" and "wisdom is Yemen").

18. With the weakening of the 'Abbasids, the Isma'ilis became active in the Yemen in 879–80. As well as being opposed by local powers, the Ziyadis and Ya'furids, they had to reckon with Yahya b. al-Husayn (Rentz 1960: 550–1). In the Yemen the Isma'ilis constitute the second largest Shi'i community.

19. On this issue, see Gochenour (1984: 204); Meissner (1987: 77–8); Dresch (1989: 188).

20. Shaykhs may also be hijrah. Bayt al-Ahmar have been hijrah among both Hashid and Bakil for over 300 years. On the subject of donations compare Gochenour (1984: 167).

21. It appears that in the Yemen the notion of the *sadah* as providers of baraka was much weaker than in countries such as Morocco. In conversation, the *sadah* never claimed that baraka passes through their patriline (e.g., Crapanzano 1973: 52, 48, 73). My own experience is commensurate with Caton's (1986: 294) who writes that the *sadah* of Hijrat Kibs (and more generally in the Yemen) do not subscribe to the belief that "magical power" is inherent in the person and deeds of the Prophet's descendants. "Instead, they place great emphasis on piety, which might be defined as an attitude of reverence for God demonstrated in the performance of Islamic ritual and strict adherence to Islamic credo as defined by the Zaidi (Shi'a) sect."

22. Lewcock et al. (1983: 138); Saqqaf (1987: 116); Frankl (1990: 21–2).

23. At that time about 35,000 people lived within the walls of San'a. Around 1980 only 10% of the population of San'a lived in the old town (Saqqaf 1987: 118–19; see also Lewcock et al. 1983: 138).

24. Between 1962 and 1984 the city's population rose from 35,000 to 284,910 (Saqqaf 1987: 120).

25. The motivations behind the move to the suburbs are manifold. Noise and pollution levels have risen, the old quarters have lost their intimacy and warmth, and few of the new migrants were easily absorbed into the personal networks of the San'anis.

2   The Zaydi Elite during the Twentieth-Century Imamate

1. Their situation was similar to that of the Damascene elite of the eighteenth and early nineteenth century (Schatkowski Schilcher 1985: 131).

2. "Houses" (*buyut*, sing. *bayt*) are patronymic descent categories (see chapter 3). The term *wujaha'* relates to *wajh*, "face," pre-eminence, nobility, pride (see Rassam 1977: 158). On *al-nas*, a term which conveys respectability, see Meneley (1996: 12–13).

3. In other parts of Yemen, the *sadah* worked in these professions. See, for example, Meissner's (1987: 201) work on Sharafayn. Although the *sadah* can be found in almost any occupation, including begging (e.g., E. Burke 1972: 98 on Morocco), the case of the Indian *sadah* of Mewat is exceptional (see introduction, n. 17). There they are classed collectively with the lower castes (Jamous 1996). In contrast, in Kerala the *sadah* received the honorific Malayali title *tannal*, equivalent to the Malay honorific *tengku*. Kerala Muslims performed pilgrimages to Tirurangadi where they kissed the hands or knees of the *sadah* (Dale 1997: 177–8).

4. Qat (*Catha edulis Forsk*) is a cultivated shrub the leaves of which are chewed regularly as a mild stimulant by the majority of Yemeni adults (see Schopen 1978, 1981; Weir 1985a, b; Kennedy 1987).

5. 'Abd al-Rahman b. Yahya Hamid al-Din, personal communication. The *sadah* I questioned on this matter had never received the *khums*, nor did their personal documents reveal that their forebears did.

6. Mufti Ahmad b. Muhammad Zabarah, personal communication. Among Zaydi scholars the idea of a *sayyid*'s acceptance of the zakah causes revulsion. Sayyid Muhammad al-Mansur told me that he would rather "eat a dead animal than the zakah." Assuming that obtaining the zakah paid by the *sadah* was less tainted than that of non-*sadah*, some Zaydi authorities such as Zayd b. 'Ali and al-Murtada b. al-Hadi contended that needy *sadah* were entitled to the zakah disbursed by their more fortunate peers (Madelung 1992: 26). Some wealthy *sadah* made endowments (waqf, pl. awqaf) on behalf of the Prophet's deprived offspring.

7. Van Arendonk (1996: 333–4); see Batatu (1978: 158) on Iraq; Schatkowski Schilcher (1985: 124, 130, 1310) on Syria; Sebti (1986: 444) on Morocco. Elaouani-Cherif (1999) provided an account of the rise and fall of the office which was held by members of his family in Qayrawan (Tunesia) from the nineteenth to the twentieth century. Among Hadrami Hashimite communities in Indonesia and Singapore there are registrars who keep records of their genealogies. They also deal with inquiries from people living in the Hadramawt. Every *sayyid* possesses a document certifying his ancestry (Farid al-'Attas, personal communication). Ensel (1998: 92) mentions a Moroccan *Sharif* who carried his genealogical certificate, signed by the Sultan, in his wallet.

8. A man called Muhammad Husayn, a relative of al-Mahdi Sahib al-Mawahib, left the village of al-Mawahib for about twenty years. When another man arrived at the village claiming to be him, various people, including his wife, mother, and brothers, believed him. He had relations with the absentee's wife for several days until a man from Zabid came and told the people from Dhamar and its governor that this was not the man who had disappeared earlier. He claimed that he was a *muzayyin* from the Sa'sa'ah family and a vagabond. When he (the visitor) arrived, he found the man wearing the special dress which is worn by members of the House of the Imam. The governor summoned the man who insisted he was Muhammad b. Husayn and whose testimony was supported by his mother, and his wife and children. Then the Imam summoned him and many witnesses who testified that he was a *muzayyin*. A verdict was issued and he was severely punished, dying soon afterwards.

9. A specific class of legal specialists emerged after the Umayyads created the office of the Islamic judge (*qadi*) a century after the time of the Prophet (Endress 1982: 73). From having been first a professional category, in the later period of the Imamate the *qudah* became more clearly a descent category and interest group. Descendants of the erstwhile judges have taken up various kinds of professions which are often unrelated to jurisprudence.

10. Men who read the Qur'an for the dead and are teachers of Qur'anic schools also fall into this category. In the past legal specialists who were not appointed by the government were also referred to as *fuqaha'*. Although the differences between the *fuqaha'* and *qudah* are often played down by people who do not fall into either category, assumptions are made about the *fuqaha'*'s lower levels of knowledge. Somebody who has difficulty in understanding a certain subject is jokingly called a *faqih*.

11. Halbwachs's (1992: 136–7) term "functionary nobility," by which he describes the new nobility in seventeenth-century France who took on offices which conferred a "nobility of dignity" is quite appropriate in this context.

12. Al-Shawkani was one of the few Zaydi scholars who advocated a literal interpretation of the Qur'an and the Prophetic traditions, and condemned the practice of adhering to the *taqlid ahl al-bayt* which was officially endorsed by other Imams. In doing so, he lent his support to the dynastic ambitions of the Qasimi Imams (Haykel 1997: chapter 7, 8; 1999: 194–5; 2001: 19).

13. On these category labels, compare Mottahedeh (1980: 97–104) and Abu Lughod (1986: 41, 87).

14. On the terminology used in the highlands see Gerholm (1977: 130–41); Weir (1985a: 23); Gingrich (1986: 46–7); Dresch (1989: 118–20). On the Hadramaut see Bujra (1966: 366; 1971: 14); Camelin (1995: 39). For a historical perspective see Gochenour (1984: 12, 83). The 'ulama discourage the use of the term *al-tabaqah al-wadi'ah* (the lowest class), arguing that they are neither polluted like the Indian untouchables nor lacking dignity (*sharaf*). They give preference to labels such as *al-mutawadi'un*, humble, unassuming people. One argued that Imam 'Ali had never referred to theft as *'amal al-muzayyin* (a *muzayyin*'s deed) but as *qalil al-din* (lacking piety).

15. Assistance offered at ritual occasions by friends, relatives, and neighbors is desirable and honorable as long as it is devoid of a material dimension other than reciprocity on equal terms. Indeed the Berti of Sudan despise the killing of animals if only carried out for profit (Holy 1991: 63). The Sharh al-azhar (Vol. 2: 301–2) describes both the patrilineages and occupations held by members of these professional classes as *dani'* (low, inferior). Razih in north-west Yemen provides an exceptional case. Neither small-scale market trading nor blacksmithing are considered demeaning occupations by either the *qaba'il* or the *sadah* (Weir 1986: 227; compare chapter 8, p. 165).

16. For example, 'Abdullah al-Hammami attended a diplomatic school in Cairo in the 1960s and was later given the post of Minister Plenipotentiary at the Yemeni embassy in London. (He was assassinated with 'Abdullah al-Hajri, the deputy president of the High Court, in April 1977.) The son of another barber, Husayn al-Maswari, was trained at the Police College during the time of Imam Ahmad. In al-Iryani's government (1967–74) he was Head of the Armed Forces (*ra'is arkan*). Later he became mayor of San'a and had business interests. One way of elevating these men's status is to render their descent affiliation ambiguous. In the 1980s, no one expressed doubt about al-Maswari's descent. In 1990, a powerful shaykh argued that his father was not a proper *muzayyin* because he had worked as Imam Ahmad's personal barber. A few years later, rumors spread that he was a member of the Sharaf al-Din family who had changed his name in order to escape Turkish persecution (the Sharaf al-Din are a prominent 'Alid house which fought Ottoman occupation). In 2001 I was told by a *sayyid* that the man's patronymic refers to the area of Jabal Maswar where there are *qudah*, *mazayinah*, and *sadah*. He had heard that al-Maswari might even be of 'Alid origin. According to his brother, the former barber of San'a who carried that name performed all jobs from circumcising to shaving. However, he was not the mayor's father but belonged to the same family (compare vom Bruck 1996: 156, 158). Mermier (1985) has documented name changes among those labeled *qalil 'asl*.

17. Imam Ahmad had a flag of black color with an insignia which symbolized only the office of the Imam. The army had a green flag.

18. E. Burke (1993: 10) identifies the *tanzimat* reforms carried out by the Ottomans as a major source of disruption of the lives of Middle Easterners during the nineteenth century. (On the Ottoman *tanzimat* reforms, see Yapp 1987: 108–14.) In Yemen these reforms included the creation of a gendarmerie, the establishment of schools for the training of officers and students, the inauguration of a bilingual newspaper, and the reorganization of the administration.

19. Elsewhere wife-taking symbolized the superiority of the Ottoman rulers. For example, in the fifteenth century, marriage marked the submission of Anatolian princely houses to the Ottomans (Peirce 1993: 29). One of the Turkish women who had married into the Yemeni nobility was Sayf al-Islam 'Ali's wife Bint Rajab, who died in December 2002. Her husband was executed in 1962.

20. By 1944 all provinces were governed by the Imam and his sons (M.A. Zabarah 1982: 24).

21. They continued their studies in Egypt and non-Arab countries such as France. Compare Douglas (1987: 24–6; 110–11, 172); Wenner (1991: 129).

22. 'Abd al-Rahman b. Yahya Hamid al-Din, personal communication.

23. Similar processes occurred in other Islamic countries. On the Ottoman state see Gilsenan (1982: 40); on India, Metcalf (1978: 111); on Morocco, Eickelman (1978: 488); on Iran, Fischer (1980: 109, 117).

24. The madrasas retained the character of the previous teaching mosques in which teaching was pursued as an honorary task (Madelung 1987: 176).

25. According to 'A. al-Iryani (n.d.: 10–11), a former judge in Ibb and a reformer, the waqf for Hijrat Iryan had been administered by his father and grandfather until the funds were channeled to the *wizarat al- awqaf* (Ministry of Endowments) on the order of the Imam. Iryani claims that these funds were spent on the *Madrasah al-'ilmiyyah*. His relatives interpreted the Imam's policy as an attempt to weaken the learned houses outside the capital and to force their young men to study under the government's eye. However, rather than alienating a great number of learned families the Imam strengthened some favored Zaydi enclaves. The 'ulama of other areas who complained about the limited funding of the mosques were advised to send their students to the *Madrasah al-'ilmiyyah*. Some students left the hijrahs because they had been severely weakened as a result of the new decree which limited their autonomy and funds. In some mosques teaching had to be abandoned altogether. Another factor which contributed to the weakening of the hijrahs was the considerable prestige the *Madrasah 'ilmiyyah* had acquired. San'ani students were keen to be enrolled in the School and few chose to study in the hijrahs. From a purely economic perspective, the School was likely to offer more financial security to students.

26. For example, in the 1930s the son of the governor of Ta'izz, 'Abdullah b. 'Ali al-Wazir, studied Arabic and literature for nine years at the Dar al-'Ulum in Cairo where he had contact with several Arab reformers. In the late 1930s/early 1940s Husayn al-Kibsi, a scholar and diplomat, went to Cairo, London, and Japan on behalf of his government. He discussed the border problem with the British government and attended the coronation of King George. Ahmad Nu'man, studied at al-Azhr in Cairo. Even before the 1930s (northern) Yemenis had studied in Mecca and Cairo.

27. A translation of the text can be found in Douglas (1987: 251–8). For a commentary on the charter see al-'Amri (2000: 1–3). According to Obermeyer (1981: 191) and Detalle (1996: 332), the reformers advocated a constitutional monarchy. Although some of the reformers held this concept (Ahmad al-Marwani quoted in

Nu'man 1963: 112), others, notably Bayt al-Wazir, repudiated the monarchic principle (Zayd b. 'Ali al-Wazir, personal communication).

28. On this issue, see Bujra (1966: 359–61); Clarence-Smith (1997: 7); Mobini-Kesheh (1997).

29. Quoted in al-Akwa' (1987: 216, 225, 228).The first intermediate school which was established in the San'ani quarter of Bustan Sharib after the revolution was called after Nashwan.

30. Ahmad Nu'man, personal communication; Alaini 2004. There was also the idea of a sultanate which was to replace the Imamate. For example, Hamid b. Husayn al-Ahmar, son of the paramount shaykh of Hashid who was taught by Nu'man and Zubayri who at one time pinned their hopes on the tribal elite, was addressed as Sultan Hamid.

31. M.A. Zabarah (1982: 41); Zayd al-Wazir, personal communication.

32. These men furnished the middle ranks in the civil administration and the army. The army continued to have little prestige, and although the Imam doubled their salaries, the young officers who had attended military academies abroad sought greater influence and authority. Because the Egyptian President Nasir was a graduate of the Military Academy who had few of the credentials required of a Zaydi leader, the Imam spoke of him somewhat contemptuously as Bimbashi (a Turkish military rank) and did not consider him an equal (Serjeant 1979: 91). For many of the former students, the revolution yielded rewards. Men like Muhsin al-'Ayni who had attended the orphan school established by Imam Yahya and was sent to Lebanon in 1947, became the headmaster of a secondary school in San'a after his return. Under the republican president Sallal (1962–67) he headed the Foreign Office and later became prime minister and ambassador to Germany and the United States. 'Ali b. Sayf al-Khawlani who also studied in Lebanon became Chief of Staff of the Armed Forces in 1968 and Minister of Interior in 1970; 'Abd al-Latif Dayf Allah who trained at the Military College in Cairo was a member of the Revolutionary Command Council and Minister of Interior under Sallal.

33. On the contentious issue of the zakah, see Dresch (2000: 49). It was common practice for all major tribal leaders to send one or two boys, a son or a nephew, to the Imams as hostages in order to prevent their kin from engaging in hostile acts. The Imams justified hostage-taking on the grounds that the children were offered a basic education during their custody (Mufti Ahmad Zabarah, personal communication).

34. Among them were the Imam's personal secretary, 'Abdullah al-Dabbi, and the Iraqi-trained head of the Royal Guard, 'Abdullah al-Sallal, who became the first President of the Republic (Douglas 1987: 238). Imam al-Badr, a descendant of the Prophet in the thirty-seventh generation, died in his British exile in August 1996.

35. Ironically, following the 1958 pact between Nasir and the Imam, al-Badr demanded that Egyptian troops be send to Yemen to protect it against British-inspired assaults and to help liberate South Yemen (*Der Spiegel*, April 30, 1958: 40).

36. Those Yemeni tribesmen who delighted in Egypt's defeat were among the few Arabs who did not experience a sense of humiliation in 1967. One of the Hamid al-Din princes described his encounter with those men, admitting to his own sense of shame at the defeat.

37. It was said that one of the officers, 'Abd al-Mughni, told one of the former secretaries of the Imam, Ahmad Afif Jabar, who sided with the revolution, that they had to kill some *qudah* so as not to give the impression that the revolution was exclusively directed against the *sadah*. The person who gave me this information claimed that Qadi 'Abd al-Rahman al-Sayaghi, the governor of Sa'dah, was killed for this reason only.

38. Puin's (1984: 488) assertion that the *qudah* did not share the fate of the *sadah* because they had participated in the opposition against the Hamid al-Din is doubtful. As this chapter demonstrates, from the late 1920s onwards, the *sadah* played a significant role in the opposition, and a great number took part in the revolution. I am inclined to agree with Keddie (1988: 8) who argues that the *qudah* were identified more with public service than with the *ancien régime*. Note, for example, that Qadi Yahya b. Ahmad b. Husayn al- 'Amri who governed San'a until 1960 spent only one year in al-Qal'ah prison. The land of Bayt al-'Amri was confiscated but no one was executed.

39. About twenty of those who were killed belonged to the government. Those of Bayt Hamid al-Din who were executed were Sayf al-Islam 'Ali b. Yahya, his son al-Hasan, and his brother Isma'il. Others were killed during the war, among them 'Abdullah b. Hasan, Radiyyah bt. 'Ali, and 'Ali b. al-Husayn. Most of those killed were executed without trial, their bodies being left in the street. Some were buried at night or eaten by dogs, others were thrown into a pit were their bodies were left for passers-by to spit on.

40. The most prominent ones were Yahya and Ahmad al-Mutawakkil, 'Ali al-Mu'ayyad, 'Ali 'Abd al-Mughni, Hashim al-Huthi, Muhammad Mutahhar, and Muhammad al-Shami. One of them, Ghalib al-Shara'ay, was personally responsible for the execution of several *sadah*. Some foreign commentators have even argued that the revolution was led by the *sadah* (J. Hooper, "In nomine Muhammad," *Observer*, March 1, 1998). On this issue compare Wachowski (2004: 87).

41. In 1959 Shaykh Husayn al-Ahmar of Hashid announced that he sought a replacement for Imam Ahmad and the Hamid al-Din (Douglas 1987: 226).

42. I am grateful to Lucine Taminian for providing me with the poem.

43. During *mawlid* rituals the Prophet's life is narrated. In Yemen, women celebrate *mawlids* not only on the day of the Prophet's birthday but also during the forty-day celebrations on the birth of a child and after recovery from a serious illness. Compare Ben Achour's (1999: 349) study of Tunisian *sadah* during the seventeenth–nineteenth century where he points out that that the *mawlid* amounted to an "official veneration" of the *ahl al-bayt*. One of the earlier anthropological works on this institution is Tapper and Tapper (1987). On *mawlids* in Morocco, see Hammoudi (1997: 69–73).

44. Combs-Schilling's argument is supported by Moroccan writers (e.g., Alaoui quoted in Hammoudi 1997: 159 n. 1). Much of her analysis rests on the symbolic link between the *spilling* of blood of animals and humans during ritual and the blood descent of Hashimite rulers. However, this link is not obvious and it is not made clear whether Moroccans themselves draw this analogy. In spite of analyzing the rituals as transformations of each other, the study conveys an image of timelessness. Factors such as internal repression and the interests of superpowers which also account for the stability of the monarchy are not given due consideration (see Munson 1993: 121, 132–48; Hammoudi 1997: 12).

3 The Anatomy of Houses

1. For example, according to Bayt al-Daylami, there are thirty individual houses in the southern town of Dhamar, the historical seat of this house where many of its members live. In the country as a whole, about 900 people carry the patronymic al-Daylami.

2. This term, which Eickelman (1976: 183) adopted from Scheffler, seems apt in this context. By stressing descent, this definition is perhaps more precise than H. Geertz's (1979: 343) "patronymic association" which indicates "the entire aggregation of persons who can be called by the same patronym."

3. Like H. Geertz (1979: 316), I use the term "family" to indicate a variety of distinct referents like the "conjugal family" or "personal kindred."

4. Compare Waterbury (1970: 97) on Morocco; Khoury (1983: 73) on Syria. A *sayyid* who belongs to a house of great 'ulama described his house as *lu'lu' al-sadah* ("the pearl of the *sadah*").

5. The case of the Yemeni *sadah* differs from that of the rural Lebanese "Learned Families" studied by Peters. He refers to them as a "stock group" although membership is not strictly defined in genealogical terms. Irrespective of their descent, learned men are counted in as members of the "stock group" (Peters 1963: 178, 181–3, 195).

6. Islamic law prohibits adoption. However, in Yemen unwanted children quickly find a home. One of the better known cases is that of one of Imam Ahmad's servants, Salih Muhsin, whom the Imam nicknamed Sharaf al-Din. (Among the *buyut al-'ilm*, men who carry the name Muhsin may be addressed by the honorific Sharaf al-Din.) Some *sadah* speculated that the Imam named his servant thus in order to tease Bayt Sharaf al-Din during a period of tension between them. Salih Muhsin introduced himself as a *sayyid* to me but was denied recognition by other *sadah*.

7. Others refer to Imam 'Ali as *jadd al-'Alawiyyun* (ancestor of the 'Alids).

8. For example, Bayt al-Daylami (see n. 1).

9. Most *sadah* do not accept an obligation to support the less fortunate members of their houses, but are prepared to provide help in specific cases.

10. I have not met any who did not possess a written genealogy, but according to Mundy (1995: 40), in Wadi Dahr poor *sadah* do not keep them. Regarding the *sadah* of Khawlan bin 'Amir in north-west Yemen, Gingrich (1989a: 76) notes that "any adult *Sayyid* . . . is able to spell his (allegedly) complete list of ancestors, his *nisba*."

11. On this subject, see Gilsenan (1973: 11, 33); Christelow (1980: 142); Sebti (1986: 446); Valensi (1990: 91); Eickelman (1998: 181); Ensel (1998: 92).

12. Bouquet (1996) may be right in arguing that anthropological kinship diagrams are modeled on European genealogical representations, but she conveniently overlooks that non-Europeans make frequent use of the tree metaphor (see appendix III.3 and 4).

13. See Halbwachs (1992: 128) for similar observations regarding the old French nobility. "Two nobles who meet each other for the first time should be able, after exchanging just a few remarks, to recognize themselves as two members of the same extended family that establishes their kinship link or alliance. This presupposes that, in the nobility, through the generations there continues a totality of well-linked traditions and remembrances."

14. For example, the work of 'Abbas al-Khatib (1968) favors the offspring of Imam al-Qasim. Bayt al-Khatib are Qasimiyyun.

15. M. Zabarah (1999). The inclusion of the patrilines of the *mazayinah* is especially significant because their names carry pejorative connotations. "Having identified a physical or moral feature, or simply having attracted ridicule to it by using an incongruous term, it may be transmitted to his descendants: 'little ringworm' (*al-qummali*), 'apricot' (*barquq*), 'pierced nose' (*akhram*)" (Mermier 1997: 94).

16. Mundy (1995: 93); Meneley (1996: 67–8); Ensel (1998: 54–5).

17. However, in the cities there are low status houses which trace their name several generations back (e.g., Bayt Barquq ["Apricot"]). Ahmad Barquq, who lives in the San'ani quarter of al-Fulayhi, is a well-known barber and circumciser who served the elite. He was even asked by wealthy clients to come to Ta'izz. He refused to carry out work normally done by the *mazayinah* such as looking after the pipe-bowls (*bawari*, sing. *buri*). Apparently his ancestor was a fruit dealer.

18. For a similar observation, see Dresch (1989: 231 n. 7). In an exceptional case, a descendant of Imam al-Qasim at the seventh generation (Muhammad b. Yahya b. Muhammad b. Isma'il b. Muhammad b. Husayn b. Imam al-Qasim), who carried the patronymic al-Bustan (garden), adopted his wife's patronymic Hamid al-Din (M. Zabarah 1984: 141).

19. Husayn b. 'Ali, a descendant of the Prophet in the thirty-second generation, moved to San'a from hijrat Zabar after the Ottomans had withdrawn from the Yemen. In San'a he was called Husayn Zabarah (see genealogy of Bayt Zabarah, appendix III.1).

20. Other common honorifics are al-Wajih ("leader") for Khalid and anyone who carries a name with the prefix 'Abd; al-Diya' ("light," "brightness") for al-'Abbas, Sarim al-Din (*sarim* is one of the terms for sword) for Ibrahim, and al-'Alam/'Alam al-Din ("the flag of religion") for al-Qasim. These honorifics are mainly used by the *buyut al-'ilm*. I have heard people addressing their grandsons affectionately by these titles.

21. Since the revolution these names have rarely been given to girls.

22. In conjunction with the plural form of address which is also used for older kin and high ranking people, women used to refer to their husbands or fathers as, say, Sayyid Qasim or Sidi Qasim. Some men consider this form of address as an expression of their wives' respect, affection and devotion, but it is largely out of fashion. Women who spent time in exile in Lebanon and Egypt during the civil war have begun to call their husbands by a teknonym, forms consisting of *abu* (father) or *umm* (mother) with the following names (e.g., Abu Muhammad). I have also heard women addressing their spouses as *'ayni* ("my eye") which is less formal than *sidi* (my lord), yet more respectful and affectionate than the use of first names. Men of conservative *buyut al-'ilm* address their mothers as *sitti wa 'ayni* ("my lady and my eye"). Wives are addressed by their first names or honorifics such as these.

23. The only activity which could be described as corporate is the assistance of close kin at ritual occasions such as weddings. This is provided by women and rarely includes all members of a house.

24. Collectively, the *sadah* would be misrepresented as a property holding group. In the Imamate many lived in poverty; during the revolution they were never branded as an exploitative class. One of Imam Ahmad's most often quoted rulings is that all his property, including his pen, should be transferred to the Treasury (*bayt al-mal*). However, according to Gingrich and Heiss (1986: 21), in

some regions in north-west Yemen sympathy for the revolution derived from the *sadah*'s misuse of public endowments (*mal al-muslimun*). This land was donated to a mosque and administered by a *sayyid* who either cultivated the land himself or employed a tenant farmer. Over time the property reverted to private property (compare al-Zubayri in Serjeant 1979: 103; Messick 1978: 381). Intimations of the Qahtani–'Adnani divide in anti-'Alid discourse may derive their prominence from the fact that the theme of economic exploitation is largely inappropriate.

25. A waqf is "the alienation of income-producing property in perpetuity to bene-fit, although not always immediately, a religious or pious cause" (Fay 1997: 35). The object of the waqf, often real estate, yields a usufruct of which the owner has forfeited his power of disposal. The income must be used for per-mitted good purposes (Heffening 1974: 624).

26. See, for example, Batatu (1978: 161) on Iraq.

27. Among these is the house of 'Ali b. Yahya which is now the Dar al-Hamd Hotel in San'a. The property owned by the Hamid al-Din would be a thorny issue in any negotiations about their repatriation. With the exception of a few women, the family has not been permitted to return to the Yemen.

28. In one case, 65,000 square meters were confiscated from two brothers whose mother is related to Imam Ahmad. In the 1970s they received 2,000 YR per square meter in compensation, but now the property has much greater value. 6,200 square meters of the land were taken by a neighbor, an officer, who sold it. Another case is that of the house formerly owned by Ahmad b. 'Abd al-Rahman al-Shami who served under Imam Yahya and Ahmad. He was executed in 1962 and his property confiscated. (Sayyid al-Shami worked as a secretary of Imam Yahya and was the Head of *al-Hay'at al-amr bi-'l-ma'ruf wa-'l-nahy 'an al-munkar*, a court established by Imam Ahmad which dealt with moral offences such as adultery and drinking.) The house is now a hotel (Funduq San'a) in the center of the town. Later it was recognized as Sayyid al-Shami's property, but his descendants receive very little rent.

29. Sayyid Muhammad al-Mansur, personal communication. On Ibb, see Messick (1993: 95–8).

30. Islamic law grants a woman half a man's share.

31. On the alienation of heirs for personal gratification in Wadi Dahr, see Mundy (1995: 153–60, 219 n. 11, 231 n. 52).

32. Husayn al-Hubayshi, personal communication.

33. One must also consider that unless it can be used for the cultivation of coffee and qat, land is not necessarily a valuable asset. Those who had sold their land argued that it had little value or that they had difficulty in obtaining their shares from the tenant farmers. The civil war provided an opportunity for some tenant farmers to keep all yields to themselves. Tenants often refuse to deliver the stip-ulated quantity of the harvest to the landowners, claiming that there was no rain, that animals ate the crops, or that there was armed conflict. Since large numbers of the hijrah inhabitants have migrated to the cities, they have had lit-tle control over their tenant farmers. The farmers are aware that the political will to enforce the rights of those families who served under the Imams is miss-ing, nor do they expect them to try and enforce their rights in court.

34. Note that some members of the new elite who migrated to the city after 1962 have built modern houses in their villages, thus erecting monuments of their recent success in the ancestral landscape.

35. For more details on the property of San'ani families in Wadi Dahr, see Mundy (1995: 45).

36. Some scholars have begun to edit these hand-written materials and have produced computerized versions (e.g., Z. al-Wazir, n.d.). The data below are taken from Z. al-Wazir, in press and personal communication. He conceives of Bayt al-Wazir as a scholarly house (*usrat ʿilm*) rather than one which has been engaged in politics (*usrat siyasah*). The result of the conversion of family documents into public history (Slyomovics 1998: 5) is the production of a kind of memorial book which, in spite of similarities with the earlier *safinat*, has different objectives. The editors of these books aim at the general public as well as at those young members of their houses who possess no more than a "school textbook knowledge" of the past. See the genealogy of Bayt al-Wazir (appendix III.2) which is not identical with this text; a fact which proves once again that genealogies have a life of their own.

37. Hijrat Waqash was established by Ibn Abi al-Haytham, a promoter of Mutarrifiyyah teachings (see next), in the tenth century.

38. However, there are other members of the house who are referred to as *jadd* (see appendix III.2). The Mutarifiyyah goes back to Mutarrif b. Shihab al-Shihabi (d. 1067). Its followers, most of them non-ʿAlids, claimed to adhere strictly to Imam al-Qasim b. Ibrahim and other early Imams such as al-Hadi, Muhammad al-Murtada, and Ahmad al-Nasir. The Mutarrifis called for repentance, ascetic exercises, purification, and withdrawal from the world in the hijrahs. They attributed God's intervention after the Creation to miracles and prophetic messages, and they did not conceive of God as the *primum mobile* of all things and events (Gochenour 1984: 170; Madelung 1987: 176–7; 1991: 30–1; 1999: 125; Z. al-Wazir 2000).

39. Much of the pride of Bayt al-Wazir focuses on their descent lines from Imam ʿAli Abu Talib to al-Afif via Imam al-Hadi Yahya, and from al-Afif to Sarim al-Din.

40. Most members of Bayt Uthman were cultivators. Today they are traders and government employees, pursuing a modern education (*al-dirasah al-hadithah*). Ahmad b. Muhammad Uthman (b. 1921) is the imam of one of the mosques in al-Sir and the only *ʿalim*. He is referred to as Ahmad Uthman al-Wazir by members of Byat al-Wazir because he is learned. Most members of Bayt al-Mufadhdhal are cultivators and traders. Ahmad b. Muhammad was the governor (*ʿamil*) of Bani Hushaysh until 1962, and then worked as a self-employed judge (*hakim taradi*).

41. Sometimes *diwans* extend from the house like conservatories. Note that the documents mention the various stages of the rebuilding and enlargement of the house, a fact which illustrates the expansion, both physical and social, of those branches of the house in the new areas where they had settled.

42. On ʿAli al-Wazir's biography see A. al-Wazir (1987). The governor had twenty-one children most of whom died in infancy. Four are still alive.

43. Throughout the Imamate period, the hijrahs remained important for those members of the house whose main residence was elsewhere. While holding posts in other areas, they would spend time in Bayt al-Sayyid during the harvest, and mediate in local disputes. During periods of political unrest, they took refuge there. When ʿAli al-Wazir had to give up his post as governor in favor of Crown Prince Ahmad, he first went back to al-Sir. However, the hijrah could not always offer support. In 1948, Hijrat Bayt al-Sayyid was not strong enough to help the uprising. Even today, Bayt al-Wazir are hijrah among both the major tribal confederations of Bakil and Hashid.

44. Altogether thirty-five men were executed.

45. After the house was blown up, the remaining stones were put on a truck. The women and children who had lived in the house were forced to sit on top and were taken to San'a. The Imam's brother demanded that the stones be assembled in front of his palace where they remained for years, and were only removed when the Saudi king came for a visit.

46. On the trajectories of the lives of Bayt al-Wazir since the 1962 revolution, see chapter 8 and appendix II.2.

47. At that time it was improper for women to enter the suq (see chapter 8).

4   Snapshots of Childhood

1. Rosenwald and Ochberg (1992: 2); D. Rubin (1996: 2, 12); Bruner and Fleisher Feldman (1996: 293, 295); Bal (1999: x); Sturken (1999: 245).

2. I adopt a phrase of Munn (1995: 97–8).

3. They folded their arms (*yadum*). The governor, himself a Zaydi, prayed with his arms outstretched (*yusarbil*).

4. Ibrahim's mother made reference to one of the prophets before Muhammad. She was the daughter of the paramount shaykh of Bakil whose main residence was in the Jawf in the east of the country, but he also lived in the Ta'izz region where he owned land. For more details on her life, see I. al-Wazir (1997).

5. This thinking was quite unusual by virtue of the predominance of the Zaydi-Hadawi school of thought which endorses the 'ulama exercise of political authority. On this issue, see chapter 12.

6. At this point Ibrahim quoted from the surahs al-Qasas and Ghafir.

7. This is a reference to the Zaydi idea that man has a free will and is not predestined.

8. Al-Mutanabbi refers to the poet Abu-'l-Tayyib al-Ju'fi (b. 915 in Kufa) who had Shi'i leanings during his youth (Blachère 1992).

9. During Hisham's rule unrest within the borders of the Umayyad empire anticipated the end of its rule. Among those who rebelled were the Shi'is of Iraq and the Berbers of North Africa (Busse 1984: 33).

10. Both Ibrahim's father and his older half-brother 'Abdullah were married to daughters of Imam Yahya.

11. She started wearing the *lithmah*, a scarf, inside the house when she was nine years old. The *lithmah* is made of silk georgette or nylon which is worn indoors (in order to cover most of the face when, for example, speaking to a male cousin who must not see her face). For details, see Maclagan (1993: 131).

12. In the context of nineteenth-century Egypt, Mitchell (1988: 87) argues that this type of instruction cannot be conceived of as "education" or "schooling" by reason of the absence of "organization" and discipline. Accounts of their childhood by Yemeni adults speak of a rigidly organized training and of the emphasis which was placed on cleanliness at the Qur'anic school. The rhetoric of morality which informed the children's instruction aimed partly at self-discipline. Mitchell (ibid.: 94) argues that only in Western-style schools power worked " 'from the inside out'—by shaping the individual mind." As chapter 5 also shows, this was the central goal of knowledge acquisition in the Yemen. See also Messick (1993: 75), who quotes the Yemeni scholar Muhammad al-Akwa' saying that attending the Qur'anic school was like being taken to the slaughterhouse, giving a poignant idea of the disciplinary stratagems at those schools.

5 Performing Kinship

1. Strathern (1973); Iteanu (1983); Weismantel (1995); Carsten (1995, 1997, 2000); Lambek and Walsh (1999). Geertz and Geertz (1964: 99–100) were among the first scholars emphasizing the importance of non-genealogical criteria for the continued existence of kin groups.

2. The term *sulb* is also used synonymously with *dhurriyyah* (patriline).

3. The model is said to be commensurate with the teachings of the Qur'an but differs from representations elsewhere in the Middle East (see, e.g., Crapanzano 1973: 48–9).

4. It would appear that this is one of the reasons why among the oasis community of the Moroccan Dra Valley, "the mother's milk flows only in one direction." In all cases of colactation, the low status Drawa were never nursed by sharifian women. Such arrangements were maintained between Drawi and sharifian families over several generations (Ensel 1998: 82).

5. This might explain the Islamic injunction against marriage between genealogically unrelated children who have been nurtured by the same woman.

6. On the notion of the body as a vehicle of history, see also Comaroff and Comaroff (1992: 79).

7. The notion that the *sadah* acquire knowledge through a combined process of unfolding intrinsic potentials in interaction with extrinsic sources is remarkably compatible with theories developed by writers who have been inspired by Chomsky. Notably, Sperber's (1985) theory of human concept acquisition is based on the idea that children are born with a sort of mental template which helps them to understand concepts like bird (1985: 81–2). Compare Boyer (1990: 109).

8. The Zaydis include the theory of the *imama* into the category of the *'ilm al-kalam*. Mufti Ahmad Zabarah, personal communication.

9. The proverb was quoted to me by *sadah* and non-*sadah* alike.

10. My sources for this period are Kohlberg (1988) and Daou (1996).

11. Hisham's ideas were at variance with those attributed to the Jarudiyyah, a Zaydi school which has always remained marginal and which some Zaydi scholars barely recognize as Zaydi. The followers of this school were said to have identified divine inspiration (*ilham*) as the source of the Imam's knowledge of *tafsir* (exegesis) and of his rulings. According to the ninth-century Imami scholar al-Hasan al-Nawbakhti, the editor of Hisham's descriptions of the Jarudiyyah, these thinkers held that Muhammad's Family "have complete [knowledge of] everything which was brought by the Prophet . . . the young and the old of them being equal in knowledge . . . those who are in the cradle and swaddling clothes to those who are the oldest among them . . . They . . . do not need to learn from one another or from anyone else. Knowledge grows in their breasts as rain makes the seeds grows, for God has taught them by his grace howsoever He pleased" (Daou 1996: 81).

12. Sayyid Ahmad Eshkevari (Qum), personal communication.

13. In the aftermath of the revolution, the Imams have been accused of treating knowledge as their exclusive property and of attempting to keep their subjects ignorant so that they would know nothing but the principles of the *imama* and would therefore be obedient (e.g., I. al-Akwa' 1995, Vol. 3: 1668; al-Iriyani 1987: 376). In reality this issue was slightly more complex. At the religious institutions, students inculcated the basic values that informed the state doctrine and formed an image of the Imams as guarantors of religion, but they also acquired the ideological weapons for opposing them. Even a staunch enemy of

the Imamate, Muhammad al-Zubayri, conceded that "the Yemen, in truth, even under the auspices of the Imams, was profuse in activity on an extensive scale in (the sphere of) learning, the door of independent judgment (*ijtihad*) was open, the mosques crowded with scholars (*'ulama'*) and pupils, and production in the field of (literary) composition used to excite astonishment and admiration" (quoted in Serjeant 1979: 104). In his autobiography, al-Shawkani states: "I acquired knowledge without a price and I wanted to give it thus" (quoted in Messick 1986: 110). No one was prevented from studying (which was mostly paid for by waqf), but only a few were actively encouraged; often pragmatic reasons accounted for lack of interest in advanced or even basic education. The Imams did little to help further the education of the rural population, but insisted on children's overall grounding in the basics. Imam Yahya's governors were instructed to punish parents who attempted to bribe teachers in order to keep their sons out of school. Some parents felt their sons were indispensable in agriculture (personal communication Muhammad 'Abd al-Malik al-Mutawakkil, San'a University; on this issue see also Würth [2000: 60–1]). This, as well as the limited availability of public sector jobs, were the main reasons why some people did not seek an education (on this kind of debate, see also Eickelman and Anderson 1999: 11).

14. Men who had close ties with the Imams also wrote down the names of their children in their Qur'an. Dresch (2000: 23), quoting Muhammad al-Akwa', notes that men recorded the deaths of close relatives in the margins of standard law books.

15. In Sa'dah, where men commemorate the Prophet's birthday at the al-Hadi mosque, the girl of the family I stayed with described it as *haflat al-hashimiyyin* (celebration of the Hashimites). Rituals do not seem to enter into the construction of 'Alid identity as significantly as the *taqlid ahl al-bayt*. The verdicts of the *ahl al-bayt* rather than the places where they are buried constitute a lived moral geography. Karbala and Kufa are not elaborated upon, and Zaydis neither make pilgrimages to the Shi'i shrine cities of Iraq nor to al-Hadi's tomb in Sa'dah. According to the 'ulama I consulted on this issue, the Zaydiyyah places far less emphasis on ritual than the Twelver-Shi'a because of its rational (Mu'tazili) orientation (see Gimaret 1993). (One exception is the ritual of Ghadir Khumm; see chapter 12.) Sayyid Muhammad al-Mansur argued that during the Hamid al-Din era there had been little officially orchestrated ritual because the state was poor. However, one young scholar offered another explanation. He argued that specifically Shi'i rituals had been discouraged by the Hamid al-Din Imams because they did not want to provoke the Sunnis. The Mufti had told him that when during the Friday sermon (*khutbah*) at the mosque the imam spoke about 'Ashura (the tenth of the month of Muharram, the day of the martyrdom of Imam al-Husayn), Imam Yahya had said to him "For many years I have tried to unify this country, and you are untying it again." At least during the Hamid al-Din era, 'Ashura rituals were not publicly performed. However, people used to fast on the ninth and tenth Muharram (and some still do).

16. The effects of memorization were also of utmost concern in the oral cultures of Greece where remembering was seen as a mental exercise with moral and metaphysical significance. Plato's ideas about memory formed an integral part of his general theory of knowledge. One remembers in order find truth; to remember is to know, and memory leads one beyond temporal experience (Vernant 1983: 77–110).

17. Much attention has been drawn to these formal contexts of legal practice whereas the benefits of a well-trained memory in more parochial settings has been given less consideration. In one case, a scholar's unfailing memory of the Qur'an allowed him to take control of an embarrassing situation. His brother's children had brought a dog from abroad which they kept in their house as a pet. Tribesmen who came to visit were displeased with finding the ritually unclean animal inside the house. The scholar made a speech reciting sources which depict the dog as the Prophet's loyal companion and a friend of the believer (Surah 18:18). Toward the end of the speech the tribesmen no longer disapproved of the dog's presence in the house; some were seen stroking it. The scholar's ability to spontaneously contextualize Qur'anic verses lent him moral authority and gave him the power of persuasion.

18. Sayyid 'Abd al-Qadir provided these notes in writing. Like Mufti Ahmad Zabarah and others, Sayyid 'Abd al-Qadir served in the Imamic and republican governments. In spite of striving to abolish the old institutions, the republic has sought legitimacy through assigning moderate 'ulama like these to respectable positions.

19. The Ottoman Sultan 'Abd al-Hamid began the codification in 1908. In Yemen it was first ventured during al-Iryani's presidency. Carapico (1998a: 33) asserts that "unlike the PDRY, the YAR never set out to reorganize the judiciary as an instrument of social transformation." The gradual process of codification, which took several years and has contributed to curb the power and autonomy of the 'ulama (and potentially that of the shaykhs), was surely a step in that direction (vom Bruck 1998b: 169).

20. Z. al-Wazir, n.d.

21. When used in connection with a blame, the formula means "May God help us to bear our mistakes." A person may also utter the formula and stroke his beard when admonishing a man for his "unmanly" behavior.

22. Z. al-Wazir, n.d. The editor of these documents traveled to several areas where members of his house had settled, but failed to trace these men's descendants. He met some who were living in Waqash and had maintained their patronymic, but "their dress and demeanor were those of a *qabili*." They could not list more than three to five forebears.

23. In her study of a San'ani court, Würth (2000: 201) notes that a *sayyid* from a well-known family whose wife accused him of alcohol consumption was more concerned about his reputation than other men. He presented a certificate issued by renown 'ulama confirming his irreproachable conduct.

24. A similar point was made by Barth (1990: 645, 647).

25. Non-'Alid scholars who were aware of the cultural significance of descent but unable to demonstrate their links to their ancestors created genealogical schemes of intellectual reproduction. Often people did not bother to trace their genealogies back earlier than the time their forebears first became scholars. For example, in his study of Muhammad al-Shawkani, al-'Amri provides little information about the *qadi*'s ancestors. Instead of a genealogical chart, the book presents the chains of transmission drawn up like a pedigree (al-'Amri 1985: 110).

26. In his study of Salé, Brown notes that fathers who took their children to the Qur'anic school for the first time told the teacher "*He is now your son*, his education belongs to you. Hit him. If you should kill him, I will bury him!" (Brown 1976: 107; emphasis added).

27. The study of property relations has mostly been neglected in studies of performative kinship. Many of these studies either do not address this issue or deal

with cultures where private property is insignificant (e.g., Stewart [1989]; Weismantel [1995]; Storrie [1999]; Asuti [2000]; Bodenhorn [2000]).

Analyzing inheritance patterns in Langkawi, Carsten (1990) deals only with blood relatives. While it is they who inherit, household goods must be shared by all resident members (1997: 97).

6   The Politics of Motherhood

1. In Yemen the term *'arabi* is used as a synonym of Qahtani. The term *fatimiyyah* (pl. *fatimiyyat)* rather than the more vernacular term *sharifah* (female descendant of the Prophet) is used in legal writings.

2. The *kafa'ah* is predicated on the idea that a married woman is entitled to a lifestyle equal to that granted her by her natal family. Equivalence can be defined in terms of descent, wealth, education, occupation, and piety. Because it is assumed that a man of equal status will do justice to her, she is given the right to insist that the principles of the *kafa'ah* be applied.

3. Among the various conditions for the *kafa'ah, nasab* has been adopted by the Hanafi, Shafi'i, and Hanbali schools of Islam (Mutahhar 1985: 166, 169, 171). The Hanafi scholar Muhammad b. Hasan al-Shaybani proclaimed *mal* and *nasab* to be the conditions of the *kafa'ah*. He included wealth because in his view women desire a good dowry. Besides those principles, Abu Hanifa added *din* and *sina'ah*. It is said in the Sharh al-azhar (Vol. 2: 301) that those 'ulama argued that it was important for a man to have a good occupation. Those Zaydi scholars I consulted on the issue of *nasab* held that the seventeeth-century Imams who declared it as a condition for the equivalence of spouses based their verdicts on that of Imam Ahmad b. Sulayman (d. 1171). This Imam was said to have been under the influence of Hanafi 'ulama. In the fin-de-siècle Netherlands, East Indies, Sunni 'Alid immigrants from the Hadramawt endorsed the idea that a non-'Alid must not marry a *sharifah* even if she waives the *kafa'ah* and consents. In this case, the suitors were non-Arabs. According to Ho (2001: 83), the concept of race was subsumed under a more encompassing scheme for identifying persons: genealogy (compare n. 7).

4. Here again Zaydi scholars recognize social distinctions within the Prophet's House. The *fatimiyyat* were not to be married to men who worked as barbers, butchers, weavers, shoemakers, or innkeepers (sing. *muqahwi*) who were (and still are) looked down upon. It would appear that this argument is about morality rather than social ranking per se because the Faqih does not speak about the women's economic background. I am not aware that Zaydi law prohibits the exercise of these professions for the pursuit of income, which would partly explain their low status. Eminent religious authorities adhering to the Imamiyyah consider making a living based on shaving beards to be unlawful (Al-Khoei 1992: 75, Art. 217). Butchering for a living is undesirable (*makruh*) (Al-Hilli 1896/97: 160–2). Imami Shi'is consider butchering to be a morally reprehensible occupation because it is believed to affect the butcher's character. He may become a man without mercy (Mustafa b. 'Ali al-Qazwini, personal communication). These attitudes are by no means limited to the Shi'a. The Syrian Shafi'i jurist Muhyi al-Din al-Nawawi (d. 1277), whose work was taught predominantly in lower Yemen, wrote that people of high social position who engage in blood-letting, sweeping, and tanning do not qualify as witnesses. A man carrying out one of these professions is not suitable for the daughter of a man in a more respectable profession (Messick 1993: 162; compare 1988: 650–1).

5. The advance into Hadramawt about 1660 by Ahmad b. Hasan, a nephew of the ruling Imam, is an important event in Zaydi historical consciousness. Even children told me about *Ahmad sayl al-layl* in reference to the ferocious force (called "Night Flood") he commanded.

6. The document was published in vom Bruck (1989). These concepts are reflected in common Yemeni sayings such as "a woman's honor is her husband" and "marriage is a woman's shelter."

7. The Imam's verdict was in stark contrast to that of Sunni 'ulama of the Hadramawt, many of whom lived abroad. In the 1930s, 'Umar 'Attas, who was living in Sumatra, ruled that an 'Alid *sharifah* could neither marry a non-*sayyid* nor other Hashimites (i.e., who are not descendants of al-Hasam and al-Husayn) (Bujra 1966: 357, 359; Ho 2001: 82–3). In the early twentieth century, a Hadrami *sayyid* who lived in Indonesia argued that "even when a *sharifa* and her *waliy* (guardian) accepted such a match, it was the obligation of all other *sada* to oppose it. The couple should be separated, by force if necessary, for such marriages were a humiliation for the Prophet, his daughter Fatima, and all her descendants" (Azra 1997: 256); compare Bredi (1999: 378) on Uttar Pradesh and Cole (1988: 79–80) on Lucknow. It is worth noting that European aristocratic houses held similar attitudes toward women's out-marriage. In Late Renaissance Venice, the patricians defined their purity by not letting women marry out. Rather than giving them in marriage to unworthy grooms, proud patrician parents destined them to a life in prayer and seclusion. More than 60 percent of the 2,500 female patricians living at the beginning of the seventeenth century were nuns (Sperling 1999: 22, 29).

8. As Ibrahim's biography reveals (chapter 4), some people take special pride in descending from parents who are both *sadah*. Youth who do so tease other *sadah* whose mother is not a *sharifah* by claiming superiority over them.

9. In Falih, it was *'ayb* (inappropriate) to raise questions about the bride's virginity because of the implicit assumption that she might have been dishonorable.

10. Isma'il (the biblical Ishmael) is regarded as the ancestor of the northern Arabs and the Arabized tribes which are traced back to 'Adnan (Paret 1978: 185). The Quraysh are descendants of Isma'il.

11. The notion of "arranged marriage" is not entirely correct because men and women have always found ways of knowing about each other through their siblings and friends, the secret exchange of letters, and even acquaintance.

12. In the present period, girls of the old elite tend to be at least eighteen.

13. It should be noted that there is little economic gain in these marriages as far the *mahr* is concerned. Some houses have fixed sums which are below the average paid by houses of similar rank, but in most cases these marriages (even with close kin) are not "cheaper." It is argued that the girl must be able to invest a considerable sum in gold jewellery in order to maintain her status in the eyes of other women.

14. For example, Bayt al-Shami and Bayt Hajar. On this subject, see also Bujra (1971: 20); Waterbury (1970: 103).

15. Compare al-Rasheed (1991: 192–4); Mundy (1995: 179–82). Bourdieu's (1977: 46) assertion that "it is impossible to find an informant or anthropologist who will not declare that in Arab and Berber countries every boy has a 'right' to his parallel cousin" can certainly not be substantiated by data on the old Yemeni elite.

16. Therefore, old generation 'ulama who explain the *kafa'ah* in terms of *nasab* express surprise at suggestions that this interpretation implies discrimination

(see later). Würth's data on contemporary jurisdiction in San'a undermines ideal constructions of 'Alid unions (2000: 211–13).

17. When the son of one of Imam Ahmad's senior officials lost his wife in his thirties and decided to remain single for the time being, his father insisted that he take another wife. The man obliged and remarried shortly thereafter.

18. Vom Bruck (2001: 304–5). The Imams' sons and brothers were unruly and contested their authority. Imam Ahmad killed two, possibly three of his brothers and sent others abroad. Some of the Imams' affines did not feel morally bound to them and took part in their violent removal from office.

19. On these hypergamous marriages, compare Bujra (1971: 95); Gingrich and Heiss (1986: 19); Gingrich (1989a: 80–1); Ho (2001: 92).

20. For a list of the marriages contracted by Imams Yahya and Ahmad see appendix I. In exceptional cases, 'Alid men claim that even they must adhere to strict endogamous rules (Meissner 1987: 182–3).

21. The woman, Huriyyah al-Mutawakkil, and the shaykh were nursed by the same mother. Her father, a son of Imam Muhsin al-Mutawakkil and married to the sister of Shaykh Nasir Mabkhut al-Ahmar, fought with Imam al-Mansur and Imam Yahya against the Turks. She was the mother of Umm Hani (d. 1979), al-Husayn (d. 1948), al-Mutahhar (d. about 1950), al-Hasan (d. 2003), and Amat al-Rahman.

22. Unable to consult the manuscript, I had access to a page copied from it by one of the woman's relatives.

23. Bayt 'Abd al-Qadir is a branch of Bayt Sharaf al-Din. Ibrahim 'Abd al-Qadir acted as the Head of the army (*amir al-jaysh*) under both Imam Yahya and Ahmad.

24. He served as the *hakim* (district judge) of Zabid and Kawkaban during the time of Imam Yahya.

7   Marriage in the Age of Revolution

1. He admitted that he took part in the revolution because "there was blood between them [the Hamid al-Din] and us (*kan baynana dam*)."

2. Yemenis never concretized ideas about baraka which might be derived from these unions in terms of material benefit (on this issue, see El-Zein 1974: 105; Waterbury 1970: 103).

3. The controversy over the conditions of equivalence in marriage is reflected in contemporary literature. To an extent, antagonism which is expressed through historical accounts, pamphlets, and poetry replaces personal confrontation which is frowned upon. Consider the dispute between two historians. Sayyid Ahmad al-Shami's *Jinayat al-Akwa'* (1980) is an answer to the work of Qadi Muhammad b. 'Ali al-Akwa'. Like the renowned scholar Muhammad al-Amir, al-Akwa' attributes the rule prohibiting the *shara'if* from marrying non-'Alids, to Imam Ahmad b. Sulayman (see chapter 6). However, al-Shami points out that a couple of centuries beforehand, the *sadah* had been discriminated against by the *'arab*. Men of high political standing such as Muhammad b. Yahya, a son of the first Zaydi Imam, had not been able to contract marriages with tribal women. His proposal to the Mu'aydiyyin, one of the northern tribes, was rejected because of his Qurayshi descent (al-Shami 1980: 62–3). Thus, in reply to al-Akwa''s insinuation that the *sadah* introduced the discriminatory marriage rule to Yemen, al-Shami suggests that they had merely adjusted to local practice.

4. Apparently it was common for girls to start crying when they learnt about men's proposals. In this case, the woman was uncertain about moving into an unfamiliar social space. Marrying into a polygynous household is usually a cause of distress.

5. One of the reasons given by the *sadah* in favor of endogamy was the tribesmen's adherence to the *'urf* (habitual legal codes commonly applied among the *qaba'il.* For example, Falihis argued that a *sayyid* could not have a son-in-law who subscribes to a legal code incompatible with Islam. Neither could a *sayyid* fulfill his affinal duties by fighting alongside his daughter's husband and his kin.

6. Since the late 1980s, several of these highly successful young men of shaykhly background have moved to San'a and married women of reputable houses, occasionally leaving their first wives in their villages.

7. Ideas about the *sadah*'s distinct phenotypical characteristics are rarely held. Once in Sa'dah province I was told by *qaba'il* that the *sadah* looked alike, and that they had light-colored skin because they spent more time in the house and the mosque than themselves.

8. In the Jawf, cultivation is looked down upon.

9. The term appears in the Qur'an and has been used to discredit tribal custom by Zaydi rulers (Dresch 1989: 184).

10. Debates about the "new" wife and mother who is herself educated and dedicates herself to training her children as educated and enlightened citizens were held in Egypt from the late nineteenth century onwards (Abu-Lughod 1998: 8–11). In Yemen, this debate is now in full swing.

11. In certain respects, the republican government's commitment to the abolition of social divisions based on descent is reflected in the dispensation of justice. Judges are careful to avoid references to descent when establishing the social disparity between spouses. Instead other criteria such as incompatible levels of education or moral conduct, or even the absence of love, are applied (Husayn al-Hubayshi, personal communication; for examples, see Würth 2000: 199, 212). Mufti Ahmad Zabarah (personal communication) placed emphasis on women's consent to a proposed marriage and conceived of arrangements without their approval as illicit (*zina'*). The Zaydi school reckons a woman's aversion (*karahiyah*) to her husband to be a legitimate reason for her to request a divorce, a principle which has been maintained in republican family law (Würth 2000: 107).

12. Most women take it for granted that they must obey their husbands and show deference, but the girl assumed that she need not do so were she to marry a "Qahtani." The Egyptian scholar and historian of Sufism, 'Abd al-Wahhab al-Sha'rani (1492–1565), held that a man could enter into matrimony with a *shar-ifah* if he were able to afford all that was due to her, would obey her pleasure, and consider himself her slave (quoted in van Arendonk 1996: 335). El-Zein (1974: 105) reports on the Kenyan sharifs that a "non-sharif husband could not criticize his sharif wife, even if she committed the gravest sin . . . [and] he also had to suspend his role as a man and head of his household. For all these reasons, men thought it was dangerous to marry a sharif woman."

13. From *qataba* (to gather, collect).

14. The argument of the maternal uncle never weighs as heavily as that of the paternal one who is obliged to provide for his brother's children after the death of their father and brothers.

15. On the first marriage of a *sharifah* to a *qabili* in the northern town of 'Amran, see Dorsky (1986: 45). Mundy (1981: 191) describes the attitude of a *sharifah*

who had married her fifth husband, a "Qahtani" soldier, in a rural community near San'a. "Her status was her private honour: the question of marriage within the status category—simply irrelevant." Reminiscent of the woman who cursed Imam al-Mutawakkil Isma'il because he prohibited 'Alid women from marrying non-'Alid men (chapter 6), apparently the prospect of political transformation in the twentieth century gave some 'Alid widows in the southern town of Hureidah hope that social convention would eventually be relaxed (Bujra 1971: 181).

16. On marriages of *muzayyin* men who were university students to 'Alid women in the north-western highlands, see Meissner (1987: 182, 186–7). The significance of these marriages has changed dramatically since the revolution. In one case, a *sayyid*, the grandson of a former governor, took a young girl of rural *muzayyin* background as a second wife. Her father's close ties to the regime may be profitable to his family. This connection, however, did not reconcile the man's wider family who reasoned that although Islam discourages these social divisions, Yemeni traditions could not be dismissed. However one of the man's uncles, 'Abd al-Karim al-Hani (chapter 11), said "good for him."

17. Like Bujra (1971: 98, 104), in his work on the "Learned Families" of a Shi'i Lebanese village Peters (1963: 178) argues that endogamy, especially of women, serves to maintain the coherence of the major landowning group. These families, which make up about a fifth of the inhabitants, own most of the land. Among Yemeni 'Alid houses, this has rarely been the case. With regard to the high rate of female endogamy in pre-revolutionary Yemen, I have inferred that marriages were manipulated for political rather than economic ends. Both *qadi* and 'Alid *buyut al-'ilm* argued that endogamy was not primarily motivated by property considerations because wealth differentiation between wife-givers and takers tends to be minimal (compare Bujra 1971: 97). Hence, property which is alienated by out-marrying women is "returned" by those who marry into the house, and circulates between wealthy houses. Few people stated that endogamy was aimed at preventing the alienation of property owned by their houses. In exceptional cases, wealthy houses who attach special sentiments to their *waqf dhurriyyah* which is inalienable, disapprove of their daughters' out-marriage because their portion of the waqf would be inherited by non-'Alid children. These sentiments are less frequently associated with other land which can be bought and sold, and which is inherited by non-'Alid women who are married to *sadah*.

8 "'Ulama of a Different Kind"

1. Algar (1969); Thaiss (1971, 1973); de Groot (1983); Schatkowski Schilcher (1985: 174, 177, 182); Denoeux (1993: 135–48). However, there is evidence that Sunnis also disapproved of trade. In the nineteenth-century Ottoman state, high ranking 'ulama who supported reform stressed the need for greater involvement in trade and industry among the Empire's Muslim population. They appealed to the higher strata to abandon their disdain for profit-making (Heyd 1993: 30).

2. Thaiss (ibid.); Keddie (1983: 7–8); Floor (1983: 93); Amanat (1988); Nakash (1994: 231–2); Litvak (2000: 54, 62–3).

3. Sayyid 'Abd al-Majid b. Abu'l Qasim al-Khu'i, personal communication.

4. According to leading 'ulama of San'a, in the twentieth century none of them has been engaged in trade and commerce. The historians I have consulted on this matter have not challenged these findings. With regard to central Yemen,

notably Khawlan, Steven Caton (personal communication) has come up with similar conclusions. The *sadah* have not been engaged in commerce (other than being market vendors). For the most part, Hadrami *sadah* had no reservations toward trade. According to Dale (1997), those who emigrated from the Hadramawt to Asia and Africa from the eighteenth century onward were paramount in establishing commercial and religious networks linked by kinship ties from Zanzibar to Singapore.

5. It must also be kept in mind that biographers did not always provide information about people's professions or sources of income. Because their knowledge was the main criterion for their eligibility to enter the biographical dictionaries (see chapter 5), occasionally we learn only about their studies. For example, it is said of Muhammad b. Yahya b. 'Ali Zabarah (b. 1895 in Jahana, Khawlan) that he studied grammar and Zaydi jurisprudence in Shaharah, that he memorized the Qur'an, and that he continued his studies in San'a (M. Zabarah 1979: 609–10).

6. Mufti Ahmad Zabarah, personal communication. Some governors had investments in places outside the borders of the area they administered. One of the merchants, al-Khadim al-Waji, who was executed in 1948, represented the business interests of some of these men.

7. The religiously sanctioned injunction against the economic activities of the rulers did not apply to the *sadah* collectively. On the contrary, their religious authorities advised them to engage in trade rather than to violate religious precepts. In his will, Imam al-Mutawakkil (d. 1676) advised his sons, relatives and other *sadah* to refrain from "eating the *zakat*"; to avoid the temptation of appropriating funds which were forbidden to them, they should take up a trade (Serjeant 1983a: 82). It is also worth noting that in times of need, high ranking scholars had no reservations about economic activities. For example, when Imam Yahya's foreign minister, Husayn al-Kibsi, was stranded in Japan after it had entered the war, he took up a trade in order to pay for his bills and his return to Yemen.

8. Imam al-Hadi al-Siraji (d. 1834) ordered the merchants to provide funds for the maintenance of 300 students (Zayd b. 'Ali al-Wazir, personal communication). The extent to which the Zaydi 'ulama, like their Imami counterparts, attracted merchant support and funds has not been studied yet.

9. I was unable to establish whether these individuals ran their own business or merely worked through intermediaries. It seems, however, that their economic engagement was rather limited. That, at any rate, is the view of Imam Yahya's descendants. According to one of his grandsons, al-Hasan b. al-Husayn, Muhammad b. Qasim b. Yahya ran an electricity company and Muhammad b. Yahya cooperated with the merchants of Hudaydah while he was a governor in the city. He denied that al-Hasan b. Yahya, the former governor of Ibb, ever pursued commercial activities, arguing that "he was not that kind of man."

10. One might expect that by the standards of the time the man's father must have had a relatively good income. However, according to Meissner (1987: 380), the officials' salaries were often too low to support them so that they were forced to look for alternative sources of income.

11. The *riyal fransi*, also referred to as *riyal namsawi* ("Austrian riyal"), was a Maria Theresa silver or gold coin which during the time of Imam Ahmad had the value of roughly one dollar.

12. For example, Muhammad b. 'Abdullah al-Wazir, who ruled briefly in the nineteenth century, worked as a bookbinder before he became Imam. (Ibrahim b. 'Ali al-Wazir, personal communication.)

13. Some of the reformers spent time abroad (See chapter 2, n. 26).

14. When Muhammad 'Abd al-Malik attended secondary school in Hajjah, a relative of the Imam held a speech at the school arguing that those who wore hats like the infidels would become like them. Muhammad exclaimed that Allah only looked to people's hearts and asked whether a Christian who donned an *'imamah* and a *thumah* (dagger) would become a Muslim. The Imam, who was informed about the incident, sent a telegram to Muhammad's father enquiring whether his son was still attached to the country's traditions. The boy also encouraged the servants in his father's house not to call him *sidi* and was angry that unlike his *khalah* (his father's second wife), his mother, a shaykh's daughter, was addressed by her personal name rather than by the title *sharifah*.

15. Both the 1948 revolt and the 1962 revolution owed much to the merchants' lavish funding. In the first republican government of 1962, some merchants were given ministerial posts, and today they are members of parliament.

16. For example, al-Ahmar Trading Company received exclusive contracts to operate pay phones and a planned digital cell-phone service (*Wall Street Journal*, January 2, 2001).

17. Among the new elite the woman's success provoked comments with slightly anti-'Alid undertones. When I asked a top official about his daughter's school, he answered "she is going to *Bayt sidi*," that is, an institution run by a member of the Prophet's House. (In other contexts, San'anis refer to the house of their grandfather as Bayt *sidi*).

18. This occupation was and sometimes still is looked down upon. In the countryside guesthouses and restaurants were usually under one roof. Until 1962, there were no guesthouses in San'a apart from one run by the government, and the so-called *samsarah* for peddlers. Ownership of a restaurant is more acceptable than serving in it.

19. In 1999 the monthly salary of a Head of department in San'a university was about $200. Some teachers provide translations for foreign organizations.

20. Note that the number of members of Bayt al-Wazir who live abroad (about eighty in the United States alone) is exceptionally high. Among the former ruling houses, they represent the second largest exile group after the Hamid al-Din.

21. See Rabinow (1975) for an example of *sadah* who became marginalized because they adhered to the ideology of noble descent and religious learning during a period of major economic and political change. Peters (1972: 192) writes of the Lebanese rural "Learned Families," some of whom are *sadah*, that under the impact of capitalism on the village economy in the 1950s, religious learning did not equip them to operate in the bureaucracy and the market.

9   The Moral Economy of Taste

1. I adopt a term coined by McKendrick et al. (1982).

2. 'Adnan Tarcici, Interview Geneva 1986. Tarcici was the head of a group of Lebanese experts who worked in Yemen during the time of Imam Yahya.

3. An elderly inhabitant of Hijrat Kibs in the central highlands recalled that in the days of Imam Yahya a man had to wear the *'imamah* even while at home with his family. When in the 1950s opposition to Imam Ahmad was increasing and there was fear of social and political disintegration, lapses in the observance of traditional dress codes were often not tolerated among the older generation. When in 1959 a young man, who had read law in Cairo, arrived at the airport wearing a suit, his uncle refused to take him home unless he changed his outfit.

4. Daggers are highly treasured and some are said to be as valuable as real estate. Those men who went into exile in the 1960s made sure they would not leave their daggers behind. When Imam Yahya's eldest son Muhummad al-Badr gave his famous sword which was ornamented with precious stones to King Farouk as a gift, his father was very angry.

5. Many of the shaykhs were wealthy landowners in the past. As noted in chapter 8, there was no wealthy commercial class in twentieth-century San'a prior to 1962. The Hujjari merchants who moved from Aden to San'a in the late 1950s and 1960s lost most of their property in 1967.

6. He aspired to strengthen the authority of the central government and launched several fixed term plans for development and state-building. Development cooperatives enjoyed governmental patronage (Peterson 1982: 145).

7. The first Internet café opened in San'a in 1998.

8. Meneley (1996: 48) describes how new political alliances between men are mirrored by formal visits among elite women of Zabid.

9. On Damascus, see Lindisfarne (1991, 2000) and Salamandra (2000). It must be stressed, however, that at Yemeni weddings male and female guests do not mix and there are no hired dancers.

10. Sunni-inspired religious opposition parties such as al-Islah also employ anti-consumption rhetoric focusing on wedding expenditure.

11. Each wedding is discussed in great detail by the guests; the number of sheep and cows slaughtered, the amount and kind of qat and food offered, the performance of the musicians, the bride's dress and hair-style, and other paraphernalia are scrutinized and evaluated. In the summer of 1990, the total amount of the wedding expenditures of a shaykh's son was said to have amounted to 25 million YR.

10 Defining through Defaming

1. A dish made of vegetables and *lahuh*, pancake bread of fermented millet.

2. The *sadah* are very rarely the object of prejudice during *tafritah* conversations. Women tend to refrain from discriminatory language against collective social entities. Rather, they agitate against individual men and women whom they accuse of improper conduct or those who have harmed them or their kin.

3. I should stress that relationships between individuals of different descent affiliation are inevitably multi-stranded, and that in many contexts the stereotypes discussed here are concealed. Difference, of course, is always contextual and relational (Moore 2000: 1130).

4. Just as van der Veer (1993) has argued that Dumont's reification of Indian culture as quintessentially Hindu has fed into Hindu nationalist discourse, assumptions of students of Arab cultures about the ancestry of those defining themselves as *'arab* would be received favorably by Yemeni anti-'Alid nationalists. For example, Rentz (1960: 544) writes that "in any event, Kahtan [Qahtan] clearly comes out closer than 'Adnan to genuine Arabness." Gochenour speaks of "the two racial stocks of 'Adnan and Qahtan," claiming that "the observant researcher can quickly discern the difference today between the ethnic Yemeni and Northern Arab in almost any group of Arabs in the peninsula" (1984: 153, 215 n. 8).

5. Officially Yemen supported Iraq and helped in the war effort.

6. Daylam was an 'Alid dynasty in tenth-century Iran. As noted in chapter 3, Bayt al-Daylami trace descent to Imam al-Nasir Abu'l-Fatah al-Daylami who made his *da'wah* to Yemen 1046.

7. Haydar Abu Bakr al-'Attas, the former President of the PDRY and Prime Minister of the Yemeni republic until the war of 1994, told a friend that he only became aware of being a *sayyid* when he came to the North and heard people talking about the *sadah* in official positions. In the North, the removal from power of the General Secretary of the Socialist Party, al-Bid, following the outbreak of hostilities gave rise to anti-'Alid rhetoric. "We finished with the *sadah* in the North, now we finished with the *sadah* in the South." According to Ho (1995: 2), more than half of the senior officials named by the socialist party during the war were *sadah*. "Their opponents gleefully seized upon this fact as proof that the whole sorry scheme was an attempt by anti-democratic ancien régime elements to seize power in the midst of the upheavals, with the backing of foreign conspirators."

8. Muhammad b. Isma'il al-Amir (d. 1769) is one of the scholars who deviated from classical Zaydi-Hadawi teachings. Among them were Muhammad b. Ibrahim al-Wazir, Salih al-Maqbali (see chapter 6), and Muhammad al-Shawkani.

9. The *sadah*'s descent from Fatima has also provoked polemic among Moroccan jurists. "Ibn al-Hajj does not fail to refer to the controversy among jurists regarding the sharaf from the mother's side" (Sebti 1986: 441).

10. Saddam Husayn claimed 'Alid descent and rehabilitated the Hashimite monarchy, portraying the kings as devoted nationalists (Bengio 1995: 142). According to one of Saddam's biographers, "a study of the family tree giving the descent of Saddam Hussein would show us that it goes right back to the noblest family of all, whose greatest scion was the Imam 'Ali bin Abi Talib" (Iskander 1980: 20).

11. Over a century ago, Glaser (1885: 202) claimed that the shaykhs appear to let the *sadah* take precedence only for religious reasons.

12. The Sunni Islamists comprise several different groupings. They have affinities to similar groups, especially in Egypt and the Sudan, and contain a strong Wahhabi component. Their common denominator is their claim that Islamic cultural and political authenticity is based exclusively on the *shari'ah*. Before a notable degree of liberalization was ushered in after unification, the government took no measures to suppress the Islamists in spite of their attacks on major government policies. On this topic, see Grosgurin (1994); Dresch and Haykel (1995); Detalle (1997); Zayd (1997); vom Bruck (1998b); Haykel (1999).

13. This school is based on the teachings of the Hanbali scholar Muhammad b. 'Abd al-Wahhab (1703/04–92) who saw the Qur'an and the Sunna as the only foundation of Muslim faith. He rejected both the Sunni–Hanafi orthodoxy practiced by the Ottomans during his time, and the Shi'a, as heresies. In 1745 he entered an alliance with Muhammad b. Sa'ud, the founder of the first Saudi kingdom.

14. Beliefs in the special powers of the Prophet's descendants are commonplace in the literature on the cultures of the Middle East, but they were hardly ever made explicit during my stay in Yemen. In exceptional cases soil is removed from the graves of great 'ulama and used for curing. I was told that in San'a Imam Yahya was asked to read the *fatihah*, the opening surah, on behalf of sick children. Access to the Imam's grave was restricted after the earth covering it had repeatedly been taken. According to Gingrich (1989b: 362; 1997: 164), among some northern tribes the *sadah* played an important role in bull sacrifices and rituals asking for rain (*istisqa'*). On curing procedures in the Sunni village of

Bani Ghazi in southern Yemen, Myntti (1983: 307) writes that the *sadah* "add their *baraka* to the mixture which is vital to the healing process" (ibid.: 307).

15. These deferential gestures are no longer displayed. Among the *sadah* I have only seen sons and daughters kissing their parents' knees on visits or *'id* days.

16. 'Abdullah al-Sallal, the first president of the republic, came from the despised professional class of blacksmiths (see Douglas 1987: 25). In suggesting that Sallal's presidency had motivated his father's alignment with the Imam whom he had loved and admired, the young man, who was very embarrassed about his father's role in the civil war, tried to gloss over his close ties with the former rulers.

17. Vom Bruck 1993. As shown in chapter 6, a complex set of gender attributes pertained to relations between members of different status categories, and these have come under revision (vom Bruck 1996).

18. Women deemed to be of lowly birth are said to have provided the only access to extra-marital relations prior to 1962 (on the *akhdam*, see Walters 1987: 76). Dresch (1989: 55, 119) notes that *qaba'il* flirt with these women in a way that would be insulting were they other tribesmen's sisters or daughters. Furthermore, they consider a man who fails to control his women to be deficient in *sharaf*.

11 Memory, Trauma, Self-Identification

1. For example, al Khoei (1991: 649).

2. This is a slightly edited version of the taped transcript.

3. He also admitted that excluding non-*sadah* from matrimony with 'Alid women was discriminating; yet he would have neither agreed to his sister's marriage to a Qahtani nor himself married a woman other than a *sharifah*. However, if his daughter, who was seven years old when I talked to him, wanted to marry a Qahtani, he would not stand in her way.

4. Besides its metaphorical function, here genderization seems appropriate because low status *sadah* were excluded from candidacy for the office of the Imam who must be both male and knowledgeable.

5. Any man called Yahya may be addressed by this title (see chapter 3, n. 20).

6. J. Scott (1990: 31) notes that after French revolutionaries had banned the use of the second person pronoun *vous* which had inscribed asymmetrical relationships, it has been used "*reciprocally* to express not status, but lack of close acquaintance." Among Yemeni old elite houses, the pronoun *antu* (the equivalent of *vous*) is used by junior kin who wish to communicate their respect to senior kin. However, it is also common among close friends and even brothers of similar age to express mutual respect, esteem, and affection.

7. This fact is of interest in analogy to naming practices among people of lowly birth who in the pre-revolutionary days felt discouraged from using names with explicit religious connotations such as 'Abd al-Rahman and 'Abd al-Qadir (Mermier 1997: 95).

8. When the authority of the Moroccan *shurfa'* was challenged in the 1960s, the children of their opponents were asked not to play with the children of the *wlad siyyd* (Rabinow 1975: 86).

9. Since the revolution, there has been heightened sensitivity among *sadah* who have found themselves confronted with such narratives. A few years ago, an historian's analysis of the events surrounding the destruction of a hijrah whose inhabitants were considered to be heretics by Imam 'Abdullah b. al-Hamzah

caused indignation. His description of the Imam as an unbeliever (*kafir*) and oppressor (*zalim*) angered some *sadah* who argued that he had dishonored the memory of their forebear. They accused him of tarnishing the *sadah*'s image and of playing into the hands of their enemies, and of risking to divide the Zaydis. In reply, the scholar challenged these men's definitions of kinship obligations, advising them that he had written the piece "as an historian, not a son."

10. As a child Hasan al-'Amri was sent to the *Madrasat al-aytam*, an orphan school established by Imam Yahya. In 1935 he attended the Military College at Baghdad and San'a. He took part in the 1948 uprising during which the Imam's prime minister 'Abdullah al-'Amri was killed.

11. Ideas about al-Hadi's potency are also expressed in jokes. After a storm a *qabili* assesses the damage caused by hail. He goes to the kitchen and gets a mortar (*malkad*). Then he goes outside and says to God "*Idha anta qawi, iksir zubb al-Hadi*" (if you are [that] strong, [try and] break al-Hadi's member).

12. This passage draws on J. Butler 1997a: 13–15, 158; 1997b: 104.

13. Yahya al-Nasir had already changed the names of his brother's daughters during the civil war. Both girls had names with the prefix "Amah" (see chapter 3). One of the girls, Amat al-Malik ("servant of the king"), was renamed Farida ("precious pearl"). On Farida al-Nasir see chapter 7.

14. 'A'isha, the Prophet's second wife, is not revered among Shi'is because of her enmity toward 'Ali (see Schimmel 1989: 43; Spellberg 1994: 6). It is likely that the woman's father's choice of names was motivated by his employment in the Sunni town of Ta'izz. Since I began research in the 1980s, I have only met two girls called 'A'isha who were born after the revolution.

15. The Yu'firid Dhu Hiwal, who came from Shibam to the north-west of San'a, defeated the 'Abbasid governor, Himyar b. al-Harith, in 847 (Rex Smith 1997: 130). According to Geddes (quoted in Eagle 1994: 112, 121 n. 63), Imam al-Hadi was opposed from the outset by Yu'firids who belonged to Dhu Hiwal.

16. Ahmad al-Shami, personal communication. However, after the revolution the most prominent scion of this house, President 'Abd al-Rahman al-Iryani (d. 1998), called himself al-Iryani al-Yahsubi. *Haql Yahsub*, literally "the field of Yahsub," was already mentioned in al-Hamdani's description of pre-Islamic Yemen (Hamid b. 'Ali al-Iryani, personal communication; compare Gochenour 1984: 17).

17. Baddeley quoted in Fentress and Wickham (1992: 22 n. 5); compare Baddeley (1989: 43). However, Tulving may not disagree with Baddeley's argument. In spite of postulating two functionally separate memory systems, he claims that the distinction is merely of heuristic value (Tulving 1983: 7). In his early work, Tulving (1972: 391–2) already suggested that semantic memory plays a role in the retrieval of episodic memory, and he stressed later that that episodic memory closely interacts with other systems (1983: 6, 8).

18. On this aspect of forgetting, see Bhabha (1994: 160–1); Hacking (1996: 70); Middleton (1999: 221).

12 History through the Looking Glass

1. Some influential new teachers of tribal background have also emerged, among them Muhammad Yahya 'izzan, Hasan 'Ali Fada'iq, and 'Abd al-Karim Jarban.

2. "The prospect of a Shiite-led Iraq needn't scare us," *Los Angeles Times*, April 13, 2003, p. 3.

3. The Islamists provided both ideological and military support to the government during major crises. During the 1994 war, paramilitaries fought on the side of the North (Halliday 1995: 134; 2000: 68).

4. "Politics in the name of the Prophet," *Le Monde Diplomatique*, November 2001. As noted by the *Economist* (February 14, 2002), "even today, Yemen remains one of the rare Arab states that integrates rather than suppresses its Islamists. They have their own party, Islah, the country's largest opposition party, with 64 of 301 seats in parliament."

5. One testimony to this discordance is the destruction of tombs and books by neo-Salafiyyun (sympathizers of a nineteenth-century reform movement which favors a return to the early principles of Islam as laid down by the Prophet and as practiced by the first caliphs) in Sunni areas since 1994. Similar devastation occurred during the Wahhabi invasion of the Hadramawt in the nineteenth century (Serjeant 1957: 21). Yemenis claim that before the Wahhabis reappeared in the late twentieth century, Shafi'is and Zaydis prayed in each other's mosques.

6. On the early nineteenth century, see Haykel (1997: 228).

7. Anonymous communication. Shawkani, however, maintained a critical distance toward the teachings of Muhammad b. 'Abd al-Wahhab (Shawkani 1982: 154–8).

8. According to the Ministry of Education, by 1997 between 900 and 1000 institutes were established with half a million students. Their budget for 1997 was five billion YR (100YR = 1$). Zayd (1997: 12–14) argues that after the reconciliation between Saudi Arabia and the Yemeni republic in 1972, Saudi Arabia started sponsoring the *ma'ahid*. According to the author, this was one of the reasons why the principles of the Zaydi school have not been taught at the institutes.

9. *Mujtahids* like al-Shawkani, who considered themselves to be above the various schools were, of course, a product of that line of thinking.

10. Consider also the "conversions" of Meccan Hasani *ashraf* from Zaydi to Sunni Islam once the former had begun to decline in the second half of the fourteenth century. Several became hadith scholars in the Sunni tradition and taught in Cairo and Mecca (Mortel 1987: 468).

11. *Mut'ah* marriage is discouraged by the Zaydiyyah. The boy was not accepted as a brother by the man's other sons, and was not invited to his father's funeral.

12. It would appear that its significance lay in the very fact of its existence. From the time of its inception, it has suffered from internal disputes. It never had a popular appeal and won only two seats in the parliamentary elections of 1993 and none in 1997 and 2003.

13. See Gingrich and Heiss (1986: 165); Meissner (1987: 32); Weir (1997). On the Ghadir ritual in the context of Zaydi history and its significance for the relations between the tribes and the Imams, see Mermier (1991).

14. God was invoked to bless Imam Zayd b. 'Ali, the founder of the Zaydiyyah, for the first time in the Friday prayers at the Great Mosque of San'a in 1660 ('A. al-Wazir 1985: 167–8). A foundation which edits and publishes hand-written Zaydi manuscripts has been named after the Imam (*Mu'assasat Zayd b. 'Ali*). Beyond these activities, it is concerned with the welfare of the poor.

15. I draw on al-Shami (1987, Vol. 1: 117–23) and interviews held in Yemen and Britain.

16. This is a reference to the role some tribesmen—particularly important shaykhs—played in the revolution. According to Zaydi history, they had invited Yahya b. al-Husayn to Yemen and paid allegiance to the Imams over the centuries.

17. I refer to Z. al-Wazir (1992: 30–3, 36) and interviews held during the 1990s.
18. Al-Wazir holds the view that although the shaykhs were Shi'i at heart, they could not commit themselves to Shi'i principles because they wanted to rule themselves. Hence they sought to diminish the authority of the Zaydi 'ulama by becoming active in organizations such as al-Islah.
19. On asking Sayyid al-Mansur about his opinion on al-Shami's proposition, he held that if 'ulama participation in government benefited society, they should not refrain from it; if however it caused problems, they should remain aloof.
20. Haykel (1996) draws similar conclusions from his interpretation of Hizb al-Haqq pamphlets.
21. For example, Sayyid al-Hasan b. Ahmad al-Jalal (d. 1673) held the view that any Muslim could be an Imam, whether Qurayshi, 'Alid or neither.

Conclusion

1. Here I am taking up Boyarin's (1994: 24) point that Halbwachs, in spite of socializing the concept of memory, fails to historicize it. By reason of the current shortage of *mujtahids*, the rulings of those who are dead are usually being referred to.
2. One such case is 'Abd al-Rahman Abu Talib, Yemen's ambassador to the United Nations who favored social, economic, and political change. His execution caused widespread indignation.
3. For a recent example concerning Yemen, see Blumi (2004). The assumption often made is that maintaining categories of identity is tantamount to reproducing the ideological structures underlying hierarchical difference and nationalist claims. One wonders, however, whether authors such as Feldman (1991) could have written a comprehensive study about the Northern Ireland conflict without conceiving of local labels such as "Catholic" and "Protestant" at least as heuristic categories. For an admirable attempt to arrive at representational coherence amidst manifold nationalist claims, see Karakasidou (1997).
4. I refer to different religions as well as adherence to diverse schools within Islam.
5. Furthermore, in contemporary Yemen the promotion of cultural pluralism is problematic in consideration of regional interest groups which seek to undermine such pluralism.
6. On the notion of parity of participation aiming at reciprocal recognition and status equality, see Fraser (2003). She points out rightly that affirmative strategies all too often lend themselves to separatism and repressive communitarianism (compare Fraser 2000: 108, 112).
7. Scholars must remain alert that omitting identity categories such as these from anthropological studies may aid authoritarian nationalist projects. In deconstructing these categories in the pursuit of denaturalizing them one must consider the expense of suspensions and the political goals toward which they are used or *not* used. One poignant example is Turkey where in the aftermath of the capture of the Kurdish leader 'Abdullah Oçalan in 1999, the use of the term "Kurd" and "Kurdish" was forbidden at Turkish universities (Times Higher Education Supplement, July 9, 1999).
8. Christelow's work (1980: 140, 153 n. 7) on mid-1850 Algeria, where he notes that the "traditional vocabulary" (i.e. *awlad sayyid*) was *still* in use at that time, conveys a sense of the significance of stereotyping in contemporary Yemen. In

Algeria too the title *sayyid* is now used as a general term of respect analogous to *monsieur*.

Appendices

1. Because women's personal names are usually not mentioned in the presence of people who are not their close kin (*mahram*), those who are not professionals or could not be consulted are not listed here (see vom Bruck 2001).

Bibliography

Abu-Lughod, Lila 1986 *Veiled sentiments: Honor and poetry in a Bedouin society.* Berkeley, CA: University of California Press.

——— 1998 "Introduction: Feminist longings and postcolonial conditions." In *Remaking women: Feminism and modernity in the Middle East.* Edited by L. Abu-Lughod, pp. 3–31. Princeton, NJ: Princeton University Press.

Abu Zahra, Muhammad 1959 *Al-Imam Zayd.* Beirut: Al-Maktabah al-islamiyyah.

Adra, Najwa 1982 *Qabyala: The tribal concept in the central highlands of the Yemen Arab Republic.* Ph.D. Thesis, Temple University.

Alaini, Mohsin 2004 50 years in shifting sands. Beirut: Dar an-Nahar.

Al-Akwaʿ, Ismaʿil 1987 "Nashwan Ibn Saʿid al-Himyari and the spiritual, religious and political conflicts of his era." In *Yemen: 3000 years of art and civilisation in Arabia Felix.* Edited by W. Daum, pp. 212–31. Innsbruck and Frankfurt: Pinguin Verlag.

——— 1995 *Hijar al-ʿilm wa maʿaqiluh fi-ʾl-yaman* (5 vols.). Beirut: Dar al-fikr al-muʿasir.

Al-Akwaʿ, Muhammad 1980 *Safhah min tarikh al-yaman al-ijtimaʿi wa qissat hayati.* Damascus: Matbaʿat al-katib al-ʿarabi.

Algar, Hamid 1969 *Religion and state in Iran 1785–1906.* Berkeley and Los Angeles, CA: University of California Press.

Alonso, Ana M. 1994 "The politics of space, time and substance: State formation, nationalism, and ethnicity." *Annual Review of Anthropology* 23: 379–405.

Amanat, Abbas 1988 "In between the madrasa and the marketplace: The designation of clerical leadership in modern Shiʿism." In *Authority and political culture in Shiʿism.* Edited by S. Arjomand, pp. 98–132. Albany, NY: State University of New York Press.

Al-ʿAmri, ʿAbdullah 1985 *The Yemen in the 18th & 19th centuries: A political and intellectual history.* Durham Middle East Monographs No.1. London: Ithaca Press.

——— 2000 "Some notes on two Yemeni contemporary documents." *New Arabian Studies* 5: 1–6.

Appadurai, Arjun 1998 "Dead certainty: Ethnic violence in the era of globalization." *Public Culture* 10(2): 225–47.

Appleby, Joyce 1993 "Consumption in early modern social thought." In *Consumption and the world of goods.* Edited by J. Brewer and R. Porter, pp. 162–73. London and New York: Routledge.

van Arendonk, C. 1996 "Sharif." *Encyclopaedia of Islam* 9: 329–37.

Arjomand, Said 1988 "Introduction: Shiʿism, authority, and political culture." In *Authority and political culture in Shiʿism.* Edited by S. Arjomand, pp. 1–22. Albany, NY: State University of New York.

Armbrust, Walter 1996 *Mass culture and modernism in Egypt*. Cambridge: Cambridge University Press.

Astuti, Rita 1995a *People of the sea: Identity and descent among the Vezo of Madagascar.* Cambridge: Cambridge University Press.

—— 1995b " 'The Vezo are not a kind of people': Identity, difference and 'ethnicity' among a fishing people of Western Madagascar." *American Ethnologist* 22(3): 464–82.

—— 2000 "Kindreds and descent groups: New perspectives from Madagascar." In *Cultures of relatedness: New approaches to the study of kinship*. Edited by J. Carsten, pp. 90–103. Cambridge: Cambridge University Press.

Al-Ayni, Muhsin 1973 Preface to *Brennpunkt Jemen* by F. Sitte. Vienna: Kremayr and Scherian.

Azra, Azyumardi 1997 "A Hadhrami religious scholar in Indonesia: Sayyid 'Uthman." In *Hadhrami traders, scholars and statesmen in the Indian Ocean, 1750s–1960s*. Edited by U. Freitag and W. Clarence-Smith, pp. 249–63. Leiden: Brill.

Baddeley, Alan 1989 "The psychology of remembering and forgetting." In *History, culture and the mind*. Edited by T. Butler, pp. 33–60. Oxford: Basil Blackwell.

Bahloul, Joëlle 1996 *The architecture of memory: A Jewish-Muslim household in colonial Algeria 1937–1962*. Cambridge: Cambridge University Press.

Bälz, Kilian 1999 "The secular reconstruction of Islamic law: The Egyptian supreme constitutional court and the 'Battle over the Veil' in state-run schools." In *Legal pluralism in the Arab world*. Edited by B. Dupret, M. Berger, and L. al-Zwaini, pp. 229–43. The Hague: Kluwer Law International.

Bammer, Angelika 1994 "Introduction." In *Displacements: Cultural identities in question*. Edited by A. Bammer, pp. xi–xx. Bloomington and Indianapolis, IN: Indiana University Press.

Barth, Fredrik 1959 *Political leadership among Swat Pathans*. London: Athlone Press.

—— 1990 "The guru and the conjurer: Transactions in knowledge and the shaping of culture in Southeast Asia and Melanesia." *Man* 25(4): 640–53.

Bartlett, Frederic 1932 *Remembering: A study of experimental and social psychology*. Cambridge: Cambridge University Press.

Battaglia, Debbora 1992 "The body in the gift: Memory and forgetting in Saberl mortuary exchange." *American Ethnologist* 19(4): 3–18.

—— 1995 "Problematizing the self: A thematic introduction." In *Rhetorics of self-making*. Edited by D. Battaglia, pp. 1–15. Berkeley and Los Angeles, CA: University of California Press.

—— 1999 "Toward an ethics of the open subject: Writing culture in good conscience." In *Anthropological theory today*. Edited by H. Moore, pp. 114–50. Oxford: Polity Press.

Batatu, Hanna 1978 *The old social classes and the revolutionary movements of Iraq: A study of Iraq's old landed and commercial classes and of its communists, Ba'thists, and Free Officers*. Princeton, NJ: Princeton University Press.

Bauman, Zygmunt 1991 *Modernity and ambivalence*. Ithaca, NY: Cornell University Press.

Ben Achour, Mohamed E. 1999 "Les Šarīfs à Tunis au temps des Deys et des Beys (XVIIe-XIXe siècle)." *Oriente Moderno* 18(2): 341–50.

Bengio, Ofra 1995 "Faysal's vision of Iraq: A retrospect." In *The Hashemites in the modern Arab world*. Edited by A. Susser and A. Shmuelevitz. London: Frank Cass.

Bergson, Henry 1970 [1911] *Matter and memory*. Translated from the French. London: George Allen and Unwin Ltd.

Berque, Jacques 1955 *Structures sociales du Haut-Atlas*. Paris: Presses Universitaires de France.

Bhabha, Homi K. 1994 *The location of culture*. New York and London: Routledge.

Bird-David, Nurit 2000 "History as performance for hunter-gatherer Nayaka." Munro Lecture, Edinburgh University, February 17.

Blachère, Régis 1992 "Al-Mutanabbi." *Encyclopaedia of Islam* 7: 769–72.

Bloch, Maurice 1995 "Malagasy kinship and kinship theory." In *L'étranger in time*. Edited by B. Champion, pp. 173–80. Réunion: Presses Universitaires de la Réunion.

—— 1996 [1992] "Internal and external memory: Different ways of being in history." In *Tense past: Cultural essays in trauma and memory*. Edited by P. Antze and M. Lambek, pp. 215–33. London: Routledge.

—— 1998 *How we think they think: Anthropological approaches to cognition, memory, and literacy*. Boulder, CO: Westview Press.

Blukacz, François 1993 "Le Yémen sous l'autorité des imams zaidites au XVIIe siècle: Une éphémère unité." *Revue des Mondes Musulmans et de la Méditerranée* 67(1): 39–51.

Blumi, Isa 2004 "Shifting loyalties and failed empire: A new look at the social history of late Ottoman Yemen, 1872–1918." In *Counter-Narratives: History, contemporary society, and politics in Saudi Arabia and Yemen*. Edited by M. al-Rasheed and R. Vitalis, pp. 103–17. New York: Palgrave.

Bodenhorn, Barbara 2000 " 'He used to be my relative': Exploring the bases of relatedness among Inupiat of northern Alaska." In *Cultures of relatedness*. Edited by J. Carsten, pp. 128–48. Cambridge: Cambridge University Press.

Bordo, Susan 1993 *Unbearable weight: Feminism, western culture, and the body*. Berkeley, CA: University of California Press.

Bosworth, Clifford 1995 "Sayyid." *Encyclopaedia of Islam* 9: 115–16.

Bouquet, Mary 1996 "Family trees and their affinities: The visual imperative of the genealogical diagram." *Journal of the Royal Anthropological Institute* 2(1): 43–66.

Bourdieu, Pierre 1977 *Outline of a theory of practice*. Translated from the French. Cambridge: Cambridge University Press.

—— 1986 *Distinction: A social critique of the judgement of taste*. Translated from the French. London: Routledge.

—— 1990 *The logic of practice*. Translated from the French. Oxford: Polity Press.

—— 1994 "Structures, habitus, power: Basis for a theory of symbolic power." In *Culture/power/history: A reader in contemporary social theory*. Edited by N. Dirks, G. Eley, and S. Ortner, pp. 155–99. Princeton, NJ: Princeton University Press.

Bowen, John 1998 "Qur'an, justice, gender: Internal debates in Indonesian Islamic jurisprudence." *History of Religions* 38(1): 52–78.

Boyarin, Jonathan 1991 *Polish Jews in Paris: The ethnography of memory*. Bloomington and Indianapolis, IN: Indiana University Press.

—— 1992 *Storm from paradise*. Minneapolis, MN: University of Minnesota Press.

—— 1994 "Space, time, and the politics of memory." In *Remapping memory: The politics of timespace*. Edited by J. Boyarin, pp. 1–37. Minneapolis, MN: University of Minnesota Press.

Boyarin, Jonathan and Daniel Boyarin 1995 "Self-exposure as theory: The double mark of the male Jew." In *Rhetorics of self-making*. Edited by D. Battaglia, pp. 16–42. Berkeley and Los Angeles, CA: University of California Press.

Boyer, Pascal 1990 *Tradition as truth and communication: A cognitive description of traditional discourse*. Cambridge: Cambridge University Press.

Bredi, Daniela 1999 "Sadat in South Asia: The case of Sayyid Abu'l-Hasan 'Ali Nadwi." *Oriente Moderno* 18(2): 375–92.

Brewer, W. 1996 "What is recollective memory?" In *Remembering our past: Studies in autobiographical memory*. Edited by D. Rubin. Cambridge: Cambridge University Press.

Brown, Kenneth 1976 *People of Salé: Tradition and change in a Moroccan city 1830–1930*. Manchester: Manchester University Press.

vom Bruck, Gabriele 1989 "Heiratspolitik der Prophetennachfahren." *Saeculum* 40(3–4): 272–95.

———— 1992–93 "Enacting tradition: The legitimation of marriage practices amongst Yemeni *sadah*." *Cambridge Anthropology* 16: 54–68. Special Issue "Islamic family law: Ideals and realities."

———— 1993 "Réconciliation ambiguë: Une perspective anthropologique sur le concept de la violence légitime dans l'Imamat du Yémen." In *La Violence et l'état: Formes et évolution d'un monopole*. Edited by E. Le Roy and T. v. Trotha, pp. 85–103. Paris: L'Harmattan.

———— 1994 "Down-playing gender: Ḥatm rituals in San'a." *Quaderni di Studi Arabi* 12: 161–82.

———— 1996 "Being worthy of protection: The dialectics of gender attributes in Yemen." *Social Anthropology* 4(2): 145–62.

———— 1997a "Elusive bodies: The politics of aesthetics among Yemeni elite women." *Signs: Journal of Women in Culture and Society* 23(1): 175–214.

———— 1997b "A house turned inside out: Inhabiting space in a Yemeni town." *Journal of Material Culture* 2(2): 139–72.

———— 1998a "Kinship and the embodiment of history." *History and Anthropology* 10(4): 263–98.

———— 1998b "Disputing descent-based authority in the idiom of religion: The case of the Republic of Yemen." *Die Welt des Islams* 38(2): 149–91.

———— 1999a "Being a Zaydi in the absence of an Imam: Doctrinal revisions, religious instruction, and the (re-) invention of ritual." In *Le Yémen contemporain*. Edited by R. Leveau, F. Mermier, and U. Steinbach, pp. 169–92. Paris: Editions Karthala.

———— 1999b "Ibrahim's childhood." *Middle Eastern Studies* 35(2): 150–71.

———— 2000 "Higher education in Yemen: Knowledge and power revisited." *International Higher Education* 18: 14–16.

———— 2001 "Le nom comme signe corporel: L'exemple des femmes de la noblesse yéménite." *Annales: Histoire, Sciences Sociales* 56(2): 283–311.

Bruner, Jerome and Carol Fleisher Feldman 1996 "Group narrative as a cultural context of autobiography." In *Remembering our past: Studies in autobiographical memory*. Edited by D. Rubin, pp. 291–317. Cambridge: Cambridge University Press.

Bujra, Abdalla S. 1966 "Political conflict and stratification in Hadramaut—I." *Middle Eastern Studies*, 3(1): 355–75.

———— 1971 *The politics of stratification: A study of political change in a South Arabian town*. Oxford: Clarendon Press.

Burke, III, Edmund 1972 "The Moroccan Ulama, 1860–1912: An introduction." In *Scholars, saints, and sufis: Muslim religious institutions since 1500*. Edited by N. Keddie, pp. 93–125. Berkeley and Los Angeles, CA: University of California Press.

———— 1993 "Middle Eastern societies and ordinary people's lives." In *Struggle and survival in the modern Middle East*. Edited by E. Burke, pp. 1–27. London: I.B. Tauris.

Burke, Peter 1989 "History as social memory." In *Memory: History, culture and the mind*. Edited by T. Butler, pp. 97–113. Oxford: Blackwell.

———— 1993 "*Res et verba*: Conspicuous consumption in the early modern world." In *Consumption and the world of goods*. Edited by J. Brewer and R. Porter, pp. 148–61. London: Routledge.

Busse, Heribert 1984 "Grundzüge der islamischen Theologie und der Geschichte des islamischen Raumes." In *Der Islam in der Gegenwart.* Edited by W. Ende and U. Steinbach, pp. 17–53. Munich: C.H. Beck.

Butler, Judith 1993 *Bodies that matter: On the discursive limits of "sex."* London: Routledge.

—— 1997a *Excitable speech: A politics of the performative.* London: Routledge.

—— 1997b *The psychic life of power: Theories in subjection.* Stanford, CA: Stanford University Press.

Butler, Thomas 1989 "Memory: A mixed blessing." In *Memory: History, culture and the mind.* Edited by T. Butler, pp. 1–31. Oxford: Basil Blackwell. Wolfson College Lectures.

Camelin, Sylvaine 1995 "Les pêcheurs de Shihr: Transmission du savoir et identité sociale." *Chroniques Yéménites* 3: 38–56.

Canard, Marius 1965 "Fatimids." *Encyclopaedia of Islam* 2: 850–62.

Carapico, Sheila 1998a *Civil society in Yemen: The political economy of activism in modern Arabia.* Cambridge: Cambridge University Press.

—— 1998b "Aid dependency, public sectors, and divestiture in Yemen." Paper presented at the Middle East Studies Association (MESA), December 3–6, Chicago.

Carruthers, Mary 1990 *The book of memory: A study of memory in medieval culture.* Cambridge: Cambridge University Press.

Carsten, Janet 1990 "Women, men, and the long and short term of inheritance in Pulau Langkawi, Malaysia." *Bijdragen tot de Taal-, Land- Volkenkunde* 146(2–3): 270–88.

—— 1995 "The substance of kinship and the heat of the hearth: Feeding, personhood and relatedness among Malays in Pulau Langkawi." *American Ethnologist* 22(2): 223–41.

—— 1997 *The heat of the hearth: The process of kinship in a Malay fishing community.* Oxford: Oxford University Press.

—— 2000 "Introduction: Cultures of relatedness." In *Cultures of relatedness: New approaches to the study of kinship.* Edited by J. Carsten, pp. 1–36. Cambridge: Cambridge University Press.

Carsten, Janet and Stephen Hugh-Jones 1995 Introduction to *About the house: Lévi-Strauss and beyond.* Cambridge: Cambridge University Press.

Caruth, Cathy 1991 "Introduction." *American Imago* 48(1): 1–12 (Special issue: Psychoanalysis, culture and trauma).

Casey, Edward 1987 *Remembering: A phenomenological study.* Bloomington, IN: Indiana University Press.

Caton, Steven C. 1986 "*Salam tahiyah*: Greetings from the highlands of Yemen." *American Ethnologist* 13(2): 290–308.

Chelkowski, Peter J. 1979 (Ed.) *Ta'ziyeh: Ritual and drama in Iran.* New York: New York University Press.

Christelow, Allan 1980 "Saintly descent and worldly affairs in mid-nineteenth century Mascara, Algeria." *International Journal of Middle East Studies* 12: 139–55.

Clarence-Smith, William G. 1997 "Hadramaut and the Hadrami diaspora in the modern colonial era: An introductory survey." In *Hadhrami traders, scholars and statesmen, 1970s–1960s.* Edited by U. Freitag and W. Clarence-Smith, pp. 1–18. Leiden: Brill.

Cohn, Bernard S. 1996 *Colonialism and its forms of knowledge: The British in India.* Princeton, NJ: Princeton University Press.

Cole, Juan R.I. 1988 *Roots of North Indian Shi'ism in Iran and Iraq: Religion and state in Awadh,1722–1859.* Berkeley, CA: University of California Press.

Coleman, Janet 1992 *Ancient and medieval memories: Studies in the reconstruction of the past.* Cambridge: Cambridge University Press.

Comaroff, John 1995 "Ethnicity, nationalism and the politics of difference in an age of revolution." In *Perspectives on nationalism and war*. Edited by J. Comaroff and P. Stern. Australia: Gordon and Breach Publishers.

Comaroff, John and Jean Comaroff 1992 *Ethnography and the historical imagination*. Boulder, CO: Westview Press.

Combs-Schilling, M. Elaine 1989 *Sacred performances: Islam, sexuality, and sacrifice*. New York: Columbia University Press.

—— 1991 "Etching patriarchal rule: Ritual dye, erotic potency, and the Moroccan monarchy." *Journal of the History of Sexuality* 1(4): 658–81.

Connerton, Paul 1989 *How societies remember*. Cambridge: Cambridge University Press.

Coussonnet, Nahida 1993 "Les assides du pouvoir zaydite au XIII siècle." *Revue du Monde Musulman et de la Méditerranée* 67: 25–37.

Crapanzano, Vincent 1973 *The Hamadsha: A study in Moroccan ethnopsychiatry*. Berkeley and Los Angeles, CA: University of California Press.

Cunningham, Karla J. 1999 "Rational statecraft in the Middle East: An analysis of Hashemite survival strategies." Paper presented at the Annual Meeting of the Middle East Studies Association.

Daftary, Farhad 1990 *The Isma'ilis: Their history and doctrines*. Cambridge: Cambridge University Press.

Dakhlia, Jocelyne 1996 "New approaches in the history of memory? A French model." Paper delivered at the Marc Bloch Centre in Berlin.

Dale, Stephen 1997 "The Hadhrami diaspora in south-western India: The role of the sayyids of the Malabar coast." In *Hadhrami traders, scholars and statesmen in the Indian Ocean, 1750s–1960s*. Edited by U. Freitag and W. Clarence-Smith, pp. 175–84. Leiden: Brill.

Daou, Tamima 1996 *The Imami Shi'i conception of knowledge of the Imam and the sources of religious doctrine in the formative period: From Hisham b. al-Hakam (d. 179 A.H.) to Kulini (d. 329 A.H.)*. Ph.D. Thesis, University of London.

Daum, Werner 1987a "From the Queen of Saba' to a modern state: 3,000 years of civilization in southern Arabia." In *Yemen*. Edited by W. Daum, pp. 9–33. Innsbruck and Frankfurt: Pinguin.

—— 1987b "From Aden to India and Cairo: Jewish world trade in the 11th and 12th centuries." In *Yemen*. Edited by W. Daum, pp. 167–77. Innsbruck and Frankfurt: Pinguin.

Dawn, Clarence E. 1971 "Hashimids" *Encyclopaedia of Islam* 3: 263–5.

Delaney, Carol 1986 "The meaning of paternity and the virgin birth debate." *Man* 21(3): 494–513.

Delgado, R. 1993 "Words that wound: A tort action for racial insults, epithets, and name calling." In *Words that wound: Critical race theory, assaultive speech, and the first amendment*. Edited by M. Matsuda et al. Boulder, CO: Westview Press.

Denoeux, Guilain 1993 *Urban unrest in the Middle East: A comparative study of informal networks in Egypt, Iran, and Lebanon*. Albany, NY: State University of New York Press.

Detalle, Renaud 1996 "Les partis politiques au Yémen: Paysage après la bataille." *Revue des mondes musulmans et de la Méditerranée* 81–2(3–4): 331–48.

—— 1997 "Les islamistes yéménites et l'Etat: Vers l'émancipation?" In *Les États arabes face à la contestation islamiste*. Edited by B. Kodmani-Darwish and M. Chartouni-Dubarry, pp. 272–98. Paris: Armand Colin.

—— 2000 "The Yemeni–Saudi conflict: Bilateral transactions and interactions." In *Tensions in Arabia: The Saudi–Yemeni fault line*. Edited by R. Detalle, pp. 52–79. Baden-Baden: Nomos Verlagsgesellschaft.

Dietrich, Albert 1982 "Autorité personnelle et autorité institutionelle dans l'islam: À propos du concept de 'sayyid' ". In *La notion d'autorité au Moyen age: Islam, Byzance, Occident*. Edited by G. Makdisi et al., pp. 83–99. Paris: Presses Universitaires de France.

Dirks, Nicholas B. 1992 "Introduction: Colonialism and culture." In *Colonialism and culture*. Edited by N. Dirks, pp. 1–25. Ann Arbor, MI: University of Michigan Press.

Dorsky, Susan 1986 *Women of `Amran: A Middle Eastern ethnographic study*. Salt Lake City, UT: University of Utah Press.

Douglas, J. Leigh 1987 *The Free Yemeni Movement 1935–1962*. Edited by G. Chimienti, American University of Beirut.

Douglas, Mary and Baron Isherwood 1978 *The world of goods*. London: Allen Lane.

Dresch, Paul 1989 *Tribes, government, and history in Yemen*. Oxford: Clarendon Press.

—— 1991 Imams and tribes: The writing and acting of history in upper Yemen. In *Tribes and state formation in the Middle East*. Edited by P. Khoury and J. Kostiner. London: I.B. Tauris

—— 1995 "The tribal factor in the Yemeni crisis." In *The Yemeni war of 1994*. Edited by J. al-Suwaidi, pp. 33–55. London: Saqi Books.

—— 2000 *A history of modern Yemen*. Cambridge: Cambridge University Press.

Dresch, Paul and Bernard Haykel 1995 "Stereotypes and political styles: Islamists and tribesfolk in Yemen." *International Journal of Middle East Studies* 27: 405–31.

Eagle, A.B.D.R. 1994 "Al-Hadi Yahya b. al-Husayn b. al-Qasim (245–98/859–911): A biographical introduction and the background and significance of his Imamate." *New Arabian Studies* 2: 103–22.

Eickelman, Dale F. 1976 *Moroccan Islam: Tradition and society in a pilgrimage center*. Austin, TX: University of Texas Press.

—— 1978 "The art of memory: Islamic education and its social reproduction." *Comparative Studies in Society and History* 20(4): 485–516.

—— 1982 "The study of Islam in local contexts." *Contributions to Asian Studies* 17: 1–16.

—— 1983 "Religion and trade in western Morocco." *Research in Economic Anthropology* 5: 335–48.

—— 1985 *Knowledge and power in Morocco: The education of a twentieth-century notable*. Princeton, NJ: Princeton University Press.

—— 1998 *The Middle East and Central Asia: An anthropological approach*. New Jersey: Prentice-Hall, Inc.

Eickelman, Dale and Jon Anderson 1999 "Redefining Muslim publics." In *New media in the Muslim world*. Edited by D. Eickelman and J. Anderson, pp. 1–18. Bloomington and Indianapolis, IN: Indiana University Press.

Elaouani-Cherif, Ahmed 1999 "La famille al-'Awani-Šarif de Qayrawan." *Oriente Moderno* 23(2): 451–7.

Ende, Werner 1984 "Der schiitische Islam." In *Der Islam in der Gegenwart*. Edited by W. Ende and U. Steinbach, pp. 70–90. Munich: C.H. Beck.

Endress, Gerhard 1982. *Einführung in die islamische Geschichte*. Munich: C.H. Beck.

Ensel, Remco 1998 *Saints and servants: Hierarchical interdependence between shurfa and haratin in the Moroccan deep south*. Amsterdam: Centrale Drukkerij.

Eshkevari, Al-Sayyid Sadiq 1999 Sugli appellativi (*alqab*) dei *sadat*. Oriente Moderno 18(2): 463–71.

Evans-Pritchard, E.E. 1949 *The Sanusi of Cyrenaica*. Oxford: Oxford University Press.

Fay, Mary Ann 1997 "Women and waqf: Toward a reconsideration of women's place in the Mamluk household." *International Journal of Middle East Studies* 29(1): 33–51.

Feldman, Alan 1991 *Formations of violence: The narrative of the body and political terror in Northern Ireland.* Chicago, IL: Chicago University Press.

———— 1994 "On cultural anesthesia: From desert storm to Rodney King." *American Ethhnologist* 21(2): 404–18.

Fentress, James and Chris Wickham 1992 *Social memory: New perspectives on the past.* Oxford: Blackwell.

Fischer, Michael 1980 *Iran: From religious dispute to revolution.* Cambridge, MA: Harvard University Press.

———— 1986 "Ethnicity and the post-modern arts of memory." In *Writing culture: The poetics and politics of ethnography.* Edited by J. Clifford and G. Marcus, pp. 194–233. Berkeley, CA: University of California Press.

———— 1990 "Legal postulates in flux: Justice, wit, and hierarchy in Iran." In *Law and Islam in the Middle East.* Edited by D. Dwyer, pp. 115–42. New York: Bergin & Garvey Publishers.

———— 1991 "The uses of life histories." *Anthropology and Humanism Quarterly* 16: 24–7.

Fischer, Michael and Mehdi Abedi 1990 *Debating Muslims: Cultural dialogues in post-modernity and tradition.* Madison, WI: University of Wisconsin Press.

Floor, Willem M. 1983 "The revolutionary character of the ulama: Wishful thinking or reality?" In *Religion and politics in Iran: Shi'ism from quietism to revolution.* Edited by N. Keddie, pp. 73–97. New Haven, CT and London: Yale University Press.

Fortes, Meyer 1969 *Kinship and the social order.* London: Routledge & Keagan Paul.

Foucault, Michel 1978 *The history of sexuality.* Vol. 1. Harmonsworth: Penguin.

———— 1994 "Two lectures." In *Culture/power/history: A reader in contemporary social theory.* Edited by N. Dirks, G. Eley, and S. Ortner, pp. 200–21. Princeton, NJ: Princeton University Press.

Fox, James J. 1987 "The house as a type of social organization on the Island of Roti, Indonesia." In *De la Hutte au palais: Sociétés 'à maison' en Asie du sud-est insulaire.* Edited by C. MacDonald. Paris: CNRS.

Frankl, P.J.L. 1990 "Robert Finlay's description of San'a in 1238–1239/1823." *British Society for Middle Eastern Studies Bulletin* 17(1): 16–32.

Fraser, Nancy 2000 "Rethinking recognition." *New Left Review* 3: 107–20.

———— 2003 "Social justice in the age of identity politics: Redistribution, recognition, and participation." In *Redistribution or recognition? A political-philosophical exchange.* N. Fraser and A. Honneth. London: Verso.

Freitag, Ulrike 1997 "Dying of enforced spinsterhood: Hadramawt through the eyes of 'Ali Ahmad Ba Kathir (1910–69)." *Die Welt des Islams* 37(1): 2–27.

———— 2000 "Scheich oder Sultan—Stamm oder Staat? Staatsbildung im Hadramaut (Jemen) im 19. und 20. Jahrhundert." *Jahrbuch des Historischen Kollegs,* pp. 165–94.

Frey, F.W. 1965 *The Turkish political elite.* Cambridge, MA: MIT Press.

Friedman, Jonathan 1994 *Consumption and identity.* Chur: Harwood Academic Publishers.

Garro, Linda C. 2001 "The remembered past in a culturally meaningful life: Remembering as cultural, social, and cognitive process." In *The psychology of cultural experience.* Edited by C. Moore and H. Mathews, pp. 105–47. Cambridge: Cambridge University Press.

Geertz, Clifford 1963 *Peddlers and princes.* Chicago, IL: Chicago University Press.

———— 1968 *Islam observed.* Chicago, IL: Chicago University Press.

Geertz, Hildred 1979 "The meanings of family ties." In *Meaning and order in Moroccan society: Three essays in cultural analysis.* Edited by C. Geertz, H. Geertz, and L. Rosen, pp. 315–91. Cambridge: Cambridge University Press.

Geertz, Hildred and Clifford Geertz 1964 "Teknonymy in Bali: Parenthood, age-grading and genealogical amnesia." *Journal of the Royal Anthropological Institute* 94 (part 2): 94–108.

Gell, Simeran 1998 "L'Inde aux deux visages: Dalip Singh et le Mahatma Gandhi." *Terrain* 31: 129–44.

Gellner, Ernest 1969 *Saints of the Atlas*. London: Weidenfeld and Nicolson.

Gerholm, Tomas 1977 *Market, mosque and mafraj: Social inequality in a Yemeni town.* Stockholm Studies in Social Anthropology 5, University of Stockholm.

———— 1984 "The making of a traditonal society: External and internal factors." In *Entwicklungsprozesse in der Arabischen Republic Jemen*. Edited by H. Kopp and G. Schweizer, pp. 87–97. Wiesbaden: Ludwig Reicher.

Gerth, H.H. and C. Wright Mills 1977 *From Max Weber: Essays in Sociology*. London: Routledge.

Gibb, H. 1960 'Ali b. Abi Talib. *Encyclopaedia of Islam* Vol. 1: 381–6.

Gilmore, David 1982 "Some notes on community nicknaming in Spain." *Man* 17(4): 686–700.

Gilsenan, Michael 1973 *Saint and sufi in modern Egypt: An essay in the sociology of religion*. Oxford: Clarendon Press.

———— 1982 *Recognising Islam*. London: Croom Helm.

———— 2003 "Out of Hadhramaut." *London Review of Books* 25(6): 1–12.

Gimaret, Daniel 1993 "Mu'tazila." *Encyclopaedia of Islam* 7: 783–93.

Gingrich, Andre 1986 " 'Iš wa milh: Brot und Salz. Vom Gastmahl bei den Ḥawlan bin 'Amir im Jemen." *Mitteilungen der Anthropologischen Gesellschaft in Wien* 116: 41–69.

———— 1989a "How the chiefs' daughters marry: Tribes, marriage patterns and hierarchies in north-western Yemen." In *Kinship, social change and evolution*. Edited by A. Gingrich, S. Haas, and G. Paleczek, pp. 75–85. Vienna: Ferdinand Berger & Söhne.

———— 1989b "Kalender, Regenzeit und Stieropfer in Nordwest-Jemen." In *Beiträge zum 2.Grazer Morgenländischen Symposion* (March 2–5, 1989). Edited by Bernhard Scholz, pp. 353–70. Grazer Morgenländische Studien.

———— 1997 "Inside an 'exhausted community': An essay on case-reconstructive research about peripheral and other moralities." In *The ethnography of moralities*. Edited by S. Howell, pp. 152–77. London and New York: Routledge.

Gingrich, Andre and Johann Heiss 1986 *Beiträge zur Ethnographie der Provinz Sa'da (Nordjemen)*. Vienna: Verlag der Österreichischen Akademie der Wissenschaften.

Glaser, Eduard 1885 "Die Kastengliederung im Jemen." *Das Ausland* 11: 201–3.

Gochenour, Thomas D. 1984 *The penetration of Zaydi Islam into medieval Yemen*. Ph.D. Thesis, Harvard University.

Goldziher, Ignáz, C. van Arendonk, and A. Tritton 1960 "Ahl al-bayt." *Encyclopaedia of Islam* 1: 257–8.

Gow, Peter 1991 *Of mixed blood*. Oxford: Oxford University Press.

Graeber, David 1999 "Painful memories." In *Ancestors, power, and history in Madagascar*. Edited by K. Middleton, pp. 319–48. Leiden: Brill.

de Groot, Joanna 1983 "Mullas and merchants: The basis of religious politics in 19th century Iran." *Mashriq* 2: 11–36.

Grosgurin, Jean-Marc 1994 "La contestation Islamiste au Yémen" In *Exils et royaumes: Les appartenances au monde arabo-musulman aujourd'hui*. Edited by G. Kepel, pp. 235–50. Paris: Presses de la Fondation Nationale des Sciences Politiques.

Gruzinski, Serge 1990 "Mutilated memory: Reconstruction of the past and the mechanisms of memory among 17th century Otomis." In *Between memory and history*. Edited by M. Bourguet, L. Valensi, and N. Wachtel, pp. 131–47. London: Harwood Academic Publishers.

Hacking, Ian 1996 "Memory sciences, memory politics." In *Tense past: Cultural essays in trauma and memory*. Edited by P. Antze and M. Lambek, pp. 67–87. London: Routledge.

Halbwachs, Maurice 1992 [1925] *On collective memory*. Translated from the French. Chicago, IL: University of Chicago Press.

Halliday, Fred 1995 "The third inter-Yemeni war and its consequences." *Asian Affairs* 26: 131–40.

——— 1996 *Islam and the myth of confrontation*. London: I.B. Tauris.

——— 2000 *Nation and religion in the Middle East*. London: Saqi Books.

Hammoudi, Abdellah 1997 *Master and disciple: The cultural foundations of Moroccan authoritarianism*. Chicago, IL: University of Chicago Press.

Harrison, Simon 2003 "Cultural difference as denied resemblance: Reconsidering nationalism and ethnicity." *Comparative Studies in Society and History* 45(2): 343–61.

Haykel, Bernard 1996 "Hizb al-Haqq and the doctrine of the Imamate." Paper presented at the Annual Meeting of the Middle East Studies Association.

——— 1997 *Order and righteousness: Muhammad 'Ali al-Shawkani and the nature of the Islamic state in Yemen*. D.Phil. Thesis, University of Oxford.

——— 1999 "Rebellion, migration or consultative democracy? The Zaydis and their detractors in Yemen." In *Le Yémen contemporain*. Edited by R. Leveau, F. Mermier, and U. Steinbach, pp. 193–201. Paris: Editions Karthala.

——— 2001 "The entrenchment of 'non-sectarian' Sunnism in Yemen." *Newsletter of the International Institute for the Study of Islam in the Modern World* 7: 19.

Heffening, W. 1974 "Wakf." *Shorter Encyclopaedia of Islam*, pp. 624–8.

Hegland, Mary E. 1995 "Shi'a women of Northwest Pakistan and agency through practice: Ritual, resistance, resilience." *PoLAR: Political and Legal Anthropology Review* 18: 65–79.

——— 1997 "A mixed blessing: The *majales*, Shi'a women's rituals of mourning in Northwest Pakistan." In *Mixed blessings: Gender and religious fundamentalism cross culturally*. Edited by J. Brink and J. Mencher, pp. 179–96. London: Routledge.

——— 1998a "The power paradox in Muslim women's *majales*: North-West Pakistani mourning rituals as sites of contestation over religious politics, ethnicity, and gender." *Signs* 23(2): 391–428.

——— 1998b "Fundamentalism and flagellation: (Trans)forming meaning, identity, and gender through Pakistani women's rituals of mourning." *American Ethnologist* 25(2): 240–66.

Heine, Peter 1984 "Das Verbreitungsgebiet der islamischen Religion: Zahlen und Informationen zur Situation in der Gegenwart." In *Der Islam in der Gegenwart*. Edited by W. Ende and U. Steinbach, pp. 132–51. Munich: C.H. Beck.

Herzfeld, Michael 1982 "When exceptions define the rules: Greek baptismal names and the negotiation of identity." *Journal of Anthropological Research* 38(3): 288–302.

——— 1997 *Cultural intimacy: Social poetics in the nation-state*. New York and London: Routledge.

Heyd, Uriel 1993 "The Ottoman 'Ulema and Westernization in the time of Selim III and Mahmud II." In *The modern Middle East*. Edited by A. Hourani, P. Khoury and M. Wilson, pp. 29–59. London: I.B. Tauris.

Al-Hilli, Hasan 1896/97 *Tahrir al-ahkam al-shar'iyyah 'ala madhhab al-Imamiyyah*. Tehran.

Ho, Engseng 1995 "The Production of Persistence." Paper presented at the panel: Time Warps in Yemen, annual meeting of the Middle East Studies Association, Washington DC.

Ho, Engseng 2001 "Le don précieux de la généalogie." In *Emirs et présidents: Figures de la parenté et du politique en islam*. Edited by P. Bonte, E. Conte, and P. Dresch, pp. 79–110. Paris: CNRS.

——— 2002 "Before parochialization: Diasporic Arabs cast in Creole waters." In *Transcending borders: Arabs, politics, trade and Islam in Southeast Asia*. Edited by H. de Jonge and N. Kaptein, pp. 11–35. Leiden: KITLV Press.

Hoffman-Ladd, Valerie J. 1992 "Devotion to the Prophet and his family in Egyptian Sufism." *International Journal of Middle East Studies* 24: 615–37.

Holy, Ladislav 1991 *Religion and custom in a Muslim society: The Berti of Sudan*. Cambridge: Cambridge University Press.

Hourani, Albert 1968 "Ottoman reform and the politics of notables." In *Beginnings of modernization in the Middle East: The nineteenth century*. Edited by W. Polk and R. Chambers, pp. 41–68. Chicago, IL and London: University of Chicago Press.

Huart, Clément 1974 " 'Ali b. Abi Talib." *Shorter Encyclopaedia of Islam*: 30–2.

Hudson, Michael 1995 "Bipolarity, rational calculation and war in Yemen." In *The Yemeni war of 1994: Causes and consequences*. Edited by J. Suwaidi. London: Saqi Books.

Imam Ali 1981 *Nahjul balagha: Sermons, letters and sayings*. Translated from the Arabic. Qum: Ansariyan Publication.

Imam Khomeini 1981 *Islam and revolution*. Translated from the Persian. Berkeley, CA: Mizan Press.

Al-Iryani, 'Abd al-Rahman n.d. *Hidayat al-mustabsirin, bi-sharh 'uddat al-husn al-hasin*. Damascus (publisher and date of publication unknown).

Al-Iriyani, Hamid 1987 "School and education—formation and development. In *Yemen*. Edited by W. Daum, pp. 375–88. Innsbruck and Frankfurt: Pinguin.

Iskander, Amir 1980 *Saddam Hussein: The fighter, the thinker and the man*. Translated from the French. Paris: Hachette Réalités.

Iteanu, André 1983 "Idéologie patrilinéaire ou idéologie de l'anthropologue?" *L'Homme* 23(2): 37–55.

Jalal, Ayesha 1995 "Conjuring Pakistan: History as official imagining." *International Journal of Middle East Studies* 27: 73–89.

Jamous, Raymond 1981 *Honneur et baraka: Les structures sociales traditionelles dans le rif*. Cambridge: Cambridge University Press.

——— 1996 "The Meo as a Rajput caste and a Muslim community." In *Caste today*. Edited by C. Fuller, pp. 180–201. Delhi: Oxford University Press.

Joffé, George 1997 "Introduction: Yemen and the contemporary Middle East." In *Yemen today: Crisis and solutions*. Edited by E. Joffé, M. Hachemi, and E. Watkins, pp. 10–20. London: Caravel.

Karakasidou, Anastasia N. 1997 *Fields of wheat, hills of blood: Passages to nationhood in Greek Macedonia 1870–1990*. Chicago, IL: Chicago University Press.

Keane, W. 1997 *Signs of recognition: Powers and hazards of representation in an Indonesian society*. Berkeley, CA: University of California Press.

Keddie, Nikki R. 1983 "Introduction." In *Religion and politics in Iran: Shi'ism from Quietism to revolution*. Edited by N. Keddie, pp. 1–18. New Haven, CT and London: Yale University Press.

——— 1988 "The Yemen Arab Republic (North Yemen): History and society." In *Sojourners and settlers: The Yemeni immigrant experience*. Edited by J. Friedlander. Salt Lake City, UT: University of Utah Press.

Kennedy, John 1987 *The flower of paradise: The institutionalized use of the drug qat in North Yemen*. Dordrecht: D. Reidel Publishing Company.

Al-Khatib, 'Abbas A. 1968 *'Alawiyyun al-Yaman* (unpublished).

Khilnani, Sunil 1997 *The idea of India*. Harmondsworth: Penguin Books.

Khoury, Philip S. 1983 *Urban notables and Arab nationalism: The politics of Damascus 1860–1920.* Cambridge: Cambridge University Press.

———— 1984. Syrian urban politics in transition: The quarters of Damascus during the French mandate. *International Journal of Middle East Studies* 16: 507–40.

Al-Khoei, Abu'l-Qasim 1991 *Imam Khoei's fatawa: Articles of Islamic faith.* London: Al-Khoei Foundation.

———— 1992 *Masa'il wa rudud.* Qum: Dar al-Hadi.

Kirmayer, Lawrence J. 1996 "Landscapes of memory: Trauma, narrative, and dissociation." In *Past tense: Cultural essays in trauma and memory.* Edited by P. Antze and M. Lambek, pp. 173–98. London: Routledge.

Klein, Kerwin 2000 "On the emergence of *Memory* in historical discourse." *Representations* 69: 127–50.

Knysh, Alexander 1993 "The cult of saints in Hadramawt: An overview." *New Arabian Studies* 1: 137–52.

Kohlberg, Etan 1976a "From Imamiyya to Ithna-'ashariyya." *Bulletin of the School of Oriental and African Studies* 39: 521–34.

———— 1976b "Some Zaydi views on the companions of the Prophet." *Bulletin of the School of Oriental and African Studies* 39: 91–8.

———— 1988 "Imam and community in the pre-Ghayba period." In *Authority and political culture in Shi'ism.* Edited by S. Arjomand, pp. 25–53. Albany, NY: State University of New York Press.

Kostiner, Joseph 1995 "Prologue of Hashimite downfall and Saudi ascendancy: A new look at the Khurma dispute, 1917–1919." In *The Hashemites in the Modern Arab World.* Edited by A. Susser and A. Shmuelevitz. London: Frank Cass.

Kramer, Martin 1991 "Introduction." In *Middle Eastern lives: The practice of biography and self-narrative.* Edited by M. Kramer. New York: Syracuse University Press.

Kristeva, Julia 1989 *Etrangers à nous-mêmes.* Paris: Fayard.

Kugelmass, Jack 1995 "Bloody memories: Encountering the past in contemporary Poland." *Cultural Anthropology* 10(3): 279–301.

Lackner, Helen 1985 *P.D.R.Yemen: Outpost of socialist development in Arabia.* London: Ithaca press.

Lambek, Michael 1993 *Knowledge and practice in Mayotte: Local discourses of Islam, sorcery, and spirit possession.* Toronto: University of Toronto Press.

———— 1995 "Choking on the Quran: And other consuming parables from the western Indian Ocean front." In *The pursuit of certainty: Religious and cultural formulations.* Edited by W. James, pp. 258–81. London: Routledge.

———— 1996 "The past imperfect: Remembering as moral practice." In *Tense Past: Cultural essays in trauma and memory.* Edited by P. Antze and M. Lambek, pp. 235–54. London: Routledge.

———— 1997a "Knowledge and practice in Mayotte: An overview." *Cultural Dynamics* 9(2): 131–48.

———— 1997b Pinching the crocodile's tongue: Affinity and the anxieties of influence in fieldwork. *Anthropology and Humanism* 22(1): 31–60.

———— 2000 "The anthropology of religion and the quarrel between poetry and philosophy." *Current Anthropology* 41(3): 309–20.

———— in press. "What's in a name? Name bestowal and the identity of spirits in Mayotte and Northwest Madagascar." In *The Anthropology of names and naming.* Edited by G. vom Bruck and B. Bodenhorn. Cambridge: Cambridge University Press.

Lambek, Michael and Andrew Walsh 1999 "The imagined community of the Antankarana: Identity, history, and ritual in Northern Madagascar."

In *Ancestors, power, and history in Madagascar*. Edited by K. Middleton, pp. 145–74. Leiden: Brill.

Lambert, Helen 2000 "Sentiment and substance in North Indian forms of relatedness." In *Cultures of relatedness*. Edited by J. Carsten, pp. 73–89. Cambridge: Cambridge University Press.

Landau-Tasseron, Ella 1990 "Zaydi Imams as restorers of religion: *Ihya'* and *tajdid* in Zaydi literature." *Journal of Near Eastern Studies* 49(4): 247–63.

Layne, Linda L. 1994 *Home and homeland: The dialogics of tribal and national identities in Jordan*. Princeton, NJ: Princeton University Press.

Lea, Vanessa 1995 "The houses of the Mebengokre (Kayapó) of Central Brazil—a new door to their social organization." In *About the house: Lévi-Strauss and beyond*. Edited by J. Carsten and S. Hugh-Jones, pp. 206–25. Cambridge: Cambridge University Press.

Lenczowski, George (Ed.) 1975 *Political elites in the Middle East*. Washington, DC: American Enterprise Institute for Public Policy Research.

Lewcock, Ronald 1986 *The old walled city of San'a*. Ghent: UNESCO.

Lewcock, Ronald, Paola Costa, Robert Serjeant, and Robert Wilson 1983 The urban development of San'a. In *San'a: An Arabian Islamic city*. Edited by R. Serjeant and R. Lewcock, pp. 122–43. The World of Islam Festival Trust.

Lewis, Herbert 1989 *After the eagles landed*. Boulder, CO: Westview Press.

Lindisfarne, Nancy 1991 In spectacular fashion: The aesthetics of privilege at wedding receptions in Damascus. Paper presented at the BRISMES Annual Conference, School of Oriental and African Studies, London.

———— 2000 *Dancing in Damascus*. New York: State University of New York Press.

Linnekin, Jocelyn and Lin Poyer 1990 "Introduction." In *Cultural identity and ethnicity in the Pacific*. Edited by J. Linnekin and L. Poyer, pp. 1–16. Honolulu: University of Hawaii Press.

Litvak, Meir 2000 "The finances of the 'ulama' communities of Najaf and Karbala', 1796–1904." *Die Welt des Islams* 40(1): 41–66.

Maclagan, Ianthe 1993 *Freedom and constraints: The world of women in a small town in Yemen*. Ph.D. Thesis, University of London.

McKendrick, Neil, John Brewer, and J.H. Plumb 1982 *The birth of a consumer society: The commercialization of eighteenth-century England*. London: Europa Publication Limited.

Maddy-Weitzman, Bruce 1990 Jordan and Iraq: Efforts at intra-Hashemite unity. *Middle Eastern Studies* 24(1): 65–72.

Madelung, Wilferd 1965 *Der Imam al-Qasim ibn Ibrahim und die Glaubenslehre der Zaiditen*. Berlin: Walter de Gruyter.

———— 1971 "Imama." *Encyclopaedia of Islam* 3: 1163–9.

———— 1982 "Authority in Twelver Shiism in the absence of the Imam." In *La notion d'autorité au Moyen Age: Islam, Byzance, Occident*. Edited by G. Makdisi et al., pp. 163–73. Paris: Presses Universitaires de France.

———— 1987 "Islam in Yemen." In *Yemen*. Edited by W. Daum, pp. 174–7. Innsbruck and Frankfurt: Pinguin Verlag.

———— 1988a "Imam al-Qasim ibn Ibrahim and Mu'tazilism." In *On both sides of al-Mandab*. Edited by U. Ehrensvard and C. Toll, pp. 39–48. Swedish Research Institutute in Istanbul: Transactions Volume 2.

———— 1988b *Religious trends in early Islamic Iran*. Albany, NY: The Persian Heritage Foundation.

———— 1991 "The origins of the Yemenite *hijra*." In *Arabicus Felix Luminosus Britannicus: Essays in Honor of A.F.L. Beeston on his Eightieth birthday*. Edited by A. Jones, pp. 25–44. Reading, MA: Ithaca Press.

—— 1992 *Religious and ethnic movements in medieval Islam.* London: Variorum.

—— 1996 "Shiʿa." *Encyclopaedia of Islam* 9: 420–4.

—— 1997 *The succession to Muhammad: A study of the early caliphate.* Cambridge: Cambridge University Press.

—— 1999 "Zaydi attitudes to Sufism." In *Islamic mysticism contested: Thirteen centuries of controversies and polemics.* Edited by F. de Jong and B. Radtke, pp. 124–44. Leiden: Brill.

—— 2002. "Zaydiyya." *Enclyclopaedia of Islam.* Vol. 11, pp. 477–81.

Magraw, Roger 1983 *France 1814–1915: The bourgeois century.* Oxford: Fontana.

Al-Mahaqiri, Muhammad n.d. *Tuluʿ al-qamar.* Date and place of publication unknown.

Mahmud, Salah R. (Ed.) 1983 *Dhikrayat al-Shawkani: rasaʾil li-ʾl-muʾarrikh al-yamani Muhammad b. ʿAli al-Shawkani.* Aden: Ministry of Culture.

Maimonides, M. 2002 *Der Brief in den Jemen: Texte zum Messias.* Edited by S. Powels-Niami. Berlin: Parerga.

Malkki, Liisa H. 1995 *Purity and exile: Violence, memory, and national cosmology among Hutu refugees in Tanzania.* Chicago, IL and London: University of Chicago Press.

Maʿoz, Moshe 1971 "The ʿulamaʾ and the process of modernization in Syria during the mid-nineteenth century." *Asian and African Studies* 7: 77–88.

Al-Maqbali, Salih 1981 *Al-ʿAlam al-shamikh fi ithar al-haqq ʿala ʾl-abaʾ waʾ l-mashayikh.* Damascus: Maktabat Dar al-bayan.

Marchand, Trevor 2001 *Minaret building and apprenticeship in Yemen.* Richmond, Surrey: Curzon Press.

Marcus, Abraham 1986 "Privacy in eighteenth-century Aleppo: The limits of cultural ideals." *International Journal of Middle East Studies* 18: 165–83.

Marcus, Michael A. 1985 " 'The saint has been stolen': Sanctity and social change in a tribe of eastern Morocco." *American Ethnologist* 12(3): 455–67.

Marriot, McKim 1976 "Hindu transactions: Diversity without dualism." In *Transaction and meaning: Directions in the anthropology of exchange and symbolic behaviour.* Edited by B. Kapferer, pp. 109–42. Philadelphia: Institute for the Study of Human Issues.

Mauss, Marcel 1950 *Essai sur le don.* Paris: Presses Universitaires de France.

Meissner, Jeffrey R. 1987 *Tribes at the core: Legitimacy, structure and power in Zaydi Yemen.* Ph.D. Thesis, Columbia University.

Meneley, Anne 1996 *Tournaments of value: Sociability and hierarchy in a Yemeni town.* Toronto: University of Toronto Press.

Mermier, Franck 1985 "Patronyme et hierarchie sociale a Sanaa." *Peuples Méditerranéens* 33: 33–41.

—— 1991 "Récit d'origine et rituel d'allégeance." *Peuples méditerranéens* 56–7: 177–80.

—— 1997 *Le cheikh de la nuit. Sanaa: Organisation des souks et société citadine.* Arles: Sindbad/Actes Sud.

—— 1999 "Yémen, les héritages d'une histoire morcelée." In *Le Yémen contemporaine.* Edited by R. Leveau, F. Mermier, and U. Steinbach, pp. 7–35. Paris: Karthala.

Messick, Brinkley 1978 *Transactions in Ibb: Economy and society in a Yemeni highland town.* Ph.D. Thesis. Princeton University.

—— 1986 "The mufti, the text and the world: Legal interpretation in Yemen." *Man* 21(1): 102–19.

—— 1988 "Kissing hands and knees: Hegemony and hierarchy in shariʿa discourse." *Law & Society Review* 22(4): 637–59.

Messick, Brinkley 1993 *The calligraphic state: Textual domination and history in a Muslim society.* Berkeley, CA and Los Angeles: University of California Press.

—— 1996 "Media muftis: Radio fatwas in Yemen." In *Islamic legal interpretation: Muftis and their fatwas.* Edited by M. Masud, B. Messick, and D. Powers, pp. 310–20. Cambridge, MA.: Harvard University Press.

Metcalf, Barbara D. 1978 "The madrasa at Deoband: A model for religious education in modern India." *Modern Asian Studies* 12(1): 111–34.

Miller, Daniel 1987 *Material culture and mass consumption.* Oxford: Blackwell.

Milton-Edwards, Beverley and Peter Hinchcliffe 2001 *Jordan: a Hashemite legacy.* London: Routledge.

Mir-Hosseini, Ziba 1993 *Marriage on trial: A study of Islamic family law.* London: I.B. Tauris.

Middleton, Karen 1999 "Circumcision, death, and strangers." In *Ancestors, power, and history in Madagascar.* Edited by K. Middleton, pp. 219–55. Leiden: Brill.

Mitchell, Timothy 1988 *Colonising Egypt.* Cairo: The American University in Cairo Press.

Mobini-Kesheh, Natalie 1997 "Islamic modernism in colonial Java: The al-Irshad movement." In *Hadhrami traders, scholars and statesmen in the Indian Ocean, 1750s–1960s.* Edited by U. Freitag and W. Clarence-Smith, pp. 231–48. Leiden: Brill.

Momen, Moojan 1985 *An introduction to Shi'i Islam: The history and doctrines of Twelver Shi'ism.* New Haven, CT and London: Yale University Press.

Moore, Henrietta L. 2000 "Difference and recognition: Postmillennial identities and social justice." *Signs: Journal of Women in Culture and Society* 25(4): 1129–32.

Moraru, Christiau 2000 "We embraced each other by our names: Lévinas, Derrida and the ethics of naming." *Names* 48(1): 49–58.

Morris, Jan 1959 *The Hashemite kings.* London: Faber & Faber.

Mortel, Richard 1987 "Zaydi Shi'ism and the Hasanid Sharifs of Mecca." *International Journal of Middle East Studies* 19:455–72.

Mottahedeh, Roy 1980 *Loyalty and leadership in an early Islamic society.* Princeton, NJ: Princeton University Press.

Mousavi, Sayed Askar 1998 *The Hazaras of Afghanistan: An historical, cultural, economic and political study.* Richmond, Surrey: Curzon Press.

Mundy, Martha 1981 *Land and family in a Yemeni Community.* Ph.D. Thesis, University of Cambridge.

—— 1983 "San'a dress, 1920–75." *San'a: An Arabian Islamic city.* Edited by R. Serjeant and R. Lewcock, pp. 529–41. London: World of Islam Festival Trust.

—— 1995 *Domestic government: Kinship, community and polity in North Yemen.* London: I.B. Tauris.

Munn, Nancy D. 1995 "An essay on the symbolic construction of memory in the Kaluli Gisalo." In *Cosmos and society in Oceania.* Edited by A. Iteanu and D. de Coppet, pp. 83–104. Oxford: Berg.

Munson, Henry 1993 *Religion and power in Morocco.* New Haven, CT and London:Yale University Press.

Musallam, Basim 1983 *Sex and society in Islam.* Cambridge: Cambridge University Press.

Al-Mutahhar, Muhammad 1985 *Ahkam al-ahwal al-shakhsiyyah min fiqh al-shari'ah al-islamiyyah.* Beirut: Dar al-kitab al-lubnani/Cairo:Dar al-kitab al-masri.

Myntti, Cynthia 1983 *Medicine and its social context: Observations from rural North Yemen.* Ph.D. Thesis, University of London.

Nagel, Tilman 1987 "Das Kalifat der Abbasiden." In *Geschichte der arabischen Welt.* Edited by U. Haarmann, pp. 101–65. Munich: C.H. Beck.

Nairn, Tom 1997 *Faces of nationalism: Janus revisited*. London and New York: Verso.

Nakash, Yitzhak 1994 *The Shi'is of Iraq*. Princeton, NJ: Princeton University Press.

Navaro-Yashin, Yael 1996 "Islamist multiculturalism: Beyond ethnographic boundaries between secularity and religion." Paper submitted at the conference on "Boundaries and Identities," October 24–26, University of Edinburgh.

Needham, Rodney 1971 "Remarks on the analysis of kinship and marriage." In *Rethinking kinship and marriage*. Edited by R. Needham, pp. 1–34. London: Tavistock.

Neisser, Ulrich 1988 "Time present and time past." In *Practical aspects of memory: Current research and issues*. Edited by M. Gruneberg, P. Morris, and R. Sykes. Vol. 2: Clinical and educational implications. Chichester: John Wiley and Sons.

Nelson, Kristina 1986 *The art of reciting the Qur'an*. Austin, TX: University of Texas Press.

Nora, Pierre 1989 "Between memory and history: Les lieux de mémoire." *Representations* 26: 7–25.

Noth, Albrecht 1987 "Früher Islam." In *Geschichte der arabischen Welt*. Edited by U. Haarmann, pp. 11–100. Munich: C.H. Beck.

Nu'man, Muhammad A. 1963 *Min wara' al-aswar*. Beirut: Dar al-kitab al-'arabi.

Obermeyer, Gerald J. 1981 "*Al-Iman* and al-Imam: Ideology and state in the Yemen, 1900–1948." In *Intellectual Life in the Arab East 1890–1939*. Edited by M. Buheiry, pp. 176–92. Beirut: American University of Beirut.

Paret, Rudi 1978 "Isma'il." *Encyclopaedia of Islam* 4: 184–5.

Parry, Jonathan 1989 "The end of the body." In *Fragments for a history of the human body*. Edited by M. Feher, pp. 490–517. (Part III) New York: Zone.

Peacock, James and Dorothy Holland 1993 "The narrated self: Life stories in process." *Ethos* 21(4): 367–83.

Peirce, Leslie P. 1993 *The imperial harem: Women and sovereignty in the Ottoman Empire*. Oxford: Oxford University Press.

Peters, Emrys L. 1963 "Aspects of rank and status among Muslims in a Lebanese village." In *Mediterranean countrymen*. Edited by J. Pitt-Rivers, pp. 159–200. Paris and The Hague: Mouton and Co.

——— 1972 "Shifts in power in a Lebanese village." In *Rural politics and social change in the Middle East*. R. Antoun and I. Harik, pp. 165–97. Bloomington, IN and London: Indiana University Press.

Peterson, John 1982 *Yemen: The search for a modern state*. London: Croom Helm.

Piepenberg, Fritz 1987 Sana'a al-qadeema: The challenges of modernization. In *The Middle East city*. Edited by A. Saqqaf, pp. 93–113. New York: Paragon House Publishers.

Pinault, David 1992 *The Shi'ites: Ritual and popular piety in a Muslim community*. London: I.B. Tauris.

Platon 1957 *Sämtliche Werke* 2. Hamburg: Rowohlt.

Podeh, Elie 1995 "Ending an age-old rivalry: The *rapprochement* between the Hashemites and the Saudis, 1956–1958." In *The Hashemites in the Modern Arab world*. Edited by A. Susser and A. Shmuelevitz, pp. 85–108. London: Frank Cass.

Puin, Gerd-Rüdiger 1984 The Yemeni *hijrah* concept of tribal protection. In *Land tenure and social transformation in the Middle East*. Beirut: American University of Beirut.

Qasim Ghalib Ahmad et al. 1983 *Ibn al-amir wa 'asruhu: Surah min kifah sha'ab al-Yaman*. San'a: Ministry of Information and Culture.

Rabinow, Paul 1975 *Symbolic domination: Cultural form and historical change in Morocco.* Chicago, IL and London: University of Chicago Press.

Al-Rasheed, Madawi 1991 *Politics in an Arabian oasis: The Rashidis of Saudi Arabia.* London: I.B. Tauris.

—— 1998 *Iraqi Assyrian Christians in London: The construction of ethnicity.* Lewiston, NY: Edwin Mellen Press.

—— 1999 "Political legitimacy and the production of history: The case of Saudi Arabia." In *New frontiers in Middle East security.* Edited by L. Martin, pp. 25–46. New York: St. Martin's Press.

Rassam, Amal 1977 "Al-Taba'iyya: Power, patronage and marginal groups in Northern Iraq." In *Patrons and clients in Mediterranean societies.* Edited by E. Gellner and J. Waterbury, pp. 157–66. London: Duckworth.

Rentz, George 1960 "Djazirat al-'Arab." *Encyclopaedia of Islam* 1: 533–56.

Rex Smith, G. 1997 "The political history of the Islamic Yemen down to the first Turkish invasion (1-945/622-1538)." In *Studies in the medieval history of the Yemen and South Arabia* by G. Rex Smith. Aldershot, Hampshire: Variorum.

Robinson, John A. 1996 "Perspectives, meaning, and remembering." In *Remembering our past.* Edited by D. Rubin, pp. 199–215. Cambridge: Cambridge University Press.

Rosenwald, George and Richard Ochberg 1992 "Introduction: Life stories, cultural politics, and self-understanding." In *Storied lives: The cultural politics of self-understanding.* Edited by G. Rosenwald and R. Ochberg, pp. 1–18. New Haven, CT and London: Yale University Press.

Rubin, David C. 1996 "Introduction." In *Remembering our past: Studies in autobiographical memory.* Edited by D. Rubin, pp. 1–15. Cambridge: Cambridge University Press.

Rubin, Gayle 1975 "The traffic in women: Notes on the 'political economy' of sex." In *Toward an anthropology of women.* Edited by R. Reiter, pp. 157–210. New York: Monthly Review Press.

Sabra, A. 1971 " 'Ilm." *Encyclopaedia of Islam* 3: 1133–4.

Salamandra, Christa 2000 "Consuming Damascus: Public culture and the construction of social identity." In *Mass mediations: New approaches to popular culture in the Middle East and beyond.* Edited by W. Armbrust, pp. 182–202. Berkeley, CA: University of California Press.

Sanadjian, Manuchehr 1996 "A public flogging in south-western Iran: Juridical rule, abolition of legality and local resistance." In *Inside and outside the law: Anthropological studies of authority and ambiguity.* Edited by O. Harris, pp. 157–83. London: Routledge.

Saqqaf, Adulaziz Y. 1987 "Sana'a: A profile of a changing city." In *The Middle East city: Ancient traditions confront a modern world.* Edited by A. Saqqaf, pp. 115–33. New York: Paragon House Publishers.

Schatkowski Schilcher, Linda 1985 *Families in politics: Damascene factions and estates of the 18th and 19th centuries.* Stuttgart: Steiner.

Schimmel, Annemarie 1989 *Islamic names.* Edinburgh: Edinburgh University Press.

Schneider, David 1964 "The nature of kinship." *Man* 64: 180–1.

—— 1984 *A critique of the study of kinship.* Ann Arbor, MI: University of Michigan Press.

Schroeter, Daniel J. 1988 *Merchants of Essaouira: Urban society and imperialism in southwestern Morocco, 1844–1886.* Cambridge: Cambridge University Press.

Schubel, Vernon J. 1993 *Religious performance in contemporary Islam: Shi'i devotional rituals in Pakistan.* Charleston, SC: University of South Carolina Press.

——— 1996 "Karbala as sacred space among North American Shi'a." In *Making Muslim space in North America and Europe.* Edited by B. Metcalf, pp. 186–203. Berkeley and Los Angeles, CA: University of California Press.

Schumacher, Ilse 1987 *Ritual devotion among Shi'i in Bahrain.* Ph.D. Thesis, University of London.

Scott, Hugh 1942 *In the high Yemen.* London: John Murray.

Scott, James C. 1990 *Domination and the arts of resistance: Hidden transcripts.* New Haven, CT and London: Yale University Press.

Schopen, Armin 1978 *Das Qat: Geschichte und Gebrauch des Genussmittels Catha edulis Forsk in der Arabischen Republik Jemen.* Wiesbaden: Franz Steiner.

——— 1981 "Das Qat in Jemen." In *Rausch und Realität.* Edited by G. Völger. Köln: Rautenstrauch-Joest-Museum.

Schwarz, Bill 1994 "Memories of empire." In *Displacements: Cultural identities in question.* Edited by A. Bammer, pp. 156–71. Bloomington and Indianapolis, IN: Indiana University Press.

Sebti, Abdelahad 1986 "Au Maroc: Sharifisme citadin, charisme et historiographie." *Annales* E.S.C. 41(2): 433–57.

Serjeant, Robert B. 1953 "A Zaidi manual of hisbah of the 3rd century (H)." *Rivista degli Studi Oriental* 28: 1–33.

——— 1957 "The saiyids of Hadramawt." Inaugural Lecture, School of Oriental and African Studies, London, 5 June 1956.

——— 1969 "The Zaydis." In *Religion in the Middle East*, Vol. 2. Edited by A. Arberry, pp. 285–301. Cambridge: Cambridge University Press.

——— 1979 "The Yemeni poet Al-Zubayri and his polemic against the Zaydi Imams." *Arabian Studies* 5: 87–130.

——— 1982 The interplay between tribal affinities and religious (Zaydi) authority in the Yemen. *Al-Abhath* 30: 11–50.

——— 1983a "The post-medieval and modern history of San'a and the Yemen, ca. 953–1382/1515–1962." In *San'a: An Arabian Islamic city.* Edited by R. Serjeant and R. Lewcock, pp. 68–107. London: The World of Islam Festival Trust.

——— 1983b "The mosques of San'a: The Yemeni Islamic setting." In *San'a: An Arabian Islamic city.* Edited by R. Serjeant and R. Lewcock, pp. 310–22. London: World of Islam Festival Trust.

Serjeant, Robert and Husayn al-'Amri 1983 "Administrative organisation." In *San'a An Arabian Islamic city.* Edited by R. Serjeant and R. Lewcock, pp. 142–60. London: World of Islam Festival Trust.

Serjeant, Robert and Ronald Lewcock (Ed.) 1983 *San'a An Arabian Islamic city.* London: The World of Islam Festival Trust.

AL-Shahari, Jamal al-Din 2001 City of divine and earthly joys: The description of San'a. Translated from the Arabic by Tim Mackintosh-Smith. San'a: American Institute for Yemeni Studies.

Shakry, Omnia 1998 "Schooled mothers and structured play: Child rearing in turn-of-the-century Egypt." In *Remaking women: Feminism and modernity in the Middle East.* Edited by L. Abu-Lughod, pp. 126–70. Princeton, NJ: Princeton University Press.

Al-Shamahi, 'Abdullah 1937 *Sirat al-'arifin ila idrak ikhtiyarat amir al-mu'minin.* San'a: Matba'at al-ma'arif.

Al-Shami, Ahmad 1965 *Imam al-Yaman Ahmad Hamid al-Din.* Beirut: Dar al-Kitab al-Jadid.

——— 1980 *Jinayat al-akwa' 'ala dhakha'ir al-hamdani.* Beirut: Dar al-Nafa'is.

Al-Shami, Ahmad 1984 *Riyah al-taghyir fi-'l-yaman*. Jiddah: Al-Matba'ah al-'Arabiyyah.

——— 1987 *Tarikh al-yaman al-fikri fi-'l-'asr al-'abbasi* (4 vols.). Beirut: Dar al-Nafa'is.

Sharh al-azhar 1980 (4 vols.) Compiled by 'Abdullah Ibn Miftah. (Place of publication unknown.)

Sharon, Moshe 1986 "Ahl al-bayt—People of the House." *Jerusalem Studies in Arabic and Islam* 8: 169–84.

Al-Shawkani, Muhammad 1979 *Adab al-talab*. San'a: Markaz al-dirasat wa-'l-buhuth al-yamaniyyah.

——— 1982 *Diwan al-Shawkani aslak al-jawhar wa-'l-hayat al-fikriyyah wa-'l-siyasiyyah fi 'asrih*. Damascus: Dar al-fikr.

——— n.d. *Al-Badr al-tali' bi-mahasin man ba'd al-qarn al-sabi'*, 2 vols., Beirut: Dar al-Ma'arifah. (Date of publication unknown.)

Shryock, Andrew 1997 *Nationalism and the genealogical imagination: Oral history and textual authority in tribal Jordan*. Berkeley and Los Angeles, CA: University of California Press.

Slyomovics, Susan 1998 *The object of memory: Arab and Jew narrate the Palestinian village*. Philadelphia, PA: University of Pennsylvania Press.

Snouck Hurgronje, Christian 1931 *Mekka in the later part of the 19th century*. Leiden: Brill.

Sourdel, Dominique 1971 "Al-Hadi ila 'l-Hakk." *Encyclopaedia of Islam* 3: 22.

Southall, Aidan W. 1986 "Common themes in Malagasy culture." In *Madagascar: Society and History*. Edited by C. Kottak et al. Durham, NC: Carolina Academic Press.

Spellberg, Denise A. 1994 *Politics, gender, and the Islamic past: The legacy of 'A'isha bint Abi Bakr*. New York: Columbia University Press.

Sperber, Dan 1985 "Anthropology and psychology: Towards an epidemology of representations." *Man* 20(1): 73–89.

Sperling, Jutta 1999 "The paradox of perfection: Reproducing the body politic in Late Renaissance Venice." *Comparative Studies in Society and History* 41(1): 3–32.

Stewart, Michael 1989 " 'True speech': Song and the moral order of a Hungarian Vlach Gypsy community." *Man* 24(1): 79–102.

Stoler, Ann and Frederick Cooper 1997 "Between metropole and colony: Rethinking a research agenda." In *Tensions of empire: Colonial cultures in a bourgeois world*. Edited by F. Cooper and A. Stoler, pp. 1–56. Berkeley, CA: University of California Press.

Stoler, Ann and Karen Strassler 2000 "Castings for the colonial: memory work in 'New Order' Java." *Comparative Study in Society and History* 42(1): 4–48.

Storrie, Robert 1999 Being human: Personhood, cosmology and subsistence for the Hoti of Venezuelan Guiana. Ph.D. Thesis, University of Manchester.

Strathern, Andrew 1973 "Kinship, descent and locality: Some New Guinea examples." In *The character of kinship*. Edited by J. Goody, pp. 21–33. Cambridge: Cambridge University Press.

Strathern, Andrew and Michael Lambek 1998 "Introduction. Embodying sociality: Africanist-Melanesianist comparisons." In *Bodies and persons: Comparative perspectives from Africa and Melanesia*. Edited by M. Lambek and A. Strathern, pp. 1–25. Cambridge: Cambridge University Press.

Strothmann, Rudolf 1912 *Das Staatsrecht der Zaiditen*. Strasbourg: Karl J. Trübner.

——— 1974 "Al-Zaidiya." *Shorter Encyclopedia of Islam*: 651–2.

Sturken, Marita 1999 "Narratives of recovery: Repressed memory as cultural memory." In *Acts of memory*. Edited by M. Bal, J. Crewe, and L. Spitzer, pp. 231–48. Hanover and London: Dartmouth College.

Subhi, Ahmad 1980 *Al-Zaydiyyah*. Alexandria: Mansha'at al-ma'arif.

Susser, Asher 1995 "Introduction." In *The Hashemites in the modern Arab world*. Edited by A. Susser and A. Shmuelevitz, pp. 1–11. London: Frank Cass.

Susser, Asher and Aryeh Shmuelevitz (Ed.) 1995 *The Hashemites in the Modern Arab World: Essays in honour of the late Professor Uriel Dann*. London: Frank Cass.

Sutton, David 1998. *Memories cast in stone: The relevance of the past in everyday life*. Oxford: Berg.

Tachau, Frank 1975 (Ed.) *Political elites and political development in the Middle East*. New York: John Wiley and Sons.

Taminian, Lucine 1999 "Persuading the monarchs: Poetry and politics in Yemen (1920–1950)." In *Le Yémen contemporain*. Edited by R. Leveau, F. Mermier, and U. Steinbach, pp. 203–19. Paris: Editions Karthala.

Tapper, Nancy 1991 *Bartered brides: Politics, gender and marriage in an Afghan tribal society*. Cambridge: Cambridge University Press.

Tapper, Nancy and Richard Tapper 1987 "The birth of the Prophet: Ritual and gender in Turkish Islam." *Man* 22(1): 69–92.

Teitelbaum, Joshua 2001 *The rise and fall of the Hashimite kingdom of Arabia*. London: Hurst & Company.

Terrace, H. 1960 " 'Alawis." *Encyclopaedia of Islam* 1: 355–8.

Thaiss, Gustav 1971 "The bazaar as a case study of religion and social change." In *Iran faces the seventies*. Edited by E. Yar-Shater. New York: Praeger.

—— 1972 "*Religious symbolism and social change: The drama of Husain*." In *Scholars, saints and sufis: Muslim religious institutions since 1500*. Edited by N. Keddie, pp. 349–66. Berkeley and Los Angeles, CA: University of California Press.

—— 1973 *Religious symbolism and social change: The drama of Husain*. Ph.D. Thesis, Washington University, St. Louis.

Thomas, Nicholas 1999 "Becoming undisciplined: Anthropology and cultural studies." In *Anthropological theory today*. Edited by H. Moore, pp. 262–79. Oxford: Polity Press.

Thompson, John 1991 "Editor's introduction." In *Language and symbolic power*. Edited by P. Bourdieu, pp. 1–31. Oxford: Polity Press.

Torab, Azam 1996 "Piety as gendered agency: A study of *jalaseh* ritual discourse in an urban neighbourhood in Iran." *Journal of the Royal Antropological Institute* 2(2): 235–51.

—— 2002 "The politicization of women's religious circles in post-revolutionary Iran." In *Women, religion and culture in Iran*. Edited by S. Ansari and V. Martin, pp. 143–68. Richmond, Surrey: Curzon.

Tripp, Charles 2000 *A history of Iraq*. Cambridge: Cambridge University Press.

Trouillot, Michel 1995 *Silencing the past: Power and the production of history*. Boston, MA: Beacon Press.

Tulving, Endel 1972 "Episodic and semantic memory." In *Organization of memory*. Edited by E. Tulving and W. Donaldson, pp. 381–403. New York and London: Academic Press.

—— 1983. *Elements of episodic memory*. Oxford: Clarendon Press.

van der Veer, Peter 1993 "The foreign hand: Orientalist discourse in sociology and communalism." In *Orientalism and the postcolonial predicament*. Edited by C. Breckenridge and P. van der Veer, pp. 23–44. Philadelphia, PA: University of Pennsylvania Press.

Veblen, Thorstein 1970 [1898] *The theory of the leisure class*. London: George Allen and Unwin.

Veccia Vaglieri, Laura 1964 "Ghadir Khumm." *Encyclopaedia of Islam* 2: 993–4.

Vernant, Jean-Pierre 1983 [1965] *Myth and thought among the Greeks.* Translated from the French. London: Routledge.

Vieille, Paul 1978 "Iranian women in family alliance and sexual politics." In *Women in the Muslim world.* Edited by L. Beck and N. Keddie, pp. 451–72. Cambridge, MA: Harvard University Press.

Voll, John 1975 "Old ' 'ulama' families and Ottoman influence in eighteenth century Damascus." *American Journal of Arabic Studies* 3: 48–59.

Wachowski, Markus 2004 *Sada in San'a: Zur Fremd-und Eigenwahrnehmung der Prophetennachkommen in der Republik Jemen.* Berlin: Klaus Schwarz.

Wachtel, Nathan 1990 "Remember and never forget." In *Between history and memory.* Edited by M. Bourguet, L. Valensi, and N. Wachtel, pp. 101–29. London: Harwood Academic Publishers.

Walters, Delores M. 1987 *Perceptions of social inequality in the Yemen Arab Republic.* Ph.D. Thesis, University of New York.

Waterbury, John 1970 *The commander of the faithful: The Moroccan political elite—a study in segmented politics.* New York: Columbia University Press.

Al-Wazir, 'Abdullah 1985 *Tarikh al-yaman khilal al-qarn al-hadi 'ashr al-hijri—al-sabi' 'ashr al-miladi 1045–1090H./1645–1680A.D.* Beirut: Dar al-Masirah.

Al-Wazir, Ahmad 1987 *Hayat al-amir 'Ali b. 'Abdullah al-Wazir.* Beirut: Manshurat al-'Asr al-hadith.

Al-Wazir, Ibrahim 1997 *Umm fi ghimar thawrah.* Bethesda: Kitab Inc.

Al-Wazir, Zayd 1992 *Muhawalah li-tashih al-masar.* Beirut: Dar al-Nafa'is.

——— 2000 Al-Mutarrifiyyah: Al-afkar *w-'l-ma'sah. Al-Masar* 1(2): 1–83.

——— in press *Tarikh al-Imam Muhammad b. 'Abdullah al-Wazir,* Vol. 1. Beirut: Dar al-'Asr al-hadith.

——— n.d. *Rihlat al-alf wa-'l-thalathma'ah 'am.*

Wedeen, Lisa 2003 "Seeing like a citizen, acting like a state: Exemplary events in unified Yemen." *Comparative Studies in Society and History* 45(4): 680–713.

Weir, Shelagh 1985a *Qat in Yemen: Consumption and social change.* Dorchester: Dorset Press.

——— 1985b "Economic aspects of the Qat industry in north-west Yemen." In *Economy, society and culture in contemporary Yemen.* Edited by B. Pridham, pp. 64–82. London: Croom Helm.

——— 1986 "Tribe, hijrah and madinah in north-west Yemen." In *Middle Eastern cities in comparative perspective.* Edited by K. Brown et al., pp. 226–39. London: Ithaca Press.

——— 1997 "A clash of fundamentalisms: Wahhabism in Yemen." *Middle East Report* 204: 22–6.

Weismantel, Mary 1995 "Making kin: Kinship theory and Zumbagua adoptions." *American Ethnologist* 22(4): 685–709.

Weiss, Brad 1997 "Objects and bodies: Some phenomenological implications of *Knowledge and Practice in Mayotte.*" *Cultural Dynamics* 9(2): 161–72.

Wenner, Manfred 1975 "Saudi Arabia: Survival of traditional elites." In *Political elites and political development in the Middle East.* Edited by F. Tachau. New York: John Wiley and Sons.

——— 1991 *The Yemen Arab Republic: Development and change in an ancient land.* Boulder, CO: Westview Press.

Wilce, James 2002 Genres of memory and the memory of genres: "forgetting" lament in Bangladesh. *Comparative Studies in Society and History* 44(1): 159–85.

Willis, John 2004 "Leaving only question-marks: Geographies of rule in modern Yemen." In *Counter-Narratives: History, contemporary society, and politics in Saudi*

Arabia and Yemen. Edited by M. al-Rasheed and R. Vitalis, pp. 119–49. New York: Palgrave.

Le Wita, Beatrix 1994 *French bourgeois culture*. Cambridge: Cambridge University Press.

Wright, Gordon 1995. *France in modern times: From the enlightenment to the present*. New York: W.W. Norton & Company.

Würth, Anna 1995 "A Sanaʻa court: The family and the ability to negotiate." *Islamic Law and Society* 2(3): 320–40.

——— 2000 *Aš-Šariʻa fi bab al-yaman: Recht, Richter und Rechtspraxis an der familienrechtlichen Kammer des Gerichts Süd-Sanaa, (Republik Jemen) 1983–1995*. Berlin: Duncker und Humblot.

Yamani, Mai 1994 "You are what you cook: Cuisine and class in Mecca." In *Culinary cultures of the Middle East*. Edited by S. Zubaida and R. Tapper, pp. 173–84. London: I.B. Tauris.

Yanagisako, Sylvia and Jane Collier 1996 Comments on "Until death do us apart." *American Ethnologist* 23(2): 235–6.

Yapp, Malcolm 1987 *The making of the modern Near East 1792–1923*. London and New York: Longman.

Yerushalmi, Yosef 1982 *Zakhor: Jewish history and Jewish memory*. Seattle and London: University of Washington Press.

Yoneyama, Lisa 1999 *Hiroshima traces: Time, space, and the dialectics of memory*. Berkeley, CA: University of California Press.

Zabarah, Muhammad A. 1982 *Yemen: Traditionalism vs. modernity*. New York: Praeger.

Zabarah, Muhammad M. 1940 *Nashr al-ʻarf li-nubalaʼ al-yaman baʻd al-alf ila sanah 1357 hijriyyah*. Sanʻa: Markaz al-dirasat wa-ʼl-buhuth al-yamaniyyah (Reprint).

——— 1979 *Nuzhat al-nazar fi rijal al-qarn al-rabiʻ ʻashar*. Sanʻa: Markaz al-dirasat wa-ʼl-buhuth al-yamaniyyah (Reprint).

——— 1984 *Al-anbaʼ ʻan dawlat bilqis wa-sabaʼ*. Sanʻa: Dar al-yamaniyyah.

——— 1999 *Khulasat al-mutun fi abnaʼ wa nubalaʼ al-yaman al-maymun* (4 vols) Richmond, VA: Yemeni Heritage and Research Centre.

Zartman, I. William 1980 "Introduction." In *Elites in the Middle East*. Edited by I.W. Zartman, pp. 1–9. New York: Praeger.

Zayd, Hasan 1997 "Qissat al-maʻahid al-ʻilmiyyah." *Maʻin*: 12–17.

El-Zein, Abdul Hamid 1974 *The sacred meadows: A structural analysis of religious symbolism in an East African town*. Evanston, IL: Northwestern University Press.

Zerubavel, Yael 1991 "New beginning, old past: The collective memory of pioneering in Israeli culture." In *New perspectives on Israeli history: The early years of the state*. Edited by L. Silberstein, pp. 193–215. New York: New York University Press.

——— 1995 *Recovered roots*. Chicago, IL: Chicago University Press.

Žižek, Slavoj 1989 *The sublime object of ideology*. London: Verso.

Zonis, Marvin 1971 *The political elite of Iran*. Princeton, NJ: Princeton University Press.

Al-Zubayri, Muhammad n.d. *Al-Imamah wa khataruha ʻala wahdat al-yaman* (date of publication unknown). Beirut: Dar al-Sahafah.

Zur, Judith 1997 "Reconstructing the self through memories of violence among Mayan Indian war widows." In *Gender and catastrophe*. Edited by R. Lentin, pp. 64–76. London and New York: Zed Books.

Glossary

'abd	*Servant of.* Affixed to one of the ninety-nine names of God
adhan	call to prayer
'Adnani	a descendant of 'Adnan, putative ancestor of the "northern Arabs"
ahl	kin group
ahl al-bayt	"People of the House," members of the household of the Prophet, including, besides Muhammad, 'Ali, Fatima, al-Hasan, al-Husayn, and their progeny
'a'ilah	"family"; in San'a also: wife
ajr	remuneration, merit
akhdam	see **khadim**
al	family of (e.g., al Muhammad); not to be confused with the Arabic article
'Alawi	ruling dynasty in Morocco from 1600s to the present
Alawiyyun	descendants of 'Ali
'Alids	those who derive descent either from 'Ali and Fatima through their sons al-Hasan and al-Husayn, or from one of 'Ali's children by other wives
'alim, pl.'ulama	scholar of the Islamic sciences
allamah	title of a religious scholar
amah	slave girl; used as prefix of one of the ninety-nine names of God
'amil, pl.'ummal	governor (during the Imamate)
amir	prince (during the Hamid al-Din era also title of a governor or military commander)
amir al-mu'minin	leader of the faithful
'arabi, pl.'arab	Arab; Yemen: Non-'Alid, Qahtani
'asabiyyah	zealous partisanship, esprit de corps
'asb	gift (usually money) given by men to their female kin on *'id* days
asl	root; origin; descent
asli	original, genuine, authentic
'Ashura	The tenth day of Muharram, the first month of the Islamic calender. Shi'is hold mourning rituals to commemorate the martyrdom of Imam al-Husayn
'ayb	inappropriate, disgraceful
balad, pl. bilad	village, countryside
Banu Hashim	the "clan" of Hashim, the Prophet's great grandfather; term used to refer to the Hashimites
baraka	blessing; the divine power emanating from a religious leader or holy man
bay'ah	oath of allegiance; homage

bayt, pl. **buyut**	house, household; a patronymic descent category
bayt al-mal	lit., the house of wealth; treasury of the Muslim state
buyut al-'ilm	"houses of learning"; families with a tradition of religious learning
bint al-nabi, pl. **banat**	female descendant of the Prophet
da'i, pl. **du'at**	"he who summons"
dani'	low, inferior
da'wah	summons, call for allegiance; invitation or call to adopt the cause of a religious leader claiming the right to the Imamate
din	religion (piety)
diwan	multi-purpose "sitting-room"; also: a collection of poetry or prose; consultative assembly
al-diwan al-malaki	royal court
du'a'	invocation of God; good wish
duwaydar	houseboy (during the Imamate)
fallah, pl. **fallahun**	peasant; farmer cultivator
faqih, pl. **fuqaha'**	an expert in fiqh (jurisprudence); Qur'anic school teacher
fatimiyyah, pl. **fatimiyyat**	female descendant of the Prophet (term used mainly in legal documents)
fay'	property acquired by the Muslims without war effort
fi'ah	group
fiqh	jurisprudence
ghaybah	lit., absence; among several Shi'i groups the term has been used to describe the condition of anyone who has been withdrawn by God from the eyes of men and whose life during that period of occultation may be miracuously prolonged
hadith	sayings and deeds attributed to the Prophet Muhammad
hakim	governor, judge; district judge during the period of the Hamid al-Din
hammami	bath attendant
haram	forbidden, unlawful
hashimi	descendants of the Hashim b.'Abd Manaf, the ancestor of the Prophet, 'Ali and al-'Abbas. The chief Hashimi branches were the 'Alids and the 'Abbasids. In Yemen the term is often used synonymously with *sayyid*
Hashimiyyah	Eighth-century movement against the Umayyads in favour of 'Alid rule
hashimiyyah	female descendant of the Prophet
Hashimiyyat	poems of al-Kumayt (d. 743)
hijrah	emigration, flight; in Yemen also a place or person which enjoys a protected status
'id	the Muslim festivals; *'id al-fitr*, the breaking of the fast of Ramadan, and *'id al-adha*, the sacrificial festival of the tenth of Dhu al-hijjah
ijtihad	interpretive judgment of religious tenets
ikhtiyarat	Zaydi Imams' verdicts on debatable legal issues
'ilm, pl. **'ulum**	knowledge; more specifically religious knowledge
Imam	leader of a group of Muslims in prayer; or the supreme leader of the Muslim community. The title was particularly used by the Shi'is in reference to the persons recognized by them as the successors of the Prophet; during the Zaydi Imamate designation of the holder of supreme authority

'imamah pl.'ama'im	male headgear predominantly worn by religious scholars
imamah	the "supreme leadership" of the Muslim community after the death of the Prophet (Imamate)
Imamiyyah	Shi'i school from which the Ithna 'Ashariyyah ("Twelver-Shi'a") developed
Ithna-'ashariyyah	the branch of the Shi'a who believe in the twelve Imams descended from 'Ali, the last of whom disappeared and went into hiding in 873 and is expected to return as the redeemer who will appear at the end of the world and usher in the millenium
jadd	grandfather, eponym
janbiyyah	dagger (traditionally worn by the *qaba'il*)
jazzar	butcher
jihad	striving; effort directed toward inner religious perfection and toward holy war of the Muslims against the infifels
jinn	spirit beings, composed of vapors or flames, who are imperceptible but may have malevolent influences
jukh	coat of woolen cloth worn by men
kafa'ah	suitability, equivalence in marriage
khadim, pl. akhdam	low status sweepers and singers
khal, pl. akhwal	maternal uncle
khalah	maternal aunt, also form of address of one's mother's co-wife
khatm	sealing. *Khatm* rituals are performed when a child has finished reading the Qur'an
khums	the fifth of war booty reserved to the Prophet
khuruj	"rising" against rulers considered to be illegitimate according to Zaydi principles
khutbah	Friday sermon
kufr	unbelief, godlessness
madhhab pl. madhahib	doctrine, school of Jurisprudence
madrasah	a college or seminary of higher Muslim learning
al-Madrasah al-'ilmiyyah	religious college inaugurated by Imam Yahya
mafraj	"reception-room" which usually occupies the topmost storey of a house where qat is chewed
ma'had, pl. ma'ahid	institute
al-ma'ahid al-'ilmiyyah	Sunni-oriented religious institutes
mahr	indirect dowry, a sum of money or any valuable which a man gives to his wife at the time of marriage (paraphernalia)
mahram	being in a degree of consanguinity precluding marriage
malakiyyun	royalists
mani	seminal fluid, vaginal humor of a female
mawla	master
mawlid	ritual performed on the Prophet's birthday and other occasions such as the birth of a child
maqyal, pl. maqayl	men's daily afternoon gathering (in San'a also referred to as *madka*; in classical Arabic *maqil*)
muhaqqiq	one who conducts research of religious matters
mujtahid	religious scholar who has achieved the level of competence necessary to practice *ijtihad*
muqahwi	keeper of caravanserai, "innkeeper"

mut'ah	usufruct marriage contracted for a specified time and for the purpose of sexual pleasure (recognized in Imami-Shi'i law only)
mutawadi'un	humble, unassuming people
muwallid	term refers to foreign slaves born and raised among Arabs or Arabs of "mixed parentage"
muwatin	fellow citizen
muzayyin, pl. mazayinah	low status barber, butcher, circumcisor (San'a)
muzayyinah	fem. form of *muzayyin* (usually a woman serving at people's houses)
Mu'tazili	adherent of a school of theology (Mu'tazila) which emphasizes the unity and justice of God and man's free will. It evolved into a theology on the basis of rationality
nabi	prophet
nasab	descent, origin
nasrani, pl. nasara	Christian
nass	explicit designation of a successor by his predecessor, particularly relating to the Imami Shi'i view of succession to the Imamate
nisbah, pl. nisab	attribution; adjective denoting descent or origin
qabaliyyah	"tribalism"
qabilah	"tribe," "tribes"
qabili, pl. qaba'il	"tribesman"
qabiliyyah, pl. at	"tribeswoman"
qadi, pl. qudah	judge (in Yemen also "from a family of judges")
Qahtani, pl. Qahtaniyyun	a descendant of Qahtan, one of the putative ancestors of the "southern Arabs"
qalil asl	those who lack authenticity, people of lowly birth
qamis	wide bodied and sleeved gown worn by men
qarawi	rustic, peasant (term used in eastern Yemen to describe a person of lowly birth)
qashsham, pl. qashshamiyyun	vegetable growers, particularly of onions garlic and radishes
qat	a mild stimulant (*Catha Edulis*)
Quraysh	the patrikin from which the Prophet Muhammad came. They dominated Mecca when he was born
ra'is al-'a'ilah	"head of the family"
raj'i	reactionary
salafiyyun	adherents of an Islamic reform movement which emerged in the nineteenth century. Insisting on a literal adaptation of the Qur'an and Sunna, they believe that the Muslim world must return to the true sources and principles of Islam as laid down by the Prophet and as practised by his companions, the *Salaf al-Salih* (venerable forefathers)
sayyid, pl. sadah	lord, master; an honorific appelation for men of authority; title of a male descendant of the Prophet
Shafi'i	a member of one of the major Sunni schools of law to which the majority of Yemenis adhere (after Muhammad b. Idris al-Shafi'i, d. 820)
sharaf	nobility, honor, dignity

sharif, pl. ashraf/shurafa'/ shurfa'	noble, eminent, honorable; title of a descendant of the Prophet predominantly in Morocco but also in the eastern provinces of the Yemen
sharifah, pl. shara'if	title of a female descendant of the Prophet (predominantly during the Imamate)
sharshaf, pl. sharashif	women's outdoor garment of four pieces
shari'ah	the divinely revealed law of Islam
shaykh	elder; the chief of a tribe (Shaykh al-Islam = title of religious dignitary)
Shi'a	*Shi'at 'Ali*, followers of 'Ali; Muslims who claim that only the direct patrilineal descendants of 'Ali, the Imams, are legitimate successors of the Prophet
shurah	consultation
sidi	"my lord," title of a male descendant of the Prophet
sitti	"my lady," title of a female descendant of the Prophet
sulb	hard, firm; spinal column
Sunna	authoritative practice of the Prophet-Muhammad; it is embodied in hadith
Sunni	an adherent of one of four standard schools of shari'ah interpretation
suq	market
tabaqah, pl.-at	rank, stratum
tafritah, pl. tafarit	women's daily afternoon gathering
taghut	idolatry
taqbil	kissing of the hand or knee
taqlid, pl. taqalid	tradition
taqlid ahl al-bayt	the verdicts and practices of the 'ulama who belong to the House of the Prophet
tarjamah	translation, biography
thumah	dagger worn predominantly by the religious elite (Imamate)
'ulama	see 'alim
ummah	nation, community; any people as followers of a particular religion or prophet; in particular, the Muslims as forming a religious community
'unsuriyyah	race, nationality; in the Yemeni context distinctions made based on perceived national difference (prejudice, discrimination, racism)
'urf	legal practice; customary law, norm
usrah	family, lineage
Wahhabi	adherent of a school of thought based on the teachings of the Hanbali scholar Muhammd b. 'Abd al-Wahhab (d. 1792); most prevalent in Saudi Arabia and Qatar
wali	legal guardian, protector
waqf, pl. awqaf	endowment
waqf 'amm	endowment made for charitable purposes
waqf al-daris	endowment in perpetuity made to support recitation of the Qur'an, often for the soul of the benefactor; also in support of students
waqf dhurriyyah	family endowment; its revenue is allocated to the descendants of the person making the endowment

waqf khas	a special, personal endowmnent according to the donor's inclination
wilayah	province (term used during the Imamate, adopted from the Ottomans); also sovereign power
wilayat al-faqih	the concept that government belongs to those who are learned in jurisprudence (revived by Khumayni)
wujaha'	those who have visibility, eminent families
zakah	alms tax
zalim	oppressor
zannah, pl. **zinnin**	a full or calf length garment worn by men
zaydiyyah	one of the Shi'i schools of shari'ah jurisprudence; after Zayd b.'Ali, d. 740
zulm	injustice, oppression

Index

Note: Page numbers in italics refer to illustrations; those in bold refer to the Appendices.